Ant Colony Optimization

Ant Colony Optimization

Marco Dorigo
Thomas Stützle

A Bradford Book
The MIT Press
Cambridge, Massachusetts
London, England

This book was set in Times New Roman on 3B2 by Asco Typesetters, Hong Kong. Printed and bound in the United States of America.

Library of Congress Cataloging-in-Publication Data

Dorigo, Marco.
Ant colony optimization / Marco Dorigo, Thomas Stützle.
  p. cm.
"A Bradford book."
Includes bibliographical references (p.).
ISBN 0-262-04219-3 (alk. paper)
1. Mathematical optimization. 2. Ants–Behavior–Mathematical models. I. Stützle, Thomas. II. Title.
QA402.5.D64 2004
519.6—dc22   2003066629

10 9 8 7 6 5 4

To Serena and Roberto
To Maria José and Alexander

Contents

Preface

Ants exhibit complex social behaviors that have long since attracted the attention of human beings. Probably one of the most noticeable behaviors visible to us is the formation of so-called ant streets. When we were young, several of us may have stepped on such an ant *highway* or may have placed some obstacle in its way just to see how the ants would react to such disturbances. We may have also wondered where these ant highways lead to or even how they are formed. This type of question may become less urgent for most of us as we grow older and go to university, studying other subjects like computer science, mathematics, and so on. However, there are a considerable number of researchers, mainly biologists, who study the behavior of ants in detail.

One of the most surprising behavioral patterns exhibited by ants is the ability of certain ant species to find what computer scientists call shortest paths. Biologists have shown experimentally that this is possible by exploiting communication based only on pheromones, an odorous chemical substance that ants may deposit and smell. It is this behavioral pattern that inspired computer scientists to develop algorithms for the solution of optimization problems. The first attempts in this direction appeared in the early '90s and can be considered as rather "toy" demonstrations, though important for indicating the general validity of the approach. Since then, these and similar ideas have attracted a steadily increasing amount of research—and ant colony optimization (ACO) is one outcome of these research efforts. In fact, ACO algorithms are the most successful and widely recognized algorithmic techniques based on ant behaviors. Their success is evidenced by the extensive array of different problems to which they have been applied, and moreover by the fact that ACO algorithms are for many problems among the currently top-performing algorithms.

Overview of the Book

This book introduces the rapidly growing field of ant colony optimization. It gives a broad overview of many aspects of ACO, ranging from a detailed description of the ideas underlying ACO, to the definition of how ACO can generally be applied to a wide range of combinatorial optimization problems, and describes many of the available ACO algorithms and their main applications. The book is divided into seven chapters and is organized as follows.

Chapter 1 explains how ants find shortest paths under controlled experimental conditions, and illustrates how the observation of this behavior has been translated into working optimization algorithms.

In chapter 2, the ACO metaheuristic is introduced and put into the general context of combinatorial optimization. Basic notions of complexity theory, such as $\mathcal{NP}$-hardness, are given and other major metaheuristics are briefly overviewed.

Chapter 3 is dedicated to the in-depth description of all the major ACO algorithms currently available in the literature. This description, which is developed using the traveling salesman problem as a running example, is completed by a guide to implementing the algorithms. A short description of a basic C implementation, as well as pointers to the public software available at www.aco-metaheuristic.org/aco-code/, is given.

Chapter 4 reports on what is currently known about the theory of ACO algorithms. In particular, we prove convergence for a specific class of ACO algorithms and we discuss the formal relation between ACO and other methods such as stochastic gradient descent, mutual-information-maximizing input clustering, and cross-entropy.

Chapter 5 is a survey of current work exploiting ACO to solve a variety of combinatorial optimization problems. We cover applications to routing, assignment, scheduling, and subset problems, as well as a number of other problems in such diverse fields as machine learning and bioinformatics. We also give a few "application principles," that is, criteria to be followed when attacking a new problem using ACO.

Chapter 6 is devoted to the detailed presentation of AntNet, an ACO algorithm especially designed for the network routing problem, that is, the problem of building and maintaining routing tables in a packet-switched telecommunication network.

Finally, chapter 7 summarizes the main achievements of the field and outlines some interesting directions for future research.

Each chapter of the book (with the exception of the last chapter) ends with the following three sections: bibliographical remarks, things to remember, and exercises.

- Bibliographical remarks, a kind of short annotated bibliography, contains pointers to further literature on the topics discussed in the chapter.

- Things to remember is a bulleted list of the important points discussed in the chapter.

- Exercises come in two forms, thought exercises and computer exercises, depending on the material presented in the chapter.

Finally, there is a long list of references about ACO algorithms that gives a lot of pointers to more in-depth literature.

Overall, this book can be read easily by anyone with a college-level scientific background. The use of mathematics is rather limited throughout, except for chapter 4, which requires some deeper knowledge of probability theory. However, we assume that the reader is familiar with some basic notions of graph theory, programming, and probabilities. The book is intended primarily for (1) academic and industry researchers in operations research, artificial intelligence, and computational intelligence; (2) practitioners willing to learn how to implement ACO algorithms to solve combinatorial optimization problems; and (3) graduate and last-year undergraduate students in computer science, management science, operations research, and artificial intelligence.

Acknowledgments

The field of ant colony optimization has been shaped by a number of people who have made valuable contributions to the development and success of the field.

First of all, we wish to acknowledge the contributions of Alberto Colorni and Vittorio Maniezzo. Alberto and Vittorio collaborated closely with Marco Dorigo in the definition of the first ACO algorithms while Marco was a doctoral student at Politecnico di Milano, in Milan, Italy. Without their contribution, there would probably be no ACO research to describe. Our thoughts turn next to Jean-Louis Deneubourg and Luca Maria Gambardella. Jean-Louis, a recognized expert in the study of social insects, provided the inspiration (as described in chapter 1 of this book) for the ACO work. Luca, a computer scientist with a strong feeling for practical applications, was the one who most helped in transforming ACO from a fascinating toy into a competitive metaheuristic.

More generally, many researchers have written papers on ACO (applications, theoretical results, and so on). This book is clearly influenced by their research and results, which are reported in chapter 5.

Several colleagues and students of ours have checked large parts of the book. We appreciated very much the comments by Maria Blesa, Christian Blum, Julia Handl, Elena Marchiori, Martin Middendorf, Michael Samples, and Tommaso Schiavinotto. In addition, we would like to thank those colleagues who checked parts of the book: Mauro Birattari, Roberto Cordone, Gianni Di Caro, Karl Dörner, Alex Freitas, Luca Maria Gambardella, Jose Antonio Gámez, Walther Gutjahr, Richard Hartl, Holger Hoos, Joshua Knowles, Guillermo Leguizamón, John Levine, Helena Lourenço, Max Manfrin, Vittorio Maniezzo, Daniel Merkle, José Miguel Puerta, Marc Reimann, Andrea Roli, Alena Shmygelska, Krzysztof Socha, Christine Solnon, and Mark Zlochin. Our special thanks goes to Cristina Versino, for providing the ant drawings used in figures 1.7 and 3.2, and to all the people at the IRIDIA and Intellectics groups, for providing a stimulating scientific and intellectual environment in which to work.

People at MIT Press, and in particular Robert Prior, have greatly helped to make this project successful. We thank all of them, and in particular Bob, for gently pressing us to deliver the draft of this book.

Final thanks go to our families, in particular to our wives Laura and Maria José, who have constantly provided the comfortable environment conducive to successfully completing this book, and to our children Luca, Alessandro, and Alexander, who give meaning to our lives.

Marco Dorigo acknowledges support from the Belgian FNRS, of which he is a senior research associate. The writing of this book has been indirectly supported by the numerous institutions who funded the research of the two authors through the years. We wish to thank the Politecnico di Milano, Milan, Italy; the International Computer Science Institute, Berkeley, California; IDSIA, Lugano, Switzerland; the Intellectics Group at Darmstadt University of Technology, Germany; the IRIDIA group at the Université Libre de Bruxelles, Brussels, Belgium; and the Improving Human Potential programme of the CEC, who supported this work through grant HPRN-CT-1999-00106 to the Research Training Network "Metaheuristics Network." The information provided is the sole responsibility of the authors and does not reflect the community's opinion. The community is not responsible for any use that might be made of data appearing in this publication.

1 From Real to Artificial Ants

I am lost! Where is the line?!
—*A Bug's Life*, Walt Disney, 1998

Ant colonies, and more generally social insect societies, are distributed systems that, in spite of the simplicity of their individuals, present a highly structured social organization. As a result of this organization, ant colonies can accomplish complex tasks that in some cases far exceed the individual capabilities of a single ant.

The field of "ant algorithms" studies models derived from the observation of real ants' behavior, and uses these models as a source of inspiration for the design of novel algorithms for the solution of optimization and distributed control problems.

The main idea is that the self-organizing principles which allow the highly coordinated behavior of real ants can be exploited to coordinate populations of artificial agents that collaborate to solve computational problems. Several different aspects of the behavior of ant colonies have inspired different kinds of ant algorithms. Examples are foraging, division of labor, brood sorting, and cooperative transport. In all these examples, ants coordinate their activities via *stigmergy*, a form of indirect communication mediated by modifications of the environment. For example, a foraging ant deposits a chemical on the ground which increases the probability that other ants will follow the same path. Biologists have shown that many colony-level behaviors observed in social insects can be explained via rather simple models in which only stigmergic communication is present. In other words, biologists have shown that it is often sufficient to consider stigmergic, indirect communication to explain how social insects can achieve self-organization. The idea behind ant algorithms is then to use a form of *artificial stigmergy* to coordinate societies of artificial agents.

One of the most successful examples of ant algorithms is known as "ant colony optimization," or ACO, and is the subject of this book. ACO is inspired by the foraging behavior of ant colonies, and targets discrete optimization problems. This introductory chapter describes how real ants have inspired the definition of artificial ants that can solve discrete optimization problems.

1.1 Ants' Foraging Behavior and Optimization

The visual perceptive faculty of many ant species is only rudimentarily developed and there are ant species that are completely blind. In fact, an important insight of early research on ants' behavior was that most of the communication among individuals, or between individuals and the environment, is based on the use of chemicals produced by the ants. These chemicals are called *pheromones*. This is different from, for example, what happens in humans and in other higher species, whose most important senses are visual or acoustic. Particularly important for the social life of some ant species is the *trail pheromone*. Trail pheromone is a specific type of pheromone that some ant species, such as *Lasius niger* or the Argentine ant *Iridomyrmex humilis* (Goss, Aron, Deneubourg, & Pasteels, 1989), use for marking paths on the ground, for example, paths from food sources to the nest. By sensing pheromone trails foragers can follow the path to food discovered by other ants. This collective trail-laying and trail-following behavior whereby an ant is influenced by a chemical trail left by other ants is the inspiring source of ACO.

1.1.1   Double Bridge Experiments

The foraging behavior of many ant species, as, for example, *I. humilis* (Goss et al., 1989), *Linepithema humile*, and *Lasius niger* (Bonabeau et al., 1997), is based on indirect communication mediated by pheromones. While walking from food sources to the nest and vice versa, ants deposit pheromones on the ground, forming in this way a pheromone trail. Ants can smell the pheromone and they tend to choose, probabilistically, paths marked by strong pheromone concentrations.

The pheromone trail-laying and -following behavior of some ant species has been investigated in controlled experiments by several researchers. One particularly brilliant experiment was designed and run by Deneubourg and colleagues (Deneubourg, Aron, Goss, & Pasteels, 1990; Goss et al., 1989), who used a double bridge connecting a nest of ants of the Argentine ant species *I. humilis* and a food source. They ran experiments varying the ratio $r = l_l / l_s$ between the length of the two branches of the double bridge, where l_l was the length of the longer branch and l_s the length of the shorter one.

In the first experiment the bridge had two branches of equal length ($r = 1$; see figure 1.1a). At the start, ants were left free to move between the nest and the food source and the percentage of ants that chose one or the other of the two branches were observed over time. The outcome was that (see also figure 1.2a), although in the initial phase random choices occurred, eventually all the ants used the same branch. This result can be explained as follows. When a trial starts there is no pheromone on the two branches. Hence, the ants do not have a preference and they select with the same probability any of the branches. Yet, because of random fluctuations, a few more ants will select one branch over the other. Because ants deposit pheromone while walking, a larger number of ants on a branch results in a larger amount of pheromone on that branch; this larger amount of pheromone in turn stimulates more ants to choose that branch again, and so on until finally the ants converge to one

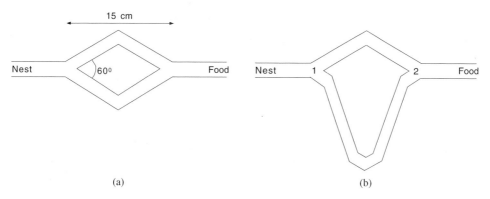

(a)   (b)

Figure 1.1
Experimental setup for the double bridge experiment. (a) Branches have equal length. (b) Branches have different length. Modified from Goss et al. (1989).

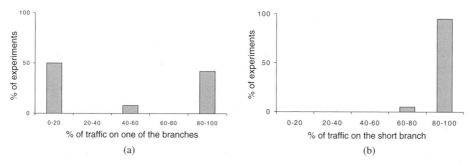

(a)   (b)

Figure 1.2
Results obtained with *Iridomyrmex humilis* ants in the double bridge experiment. (a) Results for the case in which the two branches have the same length ($r = 1$); in this case the ants use one branch or the other in approximately the same number of trials. (b) Results for the case in which one branch is twice as long as the other ($r = 2$); here in all the trials the great majority of ants chose the short branch. Modified from Goss et al. (1989).

single path. This *autocatalytic* or *positive feedback* process is, in fact, an example of a self-organizing behavior of the ants: a macroscopic pattern (corresponding to the convergence toward one branch) emerges out of processes and interactions taking place at a "microscopic" level (Camazine, Deneubourg, Franks, Sneyd, Theraulaz, & Bonabeau, 2001; Haken, 1983; Nicolis & Prigogine, 1977). In our case the convergence of the ants' paths to one branch represents the macroscopic collective behavior, which can be explained by the microscopic activity of the ants, that is, by the local interactions among the individuals of the colony. It is also an example of stigmergic communication (for a definition of *stigmergy*, see section 1.4): ants coordinate their activities, exploiting indirect communication mediated by modifications of the environment in which they move.

In the second experiment, the length ratio between the two branches was set to $r = 2$ (Goss et al., 1989), so that the long branch was twice as long as the short one (figure 1.1b shows the experimental setup). In this case, in most of the trials, after some time all the ants chose to use only the short branch (see figure 1.2b). As in the first experiment, ants leave the nest to explore the environment and arrive at a decision point where they have to choose one of the two branches. Because the two branches initially appear identical to the ants, they choose randomly. Therefore, it can be expected that, on average, half of the ants choose the short branch and the other half the long branch, although stochastic oscillations may occasionally favor one branch over the other. However, this experimental setup presents a remarkable difference with respect to the previous one: because one branch is shorter than the other (see figure 1.1b), the ants choosing the short branch are the first to reach the food and to start their return to the nest. But then, when they must make a decision between the short and the long branch, the higher level of pheromone on the short branch will bias their decision in its favor. Therefore, pheromone starts to accumulate faster on the short branch, which will eventually be used by all the ants because of the autocatalytic process described previously. When compared to the experiment with the two branches of equal length, the influence of initial random fluctuations is much reduced, and stigmergy, autocatalysis, and *differential path length* are the main mechanisms at work. Interestingly, it can be observed that, even when the long branch is twice as long as the short one, not all the ants use the short branch, but a small percentage may take the longer one. This may be interpreted as a type of "path exploration."

It is also interesting to see what happens when the ant colony is offered, after convergence, a new shorter connection between the nest and the food. This situation was studied in an additional experiment in which initially only the long branch was

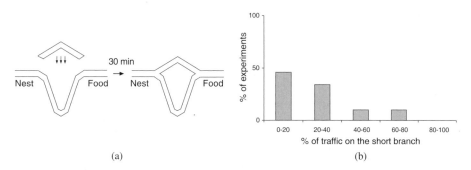

Figure 1.3
In this experiment initially only the long branch was offered to the colony. After 30 minutes, when a stable pheromone trail has formed on the only available branch, a new shorter branch is added. (a) The initial experimental setup and the new situation after 30 minutes, when the short branch was added. (b) In the great majority of the experiments, once the short branch is added the ants continue to use the long branch.

offered to the colony and after 30 minutes the short branch was added (see figure 1.3). In this case, the short branch was only selected sporadically and the colony was trapped on the long branch. This can be explained by the high pheromone concentration on the long branch and by the slow evaporation of pheromone. In fact, the great majority of ants choose the long branch because of its high pheromone concentration, and this autocatalytic behavior continues to reinforce the long branch, even if a shorter one appears. Pheromone evaporation, which could favor exploration of new paths, is too slow: the lifetime of the pheromone is comparable to the duration of a trial (Goss et al., 1989), which means that the pheromone evaporates too slowly to allow the ant colony to "forget" the suboptimal path to which they converged so that the new and shorter one can be discovered and "learned."

1.1.2 A Stochastic Model

Deneubourg and colleagues (Deneubourg et al., 1990; Goss et al., 1989) proposed a simple stochastic model that adequately describes the dynamics of the ant colony as observed in the double bridge experiment. In this model, ψ ants per second cross the bridge in each direction at a constant speed of v cm/s, depositing one unit of pheromone on the branch. Given the lengths l_s and l_l (in cm) of the short and of the long branch, an ant choosing the short branch will traverse it in $t_s = l_s/v$ seconds, while an ant choosing the long branch will use $r \cdot t_s$ seconds, where $r = l_l/l_s$.

The probability $p_{ia}(t)$ that an ant arriving at decision point $i \in \{1, 2\}$ (see figure 1.1b) selects branch $a \in \{s, l\}$, where s and l denote the short and long branch respectively, at instant t is set to be a function of the total amount of pheromone $\varphi_{ia}(t)$

on the branch, which is proportional to the number of ants that used the branch until time t. For example, the probability $p_{is}(t)$ of choosing the short branch is given by

$$p_{is}(t) = \frac{(t_s + \varphi_{is}(t))^\alpha}{(t_s + \varphi_{is}(t))^\alpha + (t_s + \varphi_{il}(t))^\alpha}, \tag{1.1}$$

where the functional form of equation (1.1), as well as the value $\alpha = 2$, was derived from experiments on trail-following (Deneubourg et al., 1990); $p_{il}(t)$ is computed similarly, with $p_{is}(t) + p_{il}(t) = 1$.

This model assumes that the amount of pheromone on a branch is proportional to the number of ants that used the branch in the past. In other words, no pheromone evaporation is considered by the model (this is in accordance with the experimental observation that the time necessary for the ants to converge to the shortest path has the same order of magnitude as the mean lifetime of the pheromone (Goss et al., 1989; Beckers, Deneubourg, & Goss, 1993)). The differential equations that describe the evolution of the stochastic system are

$$d\varphi_{is}/dt = \psi p_{js}(t - t_s) + \psi p_{is}(t), \qquad (i = 1, j = 2; i = 2, j = 1), \tag{1.2}$$

$$d\varphi_{il}/dt = \psi p_{jl}(t - r \cdot t_s) + \psi p_{il}(t), \qquad (i = 1, j = 2; i = 2, j = 1). \tag{1.3}$$

Equation (1.2) can be read as follows: the instantaneous variation, at time t, of pheromone on branch s and at decision point i is given by the ants' flow ψ, assumed constant, multiplied by the probability of choosing the short branch at decision point j at time $t - t_s$ plus the ants' flow multiplied by the probability of choosing the short branch at decision point i at time t. The constant t_s represents a time delay, that is, the time necessary for the ants to traverse the short branch. Equation (1.3) expresses the same for the long branch, except that in this case the time delay is given by $r \cdot t_s$.

The dynamic system defined by these equations was simulated using the Monte Carlo method (Liu, 2001). In figure 1.4, we show the results of two experiments consisting of 1000 simulations each and in which the branch length ratio was set to $r = 1$ and to $r = 2$. It can be observed that when the two branches have the same length ($r = 1$) the ants converge toward the use of one or the other of the branches with equal probability over the 1000 simulations. Conversely, when one branch is twice as long as the other ($r = 2$), then in the great majority of experiments most of the ants choose the short branch (Goss et al., 1989).

In this model the ants deposit pheromone both on their forward and their backward paths. It turns out that this is a necessary behavior to obtain convergence of the ant colony toward the shortest branch. In fact, if we consider a model in which ants deposit pheromone only during the forward or only during the backward trip, then

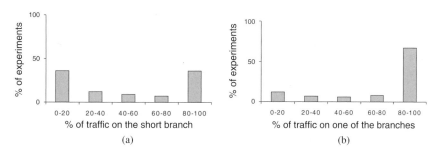

Figure 1.4
Results of 1000 Monte Carlo simulations of the model given by equations (1.1), (1.2), and (1.3), with $\psi = 0.5$ ant per second. Ants were counted between the 501st and 1000th ant crossing the bridge. (a) The ratio between the long and the short branch was set to $r = 1$. (b) The ratio between the long and the short branch was set to $r = 2$. Modified from Goss et al. (1989).

the result is that the ant colony is unable to choose the shortest branch. The observation of real ant colonies has confirmed that ants that deposit pheromone only when returning to the nest are unable to find the shortest path between their nest and the food source (Deneubourg, 2002).

1.2   Toward Artificial Ants

The double bridge experiments show clearly that ant colonies have a built-in optimization capability: by the use of probabilistic rules based on local information they can find the shortest path between two points in their environment. Interestingly, by taking inspiration from the double bridge experiments, it is possible to design artificial ants that, by moving on a graph modeling the double bridge, find the shortest path between the two nodes corresponding to the nest and to the food source.

As a first step toward the definition of artificial ants, consider the graph of figure 1.5a, which is a model of the experimental setup shown in figure 1.1b. The graph consists of two nodes (1 and 2, representing the nest and the food respectively) that are connected by a short and a long arc (in the example the long arc is r times longer than the short arc, where r is an integer number). Additionally, we assume the time to be discrete ($t = 1, 2, \ldots$) and that at each time step each ant moves toward a neighbor node at constant speed of one unit of length per time unit. By doing so, ants add one unit of pheromone to the arcs they use. Ants move on the graph by choosing the path probabilistically: $p_{is}(t)$ is the probability for an ant located in node i at time t to choose the short path, and $p_{il}(t)$ the probability to choose the long path. These probabilities are a function of the pheromone trails φ_{ia} that ants in node i ($i \in \{1, 2\}$)

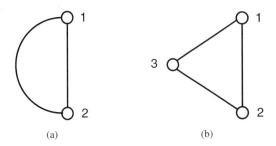

(a)   (b)

Figure 1.5
The graphs are two equivalent models of the experimental setup shown in figure 1.1b. In both cases, ants move from the nest to the food and back either via a short or via a long branch. (a) In this model the long branch is r times longer than the shorter one. An ant entering the long branch updates the pheromone on it r time units later. (b) In this model, each arc of the graph has the same length, and a longer branch is represented by a sequence of arcs. Here, for example, the long branch is twice as long as the short branch. Pheromone updates are done with one time unit delay on each arc.

encounter on the branch a, $(a \in \{s, l\})$:

$$p_{is}(t) = \frac{[\varphi_{is}(t)]^\alpha}{[\varphi_{is}(t)]^\alpha + [\varphi_{il}(t)]^\alpha}, \qquad p_{il}(t) = \frac{[\varphi_{il}(t)]^\alpha}{[\varphi_{is}(t)]^\alpha + [\varphi_{il}(t)]^\alpha}. \tag{1.4}$$

Trail update on the two branches is performed as follows:

$$\varphi_{is}(t) = \varphi_{is}(t-1) + p_{is}(t-1)m_i(t-1) + p_{js}(t-1)m_j(t-1),$$

$$(i=1, j=2; i=2, j=1), \tag{1.5}$$

$$\varphi_{il}(t) = \varphi_{il}(t-1) + p_{il}(t-1)m_i(t-1) + p_{jl}(t-r)m_j(t-r),$$

$$(i=1, j=2; i=2, j=1), \tag{1.6}$$

where $m_i(t)$, the number of ants on node i at time t, is given by

$$m_i(t) = p_{js}(t-1)m_j(t-1) + p_{jl}(t-r)m_j(t-r),$$

$$(i=1, j=2; i=2, j=1). \tag{1.7}$$

This model differs from the one presented in section 1.1.2 in two important aspects:

▪ It considers the average behavior of the system, and not the stochastic behavior of the single ants.

▪ It is a discrete time model, whereas the previous one was a continuous time model; accordingly, it uses difference instead of differential equations.

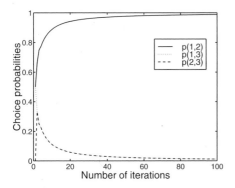

Figure 1.6
Result of the simulation of the model described by equations (1.4) through (1.7). The figure shows the probability of choosing the three branches of the graph in figure 1.5b. After a short transitory period the probabilities of choosing the long branch $((p(1,3) \equiv p_{1l}$ and $(p(2,3) \equiv p_{2l})$ become vanishingly small (in the graph they are superimposed after a few iterations from the start), while the probability of choosing the short branch $(p(1,2) \equiv p_{1s} \equiv p_{2s})$ tends to 1. Note that probabilities are symmetric: $p(i,j) = p(j,i)$. Parameter settings: $\alpha = 2$, $r = 2$, $t = 100$.

Another way of modeling the experimental apparatus of figure 1.1b with a graph is shown in figure 1.5b. In this model each arc of the graph has the same length, and a longer branch is represented by a sequence of arcs. In the figure, for example, the long branch is twice as long as the short branch. Pheromone updates are done with one time unit delay on each arc. The two models are equivalent from a computational point of view, yet the second model permits an easier algorithmic implementation when considering graphs with many nodes.

Simulations run with this discrete time model give results very similar to those obtained with the continuous time model of equations (1.1) to (1.3). For example, by setting the number of ants to twenty, the branch length ratio to $r = 2$, and the parameter α to 2, the system converges rather rapidly toward the use of the short branch (see figure 1.6).

1.3   Artificial Ants and Minimum Cost Paths

In the previous section we have shown that a set of difference equations can reproduce rather accurately the mean behavior of the continuous model of Deneubourg et al. Yet, our goal is to define algorithms that can be used to solve minimum cost problems on more complicated graphs than those representing the double bridge experiment (see, e.g., the graph in figure 1.7).

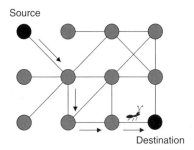

Source

Destination

Figure 1.7
Ants build solutions, that is, paths from a source to a destination node.

With this goal in mind, let us consider a static, connected graph $G = (N, A)$, where N is the set of $n = |N|$ nodes and A is the set of undirected arcs connecting them. The two points between which we want to establish a minimum cost path are called source and destination nodes, as typically done in the literature on minimum cost path problems (when the cost of arcs is given by their length, the minimum cost path problem is the same as the shortest-path problem); sometimes, in analogy to the shortest-path–finding behavior of real ants, we will also call them *nest* and *food source*.

Unfortunately, if we try to solve the minimum cost path problem on the graph G using artificial ants whose behavior is a straightforward extension of the behavior of the ants described in the previous section, the following problem arises: the ants, while building a solution, may generate loops. As a consequence of the forward pheromone trail updating mechanism, loops tend to become more and more attractive and ants can get trapped in them. But even if an ant can escape such loops, the overall pheromone trail distribution becomes such that short paths are no longer favored and the mechanism that in the simpler double bridge situation made the ant choose the shortest path with higher probability does not work anymore. Because this problem is due to forward pheromone trail updating, it might seem that the simplest solution to this problem would be the removal of the forward updating mechanism: in this way ants would rely only on backward updating. Still, this is not a solution: as was said before (see section 1.1.2, but see also exercise 1.1 at the end of this chapter), if the forward update is removed the system does not work anymore, not even in the simple case of the double bridge experiment.

We therefore need to extend the capabilities of the artificial ants in a way that, while retaining the most important characteristics of real ants, allows them to solve minimum cost path problems on generic graphs. In particular, artificial ants are

given a limited form of memory in which they can store the partial paths they have
followed so far, as well as the cost of the links they have traversed. Via the use of
memory, the ants can implement a number of useful behaviors that allow them to
efficiently build solutions to the minimum cost path problem. These behaviors are (1)
probabilistic solution construction biased by pheromone trails, without forward
pheromone updating; (2) deterministic backward path with loop elimination and
with pheromone updating; and (3) evaluation of the quality of the solutions generated and use of the solution quality in determining the quantity of pheromone to
deposit (note that while in the simple case of minimum cost path search an estimate
of the solution quality can be made by the ant also during the solution construction,
this is not necessarily true in other problems, in which there may not exist an easy
way to evaluate partial solutions).

Additionally, we show that by taking into account pheromone evaporation,
which was not necessary to explain real ants' behavior, performance can be greatly
improved.

In the following we briefly explain how the above-mentioned ants' behavior, as
well as pheromone evaporation, is implemented in an algorithm that we call Simple-
ACO (S-ACO for short). It should be noted that, although it represents a significant
step toward the definition of an efficient algorithm for the solution of minimum cost
problems on graphs, S-ACO should be taken for what it is: a didactic tool to explain
the basic mechanisms underlying ACO algorithms.

Probabilistic forward ants and solution construction.   S-ACO ants can be thought of
as having two working modes: *forward* and *backward*. They are in forward mode
when they are moving from the nest toward the food, and they are in backward
mode when they are moving from the food back to their nest. Once an ant in forward
mode reaches its destination, it switches to backward mode and starts its travel back
to the source. In S-ACO, forward ants build a solution by choosing probabilistically
the next node to move to among those in the neighborhood of the graph node on
which they are located. (Given a graph $G = (N, A)$, two nodes $i, j \in N$ are neighbors
if there exists an arc $(i, j) \in A$.) The probabilistic choice is biased by pheromone trails
previously deposited on the graph by other ants. Forward ants do not deposit any
pheromone while moving. This, together with deterministic backward moves, helps
in avoiding the formation of loops.

Deterministic backward ants and pheromone update.   The use of an explicit memory
allows an ant to retrace the path it has followed while searching the destination
node. Moreover, S-ACO ants improve the system performance by implementing
loop elimination. In practice, before starting to move backward on the path they

memorized while searching the destination node (i.e., the forward path), S-ACO ants eliminate any loops from it. While moving backward, S-ACO ants leave pheromone on the arcs they traverse.

Pheromone updates based on solution quality. In S-ACO the ants memorize the nodes they visited during the forward path, as well as the cost of the arcs traversed if the graph is weighted. They can therefore evaluate the cost of the solutions they generate and use this evaluation to modulate the amount of pheromone they deposit while in backward mode. Making pheromone update a function of the generated solution quality can help in directing future ants more strongly toward better solutions. In fact, by letting ants deposit a higher amount of pheromone on short paths, the ants' path searching is more quickly biased toward the best solutions. Interestingly, the dependence of the amount of pheromone trail deposit on the solution quality is also present in some ant species: Beckers et al. (1993) found that in the ant species *Lasius niger* the ants returning from rich food sources tend to drop more pheromone than those returning from poorer food sources.

Pheromone evaporation. In real ant colonies, pheromone intensity decreases over time because of evaporation. In S-ACO evaporation is simulated by applying an appropriately defined pheromone evaporation rule. For example, artificial pheromone decay can be set to a constant rate. Pheromone evaporation reduces the influence of the pheromones deposited in the early stages of the search, when artificial ants can build poor-quality solutions. Although in the experiments run by Deneubourg and colleagues (Deneubourg et al., 1990; Goss et al., 1989) pheromone evaporation did not play any noticeable role, it can be very useful for artificial ant colonies, as we will show in the following sections.

1.3.1   S-ACO

We now present the details of the S-ACO algorithm which adapts the real ants' behavior to the solution of minimum cost path problems on graphs. To each arc (i, j) of the graph $G = (N, A)$ we associate a variable τ_{ij} called *artificial pheromone trail*, shortened to pheromone trail in the following. Pheromone trails are read and written by the ants. The amount (intensity) of a pheromone trail is proportional to the utility, as estimated by the ants, of using that arc to build good solutions.

Ants' Path-Searching Behavior
Each ant builds, starting from the source node, a solution to the problem by applying a step-by-step decision policy. At each node, local information stored on the node itself or on its outgoing arcs is read (sensed) by the ant and used in a stochastic way to decide which node to move to next. At the beginning of the search process, a

constant amount of pheromone (e.g., $\tau_{ij} = 1$, $\forall (i,j) \in A$) is assigned to all the arcs. When located at a node i an ant k uses the pheromone trails τ_{ij} to compute the probability of choosing j as next node:

$$
p_{ij}^k = \begin{cases} \dfrac{\tau_{ij}^\alpha}{\sum_{l \in \mathcal{N}_i^k} \tau_{il}^\alpha}, & \text{if } j \in \mathcal{N}_i^k; \\[2ex] 0, & \text{if } j \notin \mathcal{N}_i^k; \end{cases} \tag{1.8}
$$

where $\mathcal{N}_i^k$ is the neighborhood of ant k when in node i. In S-ACO the neighborhood of a node i contains all the nodes directly connected to node i in the graph $G = (N, A)$, except for the predecessor of node i (i.e., the last node the ant visited before moving to i). In this way the ants avoid returning to the same node they visited immediately before node i. Only in case $\mathcal{N}_i^k$ is empty, which corresponds to a dead end in the graph, node i's predecessor is included into $\mathcal{N}_i^k$. Note that this decision policy can easily lead to loops in the generated paths (recall the graph of figure 1.7).

An ant repeatedly hops from node to node using this decision policy until it eventually reaches the destination node. Due to differences among the ants' paths, the time step at which ants reach the destination node may differ from ant to ant (ants traveling on shorter paths will reach their destinations faster).

Path Retracing and Pheromone Update
When reaching the destination node, the ant switches from the forward mode to the backward mode and then retraces step by step the same path backward to the source node. An additional feature is that, before starting the return trip, an ant eliminates the loops it has built while searching for its destination node. The problem of loops is that they may receive pheromone several times when an ant retraces its path backward to deposit pheromone trail, leading to the problem of self-reinforcing loops. Loop elimination can be done by iteratively scanning the node identifiers position by position starting from the source node: for the node at the i-th position, the path is scanned starting from the destination node until the first occurrence of the node is encountered, say, at position j (it always holds that $i \leq j$ because the scanning process stops at position i at the latest). If we have $j > i$, the subpath from position $i + 1$ to position j corresponds to a loop and can be eliminated. The scanning process is visualized in figure 1.8. The example also shows that our loop elimination procedure does not necessarily eliminate the largest loop. In the example, the loop 3-4-5-3 of length 3 is eliminated. Yet, the longest loop in this example, the loop 5-3-2-8-5 of length 4, is not eliminated because it is no longer present after eliminating the first loop. In general, if the path contains nested loops, the final loop-free path will

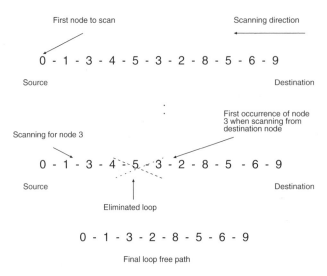

Figure 1.8
Illustration of the scanning process for loop elimination.

depend on the sequence in which the loops are eliminated. In S-ACO, loop elimination is implemented so that loops are eliminated in the same order as they are created.

During its return travel to the source the k-th ant deposits an amount $\Delta\tau^k$ of pheromone on arcs it has visited. In particular, if ant k is in the backward mode and it traverses the arc (i, j), it changes the pheromone value τ_{ij} as follows:

$$\tau_{ij} \leftarrow \tau_{ij} + \Delta\tau^k. \tag{1.9}$$

By this rule an ant using the arc connecting node i to node j increases the probability that forthcoming ants will use the same arc in the future.

An important aspect is the choice of $\Delta\tau^k$. In the simplest case, this can be the same constant value for all the ants. In this case, only the *differential path length* works in favor of the detection of short paths: ants which have detected a shorter path can deposit pheromone earlier than ants traveling on a longer path. In addition to the deterministic backward pheromone trail update, the ants may also deposit an amount of pheromone trail which is a function of the path length—the shorter the path the more pheromone is deposited by an ant. Generally, we require the amount of pheromone deposited by an ant to be a nonincreasing function of the path length.

Pheromone Trail Evaporation
Pheromone trail evaporation can be seen as an exploration mechanism that avoids quick convergence of all the ants toward a suboptimal path. In fact, the decrease in

pheromone intensity favors the exploration of different paths during the whole search process. In real ant colonies, pheromone trails also evaporate, but, as we have seen, evaporation does not play an important role in real ants' shortest-path finding. The fact that, on the contrary, pheromone evaporation seems to be important in artificial ants is probably due to the fact that the optimization problems tackled by artificial ants are much more complex than those real ants can solve. A mechanism like evaporation that, by favoring the forgetting of errors or of poor choices done in the past, allows a continuous improvement of the "learned" problem structure seems therefore to be necessary for artificial ants. Additionally, artificial pheromone evaporation also plays the important function of bounding the maximum value achievable by pheromone trails.

Evaporation decreases the pheromone trails with exponential speed. In S-ACO, the pheromone evaporation is interleaved with the pheromone deposit of the ants. After each ant k has moved to a next node according to the ants' search behavior described earlier, pheromone trails are evaporated by applying the following equation to all the arcs:

$$\tau_{ij} \leftarrow (1 - \rho)\tau_{ij}, \quad \forall(i, j) \in A, \tag{1.10}$$

where $\rho \in (0, 1]$ is a parameter. After pheromone evaporation has been applied to all arcs, the amount of pheromone $\Delta\tau^k$ is added to the arcs. We call an iteration of S-ACO a complete cycle involving ants' movement, pheromone evaporation, and pheromone deposit.

1.3.2   Experiments with S-ACO

We have run experiments to evaluate the importance of some aspects of S-ACO: evaporation, number of ants, and type of pheromone update (function of the solution quality or not).

In the experiments presented in the following the behavior of S-ACO is judged with respect to convergence toward the minimum cost (shortest) path, in a way similar to what was done for the outcome of the simulation experiments of Deneubourg et al. and for the experiments with the discrete model introduced in section 1.2. Informally, by convergence we mean that, as the algorithm runs for an increasing number of iterations, the ants' probability of following the arcs of a particular path increases—in the limit to a point where the probability of selecting the arcs of this path becomes arbitrarily close to 1 while for all the others, it becomes arbitrarily close to 0.

The experiments have been run using two simple graphs: the double bridge of figure 1.5b and the more complex graph called *extended double bridge* given in figure

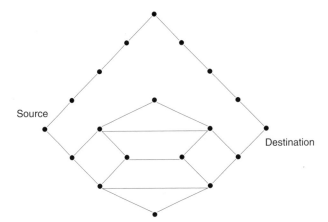

Figure 1.9
Extended double bridge. An ant starting in the source node can choose between the upper and the lower parts of the graph. The upper part consists of a single path of length 8 leading directly to the destination node. Differently, the lower part of the graph consists of a set of paths (which includes many paths shorter than eight steps) and the ant has many decisions to do before reaching the destination node. Therefore, ants choosing the upper part will always find a path of length 8, while ants choosing the lower part may find paths shorter than 8, but they may also enter loops and generate very long paths.

1.9. This second graph is designed in such a way that converging to the minimum cost path is not a trivial task for S-ACO. The difficulty of the graph is given by the fact that, in order to find the minimum cost path, an ant has to make a number of "correct" choices and if some of these choices are wrong, the ant generates suboptimal paths. To understand why, consider the graph of figure 1.9: ants exiting the source node have to choose between the loop-free, but worse than optimal, upper path of the graph, and the set of paths in the lower part of the same graph that contains two optimal paths of length 5, as well as many longer loop-free paths and infinitely many, much longer "loopy" paths. There is a trade-off between converging toward the use of an "easy" but suboptimal path, and searching for the optimal path in a region of the search space where suboptimal paths can easily be generated. In other words, to obtain convergence to the optimal solutions the ants need to choose the lower part of the graph, but then the greater number of decisions to be taken makes converging to the minimum cost path a difficult task.

Note that the choice of judging the algorithm using convergence as defined above instead of using more standard performance indices, such as the time or the number of iterations necessary to find the optimal solution, is consistent with our goals, that is, studying and understanding the relationship between design choices and the algorithm's behavior. In fact, such a study requires working on simple graphs like those discussed above so that simulation times remain reasonably short and the behavior of ants can be easily observed. But in simple graphs the shortest path is always found very quickly because of the large number of ants compared to the relatively small search space. Therefore, a performance index based on the time (or number of iterations) necessary to find the optimal solution would not be very meaningful. In fact, convergence as defined above, by requiring that all the ants do use the same path, is a more reasonable index for our purposes.

On the contrary, as we will see in the forthcoming chapters, when attacking more complex problems like $\mathcal{NP}$-hard optimization problems or routing in dynamic networks, the way experimental results are judged is different. In $\mathcal{NP}$-hard optimization problems the main goal is to find quickly very high-quality solutions and therefore we are interested mainly in the solution quality of the best solution(s) found by the ACO algorithm. In dynamic networks routing the algorithm has to be able to react rapidly to changing conditions and to maintain exploration capabilities so that it can effectively evaluate alternative paths which, due to the dynamics of the problem, may become more desirable; in both cases we will need a different definition of algorithm convergence.

Number of Ants and Types of Pheromone Update: Experiments with the Double Bridge
We ran a first set of experiments in which we studied the influence that the number of ants used and the way the amount of pheromone to be deposited is determined by ants have on the behavior of S-ACO. The experiments were run using the double bridge (see figure 1.5b). The choice of the double bridge was due to the desire of comparing the results obtained with S-ACO to those obtained with the model of real ants' behavior described in section 1.2. Note that a major difference between that model and S-ACO is that equations (1.4) through (1.7) describe the average behavior of the system, whereas in S-ACO a fixed number of ants move autonomously on the graph. Intuitively, an increasing number of ants in S-ACO should approximate better and better the average behavior given by equations (1.4) through (1.7).

In the following we report the results of two experiments:

1. Run S-ACO with different values for the number m of ants and with ants depositing a constant amount of pheromone on the visited arcs [i.e., $\Delta\tau^k = constant$ in equation (1.9)].

2. Same as in 1. above, except that the ants deposit an amount of pheromone which is inversely proportional to the length of the path they have found (i.e., $\Delta\tau^k = 1/L^k$, where L^k is the length of ant k's path).

Table 1.1
Percentage of trials in which S-ACO converged to the long path (100 independent trials for varying values of m, with $\alpha = 2$ and $\rho = 0$)

m	1	2	4	8	16	32	64	128	256	512
without path length	50	42	26	29	24	18	3	2	1	0
with path length	18	14	8	0	0	0	0	0	0	0

Column headings give the number m of ants in the colony. The first row shows results obtained performing pheromone updates without considering path length; the second row reports results obtained performing pheromone updates proportional to path length.

For each experiment we ran 100 trials and each trial was stopped after each ant had moved 1000 steps. Evaporation [see equation (1.10)] was set to $\rho = 0$, and the parameter α [see equation (1.8)] was set to 2, as in equation (1.1) of Deneubourg et al. approximating real ants' behavior. At the end of the trial we checked whether the pheromone trail was higher on the short or on the long path. In table 1.1, which gives the results of the two experiments, we report the percentage of trials in which the pheromone trail was higher on the long path. We found that, for the given parameter settings, S-ACO showed convergence behavior after 1000 ant steps so that the reported percentage is significant for understanding the algorithm behavior.

Let us focus first on the results of experiment 1. For a small number of ants (up to 32), S-ACO converged relatively often to the longer path. This is certainly due to fluctuations in the path choice in the initial iterations of the algorithm which can lead to a strong reinforcement of the long path. Yet, with an increasing number of ants, the number of times we observed this behavior decreased drastically, and for a large number of ants (here 512) we never observed convergence to the long path in any of the 100 trials. The experiments also indicate that, as could be expected, S-ACO performs poorly when only one ant is used: the number of ants has to be significantly larger than one to obtain convergence to the short path.

The results obtained in experiment 2 with pheromone updates based on solution quality are much better. As can be observed in table 1.1, S-ACO converged to the long path far less frequently than when pheromone updates were independent of the solution quality. With only one ant, S-ACO converged to the long path in only 18 out of 100 trials, which is significantly less than in experiment 1, and with eight ants or more it always converged to the short path.

In additional experiments, we examined the influence of the parameter α of equation (1.8) on the convergence behavior of S-ACO, in particular investigating the cases where α was changed in step sizes of 0.25 from 1 to 2. Again, the behavior was dependent on whether pheromone updates based on solution quality were used or not. In the first case we found that increasing α had a negative effect on the convergence behavior, while in the second case the results were rather independent of the particular value of α. In general, we found that, for a fixed number of ants, the algorithm tended to converge to the shortest path more often when α was close to 1. This is intuitively clear because large values of α tend to amplify the influence of initial random fluctuations. If, by chance, the long path is initially selected by the majority of ants, then the search of the whole colony is quickly biased toward it. This happens to a lower extent when the value of α is close to 1.

These results show that, as in the case of real ants, in S-ACO both *autocatalysis* and *differential path length* are at work to favor the emergence of short paths. While the results with S-ACO indicate that differential path length alone can be enough to let S-ACO converge to the optimal solution on small graphs, they also show that relying on this effect as the main driving force of the algorithm comes at the price of having to use large colony sizes, which results in long simulation times. In addition, the effectiveness of the differential path length effect strongly decreases with increasing problem complexity. This is what is shown by the experiments reported in the next subsection.

Pheromone Evaporation: Experiments with the Extended Double Bridge
In a second set of experiments, we studied the influence that pheromone trail evaporation has on the convergence behavior of S-ACO. Experiments were run using the extended double bridge graph (see figure 1.9).

In these experiments the ants deposit an amount of pheromone that is the inverse of their path length (i.e., $\Delta\tau^k = 1/L^k$); also, before depositing it, they eliminate loops using the procedure described in figure 1.8.

To evaluate the behavior of the algorithm we observe the development of the path lengths found by the ants. In particular, we plot the moving averages of the path lengths after loop elimination (moving averages are calculated using the $4 \cdot m$ most recent paths found by the ants, where m is the number of ants). In other words, in the graph of figure 1.10 a point is plotted each time an ant has completed a journey from the source to the destination and back (the number of journeys is on the x-axis), and the corresponding value on the y-axis is given by the length of the above-mentioned moving average after loop elimination.

We ran experiments with S-ACO and different settings for the evaporation rate $\rho \in \{0, 0.01, 0.1\}$ ($\alpha = 1$ and $m = 128$ in all experiments). If $\rho = 0$, no pheromone evaporation takes place. Note that an evaporation rate of $\rho = 0.1$ is rather large, because evaporation takes place at each iteration of the S-ACO algorithm: after ten iterations, which corresponds to the smallest number of steps that an ant needs to

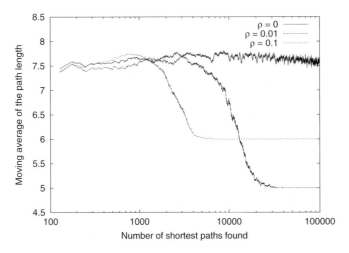

Figure 1.10
The graph plots the moving averages (given on the *y*-axis) of the ants' path length for the graph of figure 1.9 as a function of the number of completed paths (given on the *x*-axis). We give plots for not using evaporation ($\rho = 0$), low evaporation ($\rho = 0.01$), and high evaporation ($\rho = 0.1$). The trials were stopped after 5000 iterations; $\alpha = 1$ and $m = 128$.

build the shortest path and to come back to the source, roughly 65% of the pheromone on each arc evaporates, while with $\rho = 0.01$ this evaporation is reduced to around 10%.

Figure 1.10 gives the observed moving averages. Although the graphs only show results of a single run of the algorithm, they are representative of the typical algorithm behavior. If no evaporation is used, the algorithm does not converge, which can be seen by the fact that the moving average has approximately the value 7.5, which does not correspond to the length of any path (with these parameter settings, this result typically does not change if the run lasts a much higher number of iterations). With pheromone evaporation, the behavior of S-ACO is significantly different. After a short transitory phase, S-ACO converges to a single path: either the shortest one (the moving average takes the value 5 for $\rho = 0.01$) or the path of length 6 for $\rho = 0.1$. A closer examination of the results revealed that in both cases at convergence all the ants had built loop-free paths of the indicated length.

In further experiments with S-ACO on this graph we made the following general observations:

▪ Without pheromone updates based on solution quality, S-ACO performance is much worse. In particular, the algorithm converges very often to the suboptimal solution of length 8; the larger the parameters α or ρ, the faster S-ACO converges to this suboptimal solution.

- The pheromone evaporation rate ρ can be critical. In particular, we observed that S-ACO often converged to suboptimal paths when evaporation was set to a value that was too high. For example, in fifteen trials with ρ set to 0.2, S-ACO converged once to a path of length 8, once to a path of length 7, and twice to a path of length 6. Setting ρ to 0.01 S-ACO converged to the shortest path in all trials.

- Large values of α generally result in a worse behavior of S-ACO because they excessively emphasize the initial random fluctuations.

Discussion

We have seen that in real ant colonies the emergence of high-level patterns like shortest paths is only possible through the interaction of a large number of individuals. It is interesting that experimental results show that the same is true to a large extent for S-ACO: the use of a colony of ants is important to exploit the differential path length effect and to increase the robustness of the algorithm and reduce its dependence on parameter settings. As we have seen, a colony size larger than one is necessary to solve even simple problems like the double bridge.

In general, we noticed that as problems become more complex, the parameter settings of S-ACO become increasingly important to obtain convergence to the optimal solution. In particular, the experimental results presented above support the following conclusions: (1) the differential path length effect, although important, is not enough to allow the effective solution of large optimization problems; (2) pheromone updates based on solution quality are important for fast convergence; (3) large values for parameter α lead to a strong emphasis of initial, random fluctuations and to bad algorithm behavior; (4) the larger the number of ants, the better the convergence behavior of the algorithm, although this comes at the cost of longer simulation times; and (5) pheromone evaporation is important when trying to solve more complex problems. These observations will be of importance in the following chapters, where design decisions will be made both to define the ACO metaheuristic and to apply it to a multitude of different discrete optimization problems.

1.4 Bibliographical Remarks

The term *stigmergy* was introduced by Grassé to describe a form of indirect communication mediated by modifications of the environment that he observed in the workers caste of two species of termites, *Bellicositermes natalensis* and *Cubitermes* sp.

The original definition of stigmergy (see Grassé, 1959, p. 79), was: "Stimulation of workers by the performance they have achieved."

Termite nest building is the typical example of stigmergy, and is also the original example used by Grassé to introduce the concept. Termite workers build their nest using soil pellets, which they impregnate with a diffusing chemical substance called pheromone. They start nest construction (Grassé, 1959) by randomly depositing pellets on the ground. The deposits of soil pellets stimulate workers to accumulate more material on top of them through a positive feedback mechanism, since the accumulation of material reinforces the attraction of deposits by means of the diffusing pheromone emitted by the pellets (Bruinsma, 1979). This process works only if the density of the termites is above a given threshold. In fact, if the density is too low, pheromones are not added quickly enough and the positive feedback mechanism is inhibited by pheromone evaporation.

Although Grassé introduced the term stigmergy to explain the behavior of termite societies, the same term has later been used to indicate indirect communication mediated by modifications of the environment that can be observed also in other social insects. As we have seen, the foraging behavior of ant colonies described in this chapter is an example of stigmergy: ants stimulate other ants by modifying the environment via pheromone trail updating. A brief history of the notion of stigmergy can be found in Theraulaz & Bonabeau (1999).

1.5   Things to Remember

- Deneubourg and colleagues (Deneubourg et al., 1990; Goss et al., 1989) have shown in controlled experimental conditions that foraging ants can find the shortest path between their nest and a food source by marking the path they follow with a chemical called pheromone.

- The foraging behavior of ant colonies can be replicated in simulation and inspires a class of ant algorithms known as "ant colony optimization" (ACO). ACO, the subject of this book, is currently one of the most successful examples of ant algorithms.

- In experiments with foraging ants, it was shown that the pheromone evaporation rate is so slow compared to the time necessary for the ant colony to converge to the short path that, for modeling purposes, it can be neglected. When considering artificial ants things are different. Experimental results show that on very simple graphs, like the ones modeling the double bridge or the extended double bridge setups, pheromone evaporation is also not necessary. On the contrary, it improves the algorithm's performance in finding good solutions to the minimum cost path problem on more complex graphs.

■ Biologists have found that stigmergy is a useful concept to help explain the self-organizing capabilities of social insects (Theraulaz & Bonabeau, 1999; Dorigo, Bonabeau, & Theraulaz, 2000a).

1.6   Thought and Computer Exercises

Exercise 1.1   Prove by hand calculation that artificial ants using only forward (or only backward) pheromone update do not converge toward the common use of the minimum cost path in the double bridge experiment.

Exercise 1.2   Prove by hand calculation that, if artificial ants are given the capability (through the use of memory) to retrace their path to the destination node (recall section 1.3), then they are able to find the minimum cost path in the double bridge experiment even when they use only backward pheromone update.

Exercise 1.3   Implement a computer program that simulates the artificial ants in the double bridge experiment. You can do this in two ways: either by numerically solving equations (1.4) through (1.7), in this way obtaining the expected behavior of the system, or by running simulations. Is there any difference in the results? What happens if you only use a few ants in the simulation?

Exercise 1.4   Using the program above, study what happens when you change the α and r parameters. In particular, if you set $\alpha = 1$, does the probability of choosing the short branch still converge to 1? And how do the convergence properties of the algorithm change when increasing the branch length ratio r?

Exercise 1.5   An alternative model of the double bridge experiment to the one presented in section 1.2 [equations (1.4)–(1.7)] is the following. Let the amount of pheromone on a branch be proportional to the number of ants that used the branch in the past and let $m_s(t)$ and $m_l(t)$ be the numbers of ants that have used the short and the long branches after a total of t ants have crossed the bridge, with $m_s(t) + m_l(t) = t$. The probability $p_s(t)$ with which the $(t+1)$-th ant chooses the short branch can then be written as

$$p_s(t) = \frac{m_s(t)^{\alpha}}{m_s(t)^{\alpha} + m_l(t)^{\alpha}} = 1 - p_l(t). \tag{1.11}$$

The number of ants choosing the short branch is given by

$$m_s(t+1) = \begin{cases} m_s(t) + 1, & \text{if } q \le p_s(t); \\ m_s(t), & \text{otherwise;} \end{cases} \tag{1.12}$$

and the number of ants choosing the long branch by

$$m_l(t+1) = \begin{cases} m_l(t) + 1, & \text{if } q > p_l(t); \\ m_l(t), & \text{otherwise}; \end{cases} \quad (1.13)$$

where q is a uniform random number drawn from the interval $[0, 1]$.

Run Monte Carlo simulations of the dynamic system defined by the above equations and compare the results with those obtained in the first and second computer exercise.

Exercise 1.6   The ants' path-marking and foraging behavior can also be studied in unconstrained settings. Consider the following experimental setup: a squared environment contains three food sources and one nest. Ants leave the nest to search for food and, once food has been found, they go back to the nest depositing a pheromone trail on the ground. When they are looking for food, ants move stochastically using a probabilistic rule biased by pheromones they sense in the environment (see also Resnick, 1994). Implement a program which simulates the system described above and study how the ants' performance changes for different implementation choices. For example, you can study different forms of the probabilistic rules used by the ants, different ways of depositing pheromone on the ground (only while searching for food, only when going back to the nest, in both cases), different pheromone evaporation rates, and so on. (*Hint*: You may want to use Mitchel Resnick's Star-Logo programming language, available at education.mit.edu/starlogo/).

Exercise 1.7   Develop an outline for the implementation of S-ACO (section 1.3.1). Consider the following issues:

- How do you build a structure which represents the individual ants?
- How do you represent the graph, the pheromone trails, and the heuristic information?
- How do you implement the solution construction policy?
- How do you implement loop elimination?
- How do you implement pheromone update?

Once you have implemented the algorithm, run it on a number of graphs. What are your experiences with the algorithm? How do you judge the quality and the convergence of the algorithm? Would you use this algorithm for attacking large minimum cost path problems? (Consider that there exist algorithms, such as the one proposed by Dijkstra [1959], that solve the minimum cost (shortest) path problem in $\mathcal{O}(n^2)$).

2 The Ant Colony Optimization Metaheuristic

A metaheuristic refers to a master strategy that guides and modifies other heuristics to produce solutions beyond those that are normally generated in a quest for local optimality.
—*Tabu Search*, Fred Glover and Manuel Laguna, 1998

Combinatorial optimization problems are intriguing because they are often easy to state but very difficult to solve. Many of the problems arising in applications are $\mathcal{N P}$-hard, that is, it is strongly believed that they cannot be solved to optimality within polynomially bounded computation time. Hence, to practically solve large instances one often has to use approximate methods which return near-optimal solutions in a relatively short time. Algorithms of this type are loosely called *heuristics*. They often use some problem-specific knowledge to either build or improve solutions.

Recently, many researchers have focused their attention on a new class of algorithms, called metaheuristics. A *metaheuristic* is a set of algorithmic concepts that can be used to define heuristic methods applicable to a wide set of different problems. The use of metaheuristics has significantly increased the ability of finding very high-quality solutions to hard, practically relevant combinatorial optimization problems in a reasonable time.

A particularly successful metaheuristic is inspired by the behavior of real ants. Starting with Ant System, a number of algorithmic approaches based on the very same ideas were developed and applied with considerable success to a variety of combinatorial optimization problems from academic as well as from real-world applications. In this chapter we introduce ant colony optimization, a metaheuristic framework which covers the algorithmic approach mentioned above. The ACO metaheuristic has been proposed as a common framework for the existing applications and algorithmic variants of a variety of ant algorithms. Algorithms that fit into the ACO metaheuristic framework will be called in the following ACO algorithms.

2.1 Combinatorial Optimization

Combinatorial optimization problems involve finding values for discrete variables such that the optimal solution with respect to a given objective function is found. Many optimization problems of practical and theoretical importance are of combinatorial nature. Examples are the shortest-path problems described in the previous chapter, as well as many other important real-world problems like finding a minimum cost plan to deliver goods to customers, an optimal assignment of employees to tasks to be performed, a best routing scheme for data packets in the Internet, an optimal sequence of jobs which are to be processed in a production line, an allocation of flight crews to airplanes, and many more.

A combinatorial optimization problem is either a *maximization* or a *minimization* problem which has associated a set of problem instances. The term *problem* refers to the general question to be answered, usually having several parameters or variables with unspecified values. The term *instance* refers to a problem with specified values for all the parameters. For example, the traveling salesman problem (TSP), defined in section 2.3.1, is the general problem of finding a minimum cost Hamiltonian circuit in a weighted graph, while a particular TSP instance has a specified number of nodes and specified arc weights.

More formally, an instance of a combinatorial optimization problem Π is a triple $(\mathcal{S}, f, \Omega)$, where $\mathcal{S}$ is the *set of candidate solutions*, f is the *objective function* which assigns an objective function value $f(s)$ to each candidate solution $s \in \mathcal{S}$, and Ω is a set of constraints. The solutions belonging to the set $\tilde{\mathcal{S}} \subseteq \mathcal{S}$ of candidate solutions that satisfy the constraints Ω are called *feasible solutions*. The goal is to find a *globally optimal* feasible solution s^*. For minimization problems this consists in finding a solution $s^* \in \tilde{\mathcal{S}}$ with minimum cost, that is, a solution such that $f(s^*) \leq f(s)$ for all $s \in \tilde{\mathcal{S}}$; for maximization problems one searches for a solution with maximum objective value, that is, a solution with $f(s^*) \geq f(s)$ for all $s \in \tilde{\mathcal{S}}$. Note that in the following we focus on minimization problems and that the obvious adaptations have to be made if one considers maximization problems.

It should be noted that an instance of a combinatorial optimization problem is typically not specified explicitly by enumerating all the candidate solutions (i.e., the set $\mathcal{S}$) and the corresponding cost values, but is rather represented in a more concise mathematical form (e.g., shortest-path problems are typically defined by a weighted graph).

2.1.1   Computational Complexity

A straightforward approach to the solution of combinatorial optimization problems would be exhaustive search, that is, the enumeration of all possible solutions and the choice of the best one. Unfortunately, in most cases, such a naive approach becomes rapidly infeasible because the number of possible solutions grows exponentially with the instance size n, where the instance size can be given, for example, by the number of binary digits necessary to encode the instance. For some combinatorial optimization problems, deep insight into the problem structure and the exploitation of problem-specific characteristics allow the definition of algorithms that find an optimal solution much quicker than exhaustive search does. In other cases, even the best algorithms of this kind cannot do much better than exhaustive search.

Box 2.1
Worst-Case Time Complexity and Intractability

The *time complexity function* of an algorithm for a given problem Π indicates, for each possible input size n, the maximum time the algorithm needs to find a solution to an instance of that size. This is often called *worst-case time complexity*.

The worst-case time complexity of an algorithm is often formalized using the $\mathcal{O}(\cdot)$ notation. Let $g(n)$ and $h(n)$ be functions from the positive integers to the positive reals. A function $g(n)$ is said to be $\mathcal{O}(h(n))$ if two positive constants *const* and n_0 exist such that $g(n) \leq const \cdot h(n)$ for all $n \geq n_0$. In other words, the $\mathcal{O}(\cdot)$ notation gives asymptotic upper bounds on the worst-case time complexity of an algorithm.

An algorithm is said to be a *polynomial time algorithm* if its time complexity function is $\mathcal{O}(g(n))$ for some polynomial function $g(\cdot)$. If an algorithm has a time complexity function that cannot be bounded by a polynomial, it is called an *exponential time algorithm*. Note that this includes also functions such as $n^{\log n}$, which are sometimes referred to as subexponential; in any case, subexponential functions grow faster than any polynomial. A problem is said to be *intractable* if there is no polynomial time algorithm capable of solving it.

When attacking a combinatorial optimization problem it is useful to know how difficult it is to find an optimal solution. A way of measuring this difficulty is given by the notion of worst-case complexity. Worst-case complexity can be explained as follows (see also box 2.1): a combinatorial optimization problem Π is said to have worst-case complexity $\mathcal{O}(g(n))$ if the best algorithm known for solving Π finds an optimal solution to any instance of Π having size n in a computation time bounded from above by $const \cdot g(n)$.

In particular, we say that Π is solvable in polynomial time if the maximum amount of computing time necessary to solve any instance of size n of Π is bounded from above by a polynomial in n. If k is the largest exponent of such a polynomial, then the combinatorial optimization problem is said to be solvable in $\mathcal{O}(n^k)$ time.

Although some important combinatorial optimization problems have been shown to be solvable in polynomial time, for the great majority of combinatorial problems no polynomial bound on the worst-case solution time could be found so far. For these problems the run time of the best algorithms known increases exponentially with the instance size and, consequently, so does the time required to find an optimal solution. A notorious example of such a problem is the TSP.

An important theory that characterizes the difficulty of combinatorial problems is that of $\mathcal{NP}$-completeness. This theory classifies combinatorial problems in two main classes: those that are known to be solvable in polynomial time, and those that are not. The first are said to be *tractable*, the latter *intractable*.

Combinatorial optimization problems as defined above correspond to what are usually called *search problems*. Each combinatorial optimization problem Π has an associated *decision problem* defined as follows: given Π, that is, the triple $(\mathcal{S}, f, \Omega)$, and a parameter ϱ, does a feasible solution $s \in \tilde{\mathcal{S}}$ exist such that $f(s) \leq \varrho$, in case Π was a minimization problem? It is clear that solving the search version of a combinatorial problem implies being able to give the solution of the corresponding decision problem, while the opposite is not true in general. This means that Π is at least as hard to solve as the decision version of Π and proving that the decision version is intractable implies intractability of the original search problem.

The theory of $\mathcal{NP}$-completeness distinguishes between two classes of problems of particular interest: the class $\mathcal{P}$ for which an algorithm outputs in polynomial time the correct answer ("yes" or "no"), and the class $\mathcal{NP}$ for which an algorithm exists that verifies for every instance, independently of the way it was generated, in polynomial time whether the answer "yes" is correct. (Note that formally, the complexity classes $\mathcal{P}$ and $\mathcal{NP}$ are defined via idealized models of computation: in the theory of $\mathcal{NP}$-completeness, typically Turing machines are used. For details, see Garey & Johnson (1979).) It is clear that $\mathcal{P} \subseteq \mathcal{NP}$, while nothing can be said on the question whether $\mathcal{P} = \mathcal{NP}$ or not. Still, an answer to this question would be very useful because proving $\mathcal{P} = \mathcal{NP}$ implies proving that all problems in $\mathcal{NP}$ can be solved in polynomial time.

On this subject, a particularly important role is played by *polynomial time reductions*. Intuitively, a polynomial time reduction is a procedure that transforms a problem into another one by a polynomial time algorithm. The interesting point is that if problem Π_A can be solved in polynomial time and problem Π_B can be transformed into Π_A via a polynomial time reduction, then also the solution to Π_B can be found in polynomial time. We say that a problem is $\mathcal{NP}$-hard, if every other problem in $\mathcal{NP}$ can be transformed to it by a polynomial time reduction. Therefore, an $\mathcal{NP}$-hard problem is at least as hard as any of the other problems in $\mathcal{NP}$. However, $\mathcal{NP}$-hard problems do not necessarily belong to $\mathcal{NP}$. An $\mathcal{NP}$-hard problem that is in $\mathcal{NP}$ is said to be $\mathcal{NP}$-complete. Therefore, the $\mathcal{NP}$-complete problems are the hardest problems in $\mathcal{NP}$: if a polynomial time algorithm could be found for an $\mathcal{NP}$-complete problem, then all problems in the $\mathcal{NP}$-complete class (and consequently all the problems in $\mathcal{NP}$) could be solved in polynomial time. Because after many years of research efforts no such algorithm has been found, most scientists tend to accept the conjecture $\mathcal{P} \neq \mathcal{NP}$. Still, the "$\mathcal{P} = \mathcal{NP}$?" question remains one of the most important open questions in theoretical computer science.

Until today, a large number of problems have been proved to be $\mathcal{NP}$-complete, including the above-mentioned TSP; see Garey & Johnson (1979) for a long list of such problems.

2.1.2 Solution Methods for $\mathcal{NP}$-Hard Problems

Two classes of algorithms are available for the solution of combinatorial optimization problems: *exact* and *approximate algorithms*.

Exact algorithms are guaranteed to find the optimal solution and to prove its optimality for every finite size instance of a combinatorial optimization problem within an instance-dependent run time. In the case of $\mathcal{NP}$-hard problems, exact algorithms need, in the worst case, exponential time to find the optimum. Although for some specific problems exact algorithms have been improved significantly in recent years, obtaining at times impressive results (Applegate, Bixby, Chvátal, & Cook, 1995, 1998), for most $\mathcal{NP}$-hard problems the performance of exact algorithms is not satisfactory. So, for example, for the quadratic assignment problem (QAP) (to be discussed in chapter 5), an important problem that arises in real-world applications and whose goal is to find the optimal assignment of n items to n locations, most instances of dimension around 30 are currently the limit of what can be solved with state-of-the-art exact algorithms (Anstreicher, Brixius, Goux, & Linderoth, 2002; Hahn, Hightower, Johnson, Guignard-Spielberg, & Roucairol, 2001; Hahn & Krarup, 2001). For example, at the time of writing, the largest, nontrivial QAP instance from QAPLIB, a benchmark library for the QAP, solved to optimality has 36 locations (Brixius & Anstreicher, 2001; Nyström, 1999). Despite the small size of the instance, the computation time required to solve it is extremely high. For example, the solution of instance `ste36a` from a backboard wiring application (Steinberg, 1961) took approximately 180 hours of CPU time on a 800 MHz Pentium III PC. This is to be compared to the currently best-performing ACO algorithms (see section 5.2.1, for how to apply ACO to the QAP), which typically require an average time of about 10 seconds to find the optimal solution for this instance on a comparable machine. In addition to the exponential worst-case complexity, the application of exact algorithms to $\mathcal{NP}$-hard problems in practice also suffers from a strong rise in computation time when the problem size increases, and often their use quickly becomes infeasible.

If optimal solutions cannot be efficiently obtained in practice, the only possibility is to trade optimality for efficiency. In other words, the guarantee of finding optimal solutions can be sacrificed for the sake of getting very good solutions in polynomial time. Approximate algorithms, often also loosely called *heuristic methods* or simply *heuristics*, seek to obtain good, that is, near-optimal solutions at relatively low computational cost without being able to guarantee the optimality of solutions. Based on the underlying techniques that approximate algorithm use, they can be classified as being either *constructive* or *local search* methods (approximate methods may also be

Box 2.2
Constructive Algorithms

Constructive algorithms build a solution to a combinatorial optimization problem in an incremental way. Step by step and without backtracking, they add solution components until a complete solution is generated. Although the order in which to add components can be random, typically some kind of heuristic rule is employed. Often, greedy construction heuristics are used which at each construction step add a solution component with maximum myopic benefit as estimated by a heuristic function. An algorithmic outline of a greedy construction heuristic is given below.

procedure GreedyConstructionHeuristic
    $s_p \leftarrow$ ChooseFirstComponent
    while $(s_p$ is not a complete solution$)$ **do**
        $c \leftarrow$ GreedyComponent(s_p)
        $s_p \leftarrow s_p \otimes c$
    end-while
    $s \leftarrow s_p$
    return s
end-procedure

Here, the function ChooseFirstComponent chooses the first solution component (this is done at random or according to a greedy choice depending on the particular construction heuristic) and GreedyComponent returns a solution component c with best heuristic estimate. The addition of component c to a partial solution s_p is denoted by the operator $\otimes$. The procedure returns a complete solution s.

An example of a constructive algorithm for the TSP is the nearest-neighbor procedure, which treats the cities as components. The procedure works by randomly choosing an initial city and by iteratively adding the closest among the remaining cities to the solution under construction (ties are broken randomly).

In the example tour below the nearest-neighbor procedure starts from city a and sequentially adds cities b, c, d, e, and f.

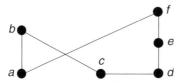

Box 2.3
Local Search

Local search is a general approach for finding high-quality solutions to hard combinatorial optimization problems in reasonable time. It is based on the iterative exploration of neighborhoods of solutions trying to improve the current solution by local changes. The types of local changes that may be applied to a solution are defined by a neighborhood structure.

Definition 2.1   *A* neighborhood structure *is a function* $\mathcal{N} : \mathcal{S} \mapsto 2^{\mathcal{S}}$ *that assigns a set of neighbors* $\mathcal{N}(s) \subseteq \mathcal{S}$ *to every* $s \in \mathcal{S}$. $\mathcal{N}(s)$ *is also called the neighborhood of s.*

The choice of an appropriate neighborhood structure is crucial for the performance of a local search algorithm and is problem-specific. The neighborhood structure defines the set of solutions that can be reached from s in one single step of a local search algorithm. Typically, a neighborhood structure is defined implicitly by defining the possible local changes that may be applied to a solution, and not by explicitly enumerating the set of all possible neighbors.

The solution found by a local search algorithm may only be guaranteed to be optimal with respect to local changes and, in general, will not be a globally optimal solution.

Definition 2.2   *A* local optimum *for a minimization problem* (a *local minimum*) *is a solution s such that* $\forall s' \in \mathcal{N}(s) : f(s) \leq f(s')$. *Similarly, a* local optimum *for a maximization problem* (a *local maximum*) *is a solution s such that* $\forall s' \in \mathcal{N}(s) : f(s) \geq f(s')$.

A local search algorithm also requires the definition of a neighborhood examination scheme that determines how the neighborhood is searched and which neighbor solutions are accepted. While the neighborhood can be searched in many different ways, in the great majority of cases the acceptance rule is either the *best-improvement* rule, which chooses the neighbor solution giving the largest improvement of the objective function, or the *first-improvement* rule, which accepts the first improved solution found.

obtained by stopping exact methods before completion (Bellman, Esogbue, & Nabeshima, 1982; Jünger, Reinelt, & Thienel, 1994), for example, after some given time bound; yet, here, this type of approximate algorithm will not be discussed further). Usually, if for an approximate algorithm it can be proved that it returns solutions that are worse than the optimal solution by at most some fixed value or factor, such an algorithm is also called an *approximation algorithm* (Hochbaum, 1997; Hromkovic, 2003; Vazirani, 2001).

Constructive algorithms (see box 2.2) generate solutions from scratch by iteratively adding solution components to an initially empty solution until the solution is complete. For example, in the TSP a solution is built by adding city after city in an incremental way. Although constructive algorithms are typically the fastest among the approximate methods, the quality of the solutions they generate is most of the time inferior to the quality of the solutions found by local search algorithms.

Local search starts from some initial solution and repeatedly tries to improve the current solution by local changes. The first step in applying local search is the definition of a *neighborhood structure* (see box 2.3) over the set of candidate solutions. In

Box 2.4
k–Exchange Neighborhoods

An important class of neighborhood structures for combinatorial optimization problems is that of *k–exchange neighborhoods.*

Definition 2.3   *The* k–exchange neighborhood *of a candidate solution s is the set of candidate solutions s' that can be obtained from s by exchanging k solution components.*

Example 2.1: The 2–exchange and k–exchange neighborhoods in the TSP   Given a candidate solution *s*, the TSP 2–exchange neighborhood of a candidate solution *s* consists of the set of all the candidate solutions *s'* that can be obtained from *s* by exchanging two pairs of arcs in all the possible ways. The figure below gives an example of one specific 2–exchange: the pair of arcs (b, c) and (a, f) is removed and replaced by the pair (a, c) and (b, f).

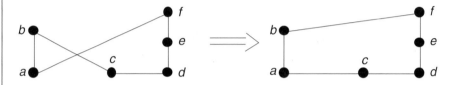

The k–exchange neighborhood is the obvious generalization in which a set of *k* arcs is replaced by a different set of *k* arcs.

practice, the neighborhood structure defines for each current solution the set of possible solutions to which the local search algorithms can move. One common way of defining neighborhoods is via k-exchange moves that exchange a set of *k* components of a solution with a different set of *k* components (see box 2.4).

In its most basic version, often called *iterative improvement*, or sometimes *hill-climbing* or *gradient-descent* for maximization and minimization problems, respectively, the local search algorithm searches for an improved solution within the neighborhood of the current solution. If an improving solution is found, it replaces the current solution and the local search is continued. These steps are repeated until no improving solution is found in the neighborhood and the algorithm terminates in a *local optimum*. A disadvantage of iterative improvement is that the algorithm may stop at very poor-quality local optima.

2.1.3   What Is a Metaheuristic?

A disadvantage of single-run algorithms like constructive methods or iterative improvement is that they either generate only a very limited number of different solutions, which is the case for greedy construction heuristics, or they stop at poor-quality local optima, which is the case for iterative improvement methods. Unfortunately, the obvious extension of local search, that is, to restart the algorithm several times from new starting solutions, does not produce significant improvements in practice (Johnson & McGeoch, 1997; Schreiber & Martin, 1999). Several general approaches, which are nowadays often called metaheuristics, have been proposed which try to bypass these problems.

A *metaheuristic* is a set of algorithmic concepts that can be used to define heuristic methods applicable to a wide set of different problems. In other words, a metaheuristic can be seen as a general-purpose heuristic method designed to guide an underlying problem-specific heuristic (e.g., a local search algorithm or a construction heuristic) toward promising regions of the search space containing high-quality solutions. A metaheuristic is therefore a general algorithmic framework which can be applied to different optimization problems with relatively few modifications to make them adapted to a specific problem.

Examples of metaheuristics include simulated annealing (Cerný, 1985; Kirkpatrick, Gelatt, & Vecchi, 1983), tabu search (Glover, 1989, 1990; Glover & Laguna, 1997), iterated local search (Lourenço, Martin, & Stützle, 2002), evolutionary computation (Fogel, Owens, & Walsh, 1966; Holland, 1975; Rechenberg, 1973; Schwefel, 1981; Goldberg, 1989), and ant colony optimization (Dorigo & Di Caro, 1999b; Dorigo, Di Caro, & Gambardella, 1999; Dorigo, Maniezzo, & Colorni, 1996; Dorigo & Stützle, 2002) (see Glover & Kochenberger [2002] for a comprehensive overview).

The use of metaheuristics has significantly increased the ability of finding very high-quality solutions to hard, practically relevant combinatorial optimization problems in a reasonable time. This is particularly true for large and poorly understood problems. A detailed description of the ant colony optimization metaheuristic is given in the next section; the other metaheuristics mentioned above are briefly described in section 2.4.

2.2   The ACO Metaheuristic

Ant colony optimization is a metaheuristic in which a colony of artificial ants cooperate in finding good solutions to difficult discrete optimization problems. Cooperation is a key design component of ACO algorithms: The choice is to allocate the computational resources to a set of relatively simple agents (artificial ants) that communicate indirectly by stigmergy, that is, by indirect communication mediated by the environment (see chapter 1, section 1.4). Good solutions are an emergent property of the agents' cooperative interaction.

ACO algorithms can be used to solve both static and dynamic combinatorial optimization problems. Static problems are those in which the characteristics of the problem are given once and for all when the problem is defined, and do not change while the problem is being solved. A paradigmatic example of such problems is the TSP (Johnson & McGeoch, 1997; Lawler, Lenstra, Rinnooy Kan, & Shmoys, 1985; Reinelt, 1994), in which city locations and their relative distances are part of the problem definition and do not change at run time. On the contrary, dynamic problems are defined as a function of some quantities whose value is set by the dynamics of an underlying system. The problem instance changes therefore at run time and the optimization algorithm must be capable of adapting online to the changing environment. An example of this situation, which we discuss at length in chapter 6, are network routing problems in which the data traffic and the network topology can vary in time.

In this section we give a formal characterization of the class of problems to which the ACO metaheuristic can be applied, of the behavior governing the artificial ants, and of the general structure of the ACO metaheuristic.

2.2.1   Problem Representation

An artificial ant in ACO is a stochastic constructive procedure that incrementally builds a solution by adding opportunely defined solution components to a partial solution under construction. Therefore, the ACO metaheuristic can be applied to any combinatorial optimization problem for which a constructive heuristic can be defined.

Although this means that the ACO metaheuristic can be applied to any interesting combinatorial optimization problems, the real issue is how to map the considered problem to a representation that can be used by the artificial ants to build solutions. In the following we give a formal characterization of the representation that the artificial ants use and of the policy they implement.

Let us consider the minimization problem $(\mathcal{S}, f, \Omega)$, where $\mathcal{S}$ is the *set of candidate solutions*, f is the *objective function* which assigns an objective function (cost) value $f(s, t)$ to each candidate solution $s \in \mathcal{S}$, and $\Omega(t)$ is a *set of constraints*. The parameter t indicates that the objective function and the constraints can be time-dependent, as is the case, for example, in applications to dynamic problems (e.g., in telecommunication network routing problems the cost of links is proportional to traffic, which is time-dependent; and constraints on the reachable nodes can also change with time: think of a network node that suddenly becomes unreachable).

The goal is to find a *globally optimal* feasible solution s^*, that is, a minimum cost feasible solution to the minimization problem.

The combinatorial optimization problem $(\mathcal{S}, f, \Omega)$ is mapped on a problem that can be characterized by the following list of items:

- A finite set $C = \{c_1, c_2, \ldots, c_{N_C}\}$ of *components* is given, where N_C is the number of components.

- The *states* of the problem are defined in terms of sequences $x = \langle c_i, c_j, \ldots, c_h, \ldots \rangle$ of finite length over the elements of C. The set of all possible states is denoted by $\mathcal{X}$. The length of a sequence x, that is, the number of components in the sequence, is expressed by $|x|$. The maximum length of a sequence is bounded by a positive constant $n < +\infty$.

- The set of (candidate) solutions $\mathcal{S}$ is a subset of $\mathcal{X}$ (i.e., $\mathcal{S} \subseteq \mathcal{X}$).

- A set of feasible states $\tilde{\mathcal{X}}$, with $\tilde{\mathcal{X}} \subseteq \mathcal{X}$, defined via a problem-dependent test that verifies that it is not impossible to complete a sequence $x \in \tilde{\mathcal{X}}$ into a solution satisfying the constraints Ω. Note that by this definition, the feasibility of a state $x \in \tilde{\mathcal{X}}$ should be interpreted in a *weak* sense. In fact, it does not guarantee that a completion s of x exists such that $s \in \tilde{\mathcal{X}}$.

- A non-empty set $\mathcal{S}^*$ of optimal solutions, with $\mathcal{S}^* \subseteq \tilde{\mathcal{X}}$ and $\mathcal{S}^* \subseteq \mathcal{S}$.

- A *cost* $g(s, t)$ is associated with each candidate solution $s \in \mathcal{S}$. In most cases $g(s, t) \equiv f(s, t)$, $\forall s \in \tilde{\mathcal{S}}$, where $\tilde{\mathcal{S}} \subseteq \mathcal{S}$ is the set of feasible candidate solutions, obtained from $\mathcal{S}$ via the constraints $\Omega(t)$.

- In some cases a cost, or the estimate of a cost, $J(x, t)$ can be associated with states other than candidate solutions. If x_j can be obtained by adding solution components to a state x_i, then $J(x_i, t) \leq J(x_j, t)$. Note that $J(s, t) \equiv g(s, t)$.

Given this formulation, artificial ants build solutions by performing randomized walks on the completely connected graph $G_C = (C, L)$ whose nodes are the components C, and the set L fully connects the components C. The graph G_C is called *construction graph* and elements of L are called *connections*.

The problem constraints $\Omega(t)$ are implemented in the policy followed by the artificial ants, as explained in the next section. The choice of implementing the constraints in the construction policy of the artificial ants allows a certain degree of flexibility. In fact, depending on the combinatorial optimization problem considered, it may be more reasonable to implement the constraints in a hard way, allowing the ants to build only feasible solutions, or in a soft way, in which case the ants can build infeasible solutions (i.e., candidate solutions in $\mathcal{S} \backslash \tilde{\mathcal{S}}$) that can be penalized as a function of their degree of infeasibility.

2.2.2   Ants' Behavior

As we said, in ACO algorithms artificial ants are stochastic constructive procedures that build solutions by moving on the construction graph $G_C = (C, L)$, where the set

L fully connects the components C. The problem constraints $\Omega(t)$ are built into the ants' constructive heuristic. In most applications, ants construct feasible solutions. However, sometimes it may be necessary or beneficial to also let them construct infeasible solutions. Components $c_i \in C$ and connections $l_{ij} \in L$ can have associated a *pheromone trail* τ (τ_i if associated with components, τ_{ij} if associated with connections), and a *heuristic value* η (η_i and η_{ij}, respectively). The pheromone trail encodes a long-term memory about the entire ant search process, and is updated by the ants themselves. Differently, the heuristic value, often called *heuristic information*, represents a priori information about the problem instance or run-time information provided by a source different from the ants. In many cases η is the cost, or an estimate of the cost, of adding the component or connection to the solution under construction. These values are used by the ants' heuristic rule to make probabilistic decisions on how to move on the graph.

More precisely, each ant k of the colony has the following properties:

- It exploits the construction graph $G_C = (C, L)$ to search for optimal solutions $s^* \in \mathcal{S}^*$.

- It has a memory $\mathcal{M}^k$ that it can use to store information about the path it followed so far. Memory can be used to (1) build feasible solutions (i.e., implement constraints Ω); (2) compute the heuristic values η; (3) evaluate the solution found; and (4) retrace the path backward.

- It has a *start state* x_s^k and one or more *termination conditions* e^k. Usually, the start state is expressed either as an empty sequence or as a unit length sequence, that is, a single component sequence.

- When in state $x_r = \langle x_{r-1}, i \rangle$, if no termination condition is satisfied, it moves to a node j in its *neighborhood* $\mathcal{N}^k(x_r)$, that is, to a state $\langle x_r, j \rangle \in \mathcal{X}$. If at least one of the termination conditions e^k is satisfied, then the ant stops. When an ant builds a candidate solution, moves to infeasible states are forbidden in most applications, either through the use of the ant's memory, or via appropriately defined heuristic values η.

- It selects a move by applying a probabilistic decision rule. The probabilistic decision rule is a function of (1) the locally available pheromone trails and heuristic values (i.e., pheromone trails and heuristic values associated with components and connections in the neighborhood of the ant's current location on graph G_C); (2) the ant's private memory storing its current state; and (3) the problem constraints.

- When adding a component c_j to the current state, it can update the pheromone trail τ associated with it or with the corresponding connection.

- Once it has built a solution, it can retrace the same path backward and update the pheromone trails of the used components.

It is important to note that ants act concurrently and independently and that although each ant is complex enough to find a (probably poor) solution to the problem under consideration, good-quality solutions can only emerge as the result of the collective interaction among the ants. This is obtained via indirect communication mediated by the information ants read or write in the variables storing pheromone trail values. In a way, this is a distributed learning process in which the single agents, the ants, are not adaptive themselves but, on the contrary, adaptively modify the way the problem is represented and perceived by other ants.

2.2.3   The Metaheuristic

Informally, an ACO algorithm can be imagined as the interplay of three procedures: ConstructAntsSolutions, UpdatePheromones, and DaemonActions.

ConstructAntsSolutions manages a colony of ants that concurrently and asynchronously visit adjacent states of the considered problem by moving through neighbor nodes of the problem's construction graph G_C. They move by applying a stochastic local decision policy that makes use of pheromone trails and heuristic information. In this way, ants incrementally build solutions to the optimization problem. Once an ant has built a solution, or while the solution is being built, the ant evaluates the (partial) solution that will be used by the UpdatePheromones procedure to decide how much pheromone to deposit.

UpdatePheromones is the process by which the pheromone trails are modified. The trails value can either increase, as ants deposit pheromone on the components or connections they use, or decrease, due to pheromone evaporation (see also section 1.3 of chapter 1). From a practical point of view, the deposit of new pheromone increases the probability that those components/connections that were either used by many ants or that were used by at least one ant and which produced a very good solution will be used again by future ants. Differently, pheromone evaporation implements a useful form of *forgetting*: it avoids a too rapid convergence of the algorithm toward a suboptimal region, therefore favoring the exploration of new areas of the search space.

Finally, the DaemonActions procedure is used to implement centralized actions which cannot be performed by single ants. Examples of daemon actions are the activation of a local optimization procedure, or the collection of global information that can be used to decide whether it is useful or not to deposit additional pheromone to bias the search process from a nonlocal perspective. As a practical example, the daemon can observe the path found by each ant in the colony and select one or a few ants (e.g., those that built the best solutions in the algorithm iteration) which are then allowed to deposit additional pheromone on the components/connections they used.

procedure ACOMetaheuristic
  ScheduleActivities
    ConstructAntsSolutions
    UpdatePheromones
    DaemonActions   % optional
  end-ScheduleActivities
end-procedure

Figure 2.1
The ACO metaheuristic in pseudo-code. The procedure DaemonActions is optional and refers to centralized actions executed by a daemon possessing global knowledge.

In figure 2.1, the ACO metaheuristic is described in pseudo-code. The main procedure of the ACO metaheuristic manages the scheduling of the three above-discussed components of ACO algorithms via the ScheduleActivities construct: (1) management of the ants' activity, (2) pheromone updating, and (3) daemon actions. The ScheduleActivities construct does not specify how these three activities are scheduled and synchronized. In other words, it does not say whether they should be executed in a completely parallel and independent way, or if some kind of synchronization among them is necessary. The designer is therefore free to specify the way these three procedures should interact, taking into account the characteristics of the considered problem.

Nowadays numerous successful implementations of the ACO metaheuristic are available and they have been applied to many different combinatorial optimization problems. These applications are summarized in table 2.1 and they are discussed in the forthcoming chapters of this book.

2.3   How Do I Apply ACO?

Probably, the best way of illustrating how the ACO metaheuristic operates is by describing how it has been applied to combinatorial optimization problems. This is done with a full and detailed description of most of the current applications of ACO in chapter 5. Here we limit ourselves to a brief description of the main points to consider when applying ACO algorithms to a few examples of problems representative of important classes of optimization problems.

First, we illustrate the application to permutation problems in their unconstrained and constrained forms: the TSP and the sequential ordering problem. Then we consider generalized assignment as an example of assignment problems, and multiple

Table 2.1
Current applications of ACO algorithms listed according to problem types and chronologically

Problem type	Problem name	Main references
Routing	Traveling salesman	Dorigo, Maniezzo, & Colorni (1991a,b, 1996) Dorigo (1992) Gambardella & Dorigo (1995) Dorigo & Gambardella (1997a,b) Stützle & Hoos (1997, 2000) Bullnheimer, Hartl, & Strauss (1999c) Cordón, de Viana, Herrera, & Morena (2000)
	Vehicle routing	Bullnheimer, Hartl, & Strauss (1999a,b) Gambardella, Taillard, & Agazzi (1999) Reimann, Stummer, & Doerner (2002)
	Sequential ordering	Gambardella & Dorigo (1997, 2000)
Assignment	Quadratic assignment	Maniezzo, Colorni, & Dorigo (1994) Stützle (1997b) Maniezzo & Colorni (1999) Maniezzo (1999) Stützle & Hoos (2000)
	Graph coloring	Costa & Hertz (1997)
	Generalized assignment	Lourenço & Serra (1998, 2002)
	Frequency assignment	Maniezzo & Carbonaro (2000)
	University course timetabling	Socha, Knowles, & Sampels (2002) Socha, Sampels, & Manfrin (2003)
Scheduling	Job shop	Colorni, Dorigo, Maniezzo, & Trubian (1994)
	Open shop	Pfahringer (1996)
	Flow shop	Stützle (1998a)
	Total tardiness	Bauer, Bullnheimer, Hartl, & Strauss (2000)
	Total weighted tardiness	den Besten, Stützle, & Dorigo (2000) Merkle & Middendorf (2000, 2003a) Gagné, Price, & Gravel (2002)
	Project scheduling	Merkle, Middendorf, & Schmeck (2000a, 2002)
	Group shop	Blum (2002a, 2003a)
Subset	Multiple knapsack	Leguizamón & Michalewicz (1999)
	Max independent set	Leguizamón & Michalewicz (2000)
	Redundancy allocation	Liang & Smith (1999)
	Set covering	Leguizamón & Michalewicz (2000) Hadji, Rahoual, Talbi, & Bachelet (2000)
	Weight constrained graph tree partition	Cordone & Maffioli (2001)
	Arc-weighted l-cardinality tree	Blum & Blesa (2003)
	Maximum clique	Fenet & Solnon (2003)

Table 2.1
(continued)

Problem type	Problem name	Main references
Other	Shortest common supersequence	Michel & Middendorf (1998, 1999)
	Constraint satisfaction	Solnon (2000, 2002)
	2D-HP protein folding	Shmygelska, Aguirre-Hernández, & Hoos (2002)
	Bin packing	Levine & Ducatelle (2003)
Machine learning	Classification rules	Parpinelli, Lopes, & Freitas (2002b)
	Bayesian networks	de Campos, Gámez, & Puerta (2002b)
	Fuzzy systems	Casillas, Cordón, & Herrera (2000)
Network routing	Connection-oriented network routing	Schoonderwoerd, Holland, Bruten, & Rothkrantz (1996) Schoonderwoerd, Holland, & Bruten (1997) White, Pagurek, & Oppacher (1998) Di Caro & Dorigo (1998d) Bonabeau, Henavy, Guérin, Snyers, Kuntz, & Theraulaz (1998)
	Connectionless network routing	Di Caro & Dorigo (1997, 1998c,f) Subramanian, Druschel, & Chen (1997) Heusse, Snyers, Guérin, & Kuntz (1998) van der Put (1998)
	Optical network routing	Navarro Varela, & Sinclair (1999)

knapsack as an example of subset problems. Finally, applications to two dynamic problems, network routing and dynamic TSP, are briefly discussed.

2.3.1   The Traveling Salesman Problem

Intuitively, the traveling salesman problem is the problem faced by a salesman who, starting from his home town, wants to find a shortest possible trip through a given set of customer cities, visiting each city once before finally returning home. The TSP can be represented by a complete weighted graph $G = (N, A)$ with N being the set of $n = |N|$ nodes (cities), A being the set of arcs fully connecting the nodes. Each arc $(i, j) \in A$ is assigned a weight d_{ij} which represents the distance between cities i and j. The TSP is the problem of finding a minimum length Hamiltonian circuit of the graph, where a Hamiltonian circuit is a closed walk (a tour) visiting each node of G exactly once. We may distinguish between symmetric TSPs, where the distances between the cities are independent of the direction of traversing the arcs, that is, $d_{ij} = d_{ji}$ for every pair of nodes, and the asymmetric TSP (ATSP), where at least for one pair of nodes i, j we have $d_{ij} \neq d_{ji}$.

A solution to an instance of the TSP can be represented as a permutation of the city indices; this permutation is cyclic, that is, the absolute position of a city in a tour is not important at all but only the relative order is important (in other words, there are n permutations that map to the same solution).

Construction graph.   The construction graph is identical to the problem graph: the set of components C corresponds to the set of nodes (i.e., $C = N$), the connections correspond to the set of arcs (i.e., $L = A$), and each connection has a weight which corresponds to the distance d_{ij} between nodes i and j. The states of the problem are the set of all possible partial tours.

Constraints.   The only constraint in the TSP is that all cities have to be visited and that each city is visited at most once. This constraint is enforced if an ant at each construction step chooses the next city only among those it has not visited yet (i.e., the feasible neighborhood $\mathcal{N}_i^k$ of an ant k in city i, where k is the ant's identifier, comprises all cities that are still unvisited).

Pheromone trails and heuristic information.   The pheromone trails τ_{ij} in the TSP refer to the desirability of visiting city j directly after i. The heuristic information η_{ij} is typically inversely proportional to the distance between cities i and j, a straightforward choice being $\eta_{ij} = 1/d_{ij}$. In fact, this is also the heuristic information used in most ACO algorithms for the TSP.

Solution construction.   Each ant is initially put on a randomly chosen start city and at each step iteratively adds one still unvisited city to its partial tour. The solution construction terminates once all cities have been visited.

General comments.   The TSP is a paradigmatic $\mathcal{NP}$-hard combinatorial optimization problem which has attracted a very significant amount of research (Johnson & McGeoch, 1997; Lawler et al., 1985; Reinelt, 1994). The TSP has played a central role in ACO, because it was the application problem chosen when proposing the first ACO algorithm called Ant System (Dorigo, 1992; Dorigo, Maniezzo, & Colorni, 1991b, 1996) and it was used as a test problem for almost all ACO algorithms proposed later. Chapter 3 gives a detailed presentation of the ACO algorithms available for the TSP.

2.3.2   The Sequential Ordering Problem

The sequential ordering problem (SOP) consists in finding a minimum weight Hamiltonian path on a directed graph with weights on the arcs and the nodes, subject to precedence constraints between nodes. It is easy to remove weights from nodes and to add them to the arcs, obtaining a kind of asymmetric traveling salesman problem in which, once all the nodes have been visited, the path is not closed (i.e., it does not become a tour as in the ATSP).

Construction graph.   Similar to the TSP, the set of components C contains all the nodes. Solutions are permutations of the elements of C, and costs (lengths) are associated with connections between nodes.

Constraints.   The only significant difference between the applications of ACO to the SOP and to the TSP is the set of constraints: while building solutions, ants choose components only among those that have not yet been used and, if possible, satisfy all precedence constraints.

Pheromone trails and heuristic information.   As in the TSP case, pheromone trails are associated with connections, and the heuristic information can, for example, be chosen as the inverse of the costs (lengths) of the connections.

Solution construction.   Ants build solutions iteratively by adding, step by step, new unvisited nodes to the partial solution under construction. They choose the new node to add by using pheromone trails, heuristic, and constraint information.

2.3.3   The Generalized Assignment Problem

In the generalized assignment problem (GAP) a set of tasks $i \in I$, has to be assigned to a set of agents $j \in J$. Each agent j has only a limited capacity a_j and each task i assigned to agent j consumes a quantity r_{ij} of the agent's capacity. Also, the cost d_{ij} of assigning task i to agent j is given. The objective then is to find a feasible task assignment with minimum cost.

Let y_{ij} be 1 if task i is assigned to agent j and 0 otherwise. Then the GAP can formally be defined as

$$\min f(y) = \sum_{j=1}^{m} \sum_{i=1}^{n} d_{ij} y_{ij} \tag{2.1}$$

subject to

$$\sum_{i=1}^{n} r_{ij} y_{ij} \leq a_j, \quad j = 1, \ldots, m, \tag{2.2}$$

$$\sum_{j=1}^{m} y_{ij} = 1, \quad i = 1, \ldots, n, \tag{2.3}$$

$$y_{ij} \in \{0, 1\}, \quad i = 1, \ldots, n, \quad j = 1, \ldots, m. \tag{2.4}$$

The constraints in equation (2.2) implement the limited resource capacity of the agents, while the constraints given by equations (2.3) and (2.4) impose that each task is assigned to exactly one agent and that a task cannot be split among several agents.

Construction graph.   The GAP can easily be cast into the framework of the ACO metaheuristic. For example, the problem could be represented on the construction graph $G_C = (C, L)$ in which the set of components comprises the set of tasks and agents, that is, $C = I \cup J$. Each assignment, which consists of n couplings (i, j) of tasks and agents, corresponds to at least one ant's walk on this graph and costs d_{ij} are associated with all possible couplings (i, j) of tasks and agents.

Constraints.   Walks on the construction graph G_C have to satisfy the constraints given by equations (2.3) and (2.4) to obtain a valid assignment. One particular way of generating such an assignment is by an ant's walk which iteratively switches from task nodes (nodes in the set I) to agent nodes (nodes in the set J) without repeating any task node but possibly using an agent node several times (several tasks may be assigned to an agent). Moreover, the GAP involves resource capacity constraints that can be enforced by an appropriately defined neighborhood. For example, for an ant k, $\mathcal{N}_i^k$ could be defined as consisting of all those agents to which task i can be assigned without violating the agents' resource capacity. If no agent meets the task's resource requirement, then the ant is forced to build an infeasible solution; in this case $\mathcal{N}_i^k$ becomes the set of all agents. Infeasibilities can then be handled, for example, by assigning penalties proportional to the amount of resource violations.

Pheromone trails and heuristic information.   During the construction of a solution, ants repeatedly have to take the following two basic decisions: (1) choose the task to assign next and (2) choose the agent the task should be assigned to. Pheromone trail information can be associated with any of the two decisions: it can be used to learn an appropriate order for task assignments or it can be associated with the desirability of assigning a task to a specific agent. In the first case, τ_{ij} represents the desirability of assigning task j directly after task i, while in the second case it represents the desirability of assigning agent j to task i.

Similarly, heuristic information can be associated with any of the two decisions. For example, heuristic information could bias task assignment toward those tasks that use more resources, and bias the choice of agents in such a way that small assignment costs are incurred and the agent only needs a relatively small amount of its available resource to perform the task.

Solution construction.   Solution construction can be performed as usual, by choosing the components to add to the partial solution from among those that, as explained above, satisfy the constraints with a probability biased by the pheromone trails and heuristic information.

2.3.4   The Multiple Knapsack Problem

Given a set of items $i \in I$ with associated a vector of resource requirements r_i and a profit b_i, the knapsack problem (KP) is the problem of selecting a subset of items from I in such a way that they fit into a knapsack of limited capacity and maximize the sum of profits of the chosen items. The multiple knapsack problem (MKP), also known as multidimensional KP, extends the single KP by considering multiple resource constraints. Let y_i be a variable associated with item i, which has value 1 if i is added to the knapsack, and 0 otherwise. Also, let r_{ij} be the resource requirement of item i with respect to resource constraint j, a_j the capacity of resource j, and m be the number of resource constraints. Then the MKP can be formulated as

$$\max f(y) = \sum_{i=1}^{n} b_i y_i, \tag{2.5}$$

subject to

$$\sum_{i=1}^{n} r_{ij} y_i \leq a_j, \quad j = 1, \ldots, m, \tag{2.6}$$

$$y_i \in \{0, 1\}, \quad i = 1, \ldots, n. \tag{2.7}$$

In the MKP, it is typically assumed that all profits b_i and all weights r_{ij} take positive values.

Construction graph.   In the construction graph $G_C = (C, L)$, the set of components C corresponds to the set of items and, as usual, the set of connections L fully connects the set of items. The profit of adding items can be associated with either the connections or the components.

Constraints.   The solution construction has to consider the resource constraints given by equation (2.6). During the solution construction process, this can be easily done by allowing ants to add only those components that, when added to their current partial solution, do not violate any resource constraint.

Pheromone trails and heuristic information.   The MKP has the particularity that pheromone trails τ_i are associated only with components and refer to the desirability of adding an item i to the current partial solution. The heuristic information, intuitively, should prefer items which have a high profit and low resource requirements. One possible choice for the heuristic information is to calculate the average resource requirement $\bar{r}_i = \sum_{j=1}^{m} r_{ij}/m$ for each item and then to define $\eta_i = b_i/\bar{r}_i$. Yet this choice has the disadvantage that it does not take into account how tight the single resource constraints are. Therefore, more information can be provided if the heuristic information is also made a function of the a_j. One such possibility is to calculate $\bar{r}_i' = 1/m \cdot \sum_{j=1}^{m} a_j/r_{ij}$ and to compute the heuristic information as $\eta_i' = b_i/\bar{r}_i'$.

Solution construction.   Each ant iteratively adds items in a probabilistic way biased by pheromone trails and heuristic information; each item can be added at most once. An ant's solution construction ends if no item can be added anymore without violating any of the resource constraints. This leads to one particularity of the ACO application to the MKP: the length of the ants' walks is not fixed in advance and different ants may have solutions of different length.

2.3.5   The Network Routing Problem

Let a telecommunications network be defined by a set of nodes N, a set of links between nodes L_{net}, and the costs d_{ij} associated with the links. Then, the network routing problem (NRP) is the problem of finding minimum cost paths among all pairs of nodes in the network. It should be noted that if the costs d_{ij} are fixed, then the NRP is reduced to a set of minimum cost path problems, each of which can be solved efficiently via a polynomial time algorithm like Dijkstra's algorithm (Dijkstra, 1959). The problem becomes interesting for heuristic approaches once, as happens in real-world applications like routing in communications networks, costs (e.g., data traffic in links) or the network topology varies in time.

Construction graph.   The construction graph is the graph $G_C = (C, L)$, where C corresponds to the set of nodes N, and L fully connects G_C. Note that $L_{net} \subseteq L$.

Constraints.   The only constraint is that ants use only connections $l_{ij} \in L_{net}$.

Pheromone trails and heuristic information.   Because the NRP is, in reality, a set of minimum cost path problems, each connection $l_{ij} \in L$ should have many different pheromone trails associated. For example, each connection l_{ij} could have associated one trail value τ_{ijd} for each possible destination node d an ant located in node i can have. Each arc can also be assigned a heuristic value η_{ij} independent of the final destination. The heuristic value η_{ij} can be set, for example, to a value inversely proportional to the amount of traffic on the link connecting nodes i and j.

Solution construction.   Solution construction is straightforward. In fact, the S-ACO algorithm presented in chapter 1, section 1.3.1, is an example of how to proceed.

Each ant has a source node s and a destination node d, and moves from s to d hopping from one node to the next, until node d has been reached. When ant k is located at node i, it chooses the next node j to move to using a probabilistic decision rule which is a function of the ant's memory, of local pheromones, and heuristic information.

2.3.6   The Dynamic Traveling Salesman Problem

The dynamic traveling salesman problem (DTSP) is a TSP in which cities can be added or removed at run time. The goal is to find as quickly as possible the new shortest tour after each transition.

Construction graph.   The same as for the TSP: $G_C = (C, L)$, where $C = C(t)$ is the set of cities and $L = L(t)$ completely connects G_C. The dependence of C and L on time is due to the dynamic nature of the problem.

Constraints.   As in the TSP, the only constraint is that a solution should contain each city once and only once.

Pheromone trails and heuristic information.   As in the TSP: pheromone trails are associated with connections, and heuristic values can be given by the inverse of the distances between cities. An important question is how to handle the problem of connections that disappear and appear in case a city is removed or a new city is added. In the first case the values no longer used can simply be removed, while in the second case the new pheromone values could be set, for example, either to values proportional to the length of the associated connections or to the average of the other pheromone values.

Solution construction.   Solution construction follows the same rules as in the TSP.

2.4   Other Metaheuristics

The world of metaheuristics is rich and multifaceted and, besides ACO, a number of other successful metaheuristics are available in the literature. Some of the best known and most widely applied metaheuristics are simulated annealing (SA) (Cerný, 1985; Kirkpatrick et al., 1983), tabu search (TS) (Glover, 1989, 1990; Glover & Laguna, 1997), guided local search (GLS) (Voudouris & Tsang, 1995; Voudouris, 1997), greedy randomized adaptive search procedures (GRASP) (Feo & Resende, 1989, 1995), iterated local search (ILS) (Lourenço et al., 2002), evolutionary computation (EC) (Fogel et al., 1966; Goldberg, 1989; Holland, 1975; Rechenberg, 1973; Schwefel, 1981), and scatter search (Glover, 1977).

All metaheuristics have in common that they try to avoid the generation of poor-quality solutions by introducing general mechanisms that extend problem-specific, single-run algorithms like greedy construction heuristics or iterative improvement local search. Differences among the available metaheuristics concern the techniques employed to avoid getting stuck in suboptimal solutions and the type of trajectory followed in the space of either partial or full solutions.

A first important distinction among metaheuristics is whether they are constructive or local search based (see boxes 2.2 and 2.3). ACO and GRASP belong to the first class; all the other metaheuristics belong to the second class. Another important distinction is whether at each iteration they manipulate a single solution or a population of solutions. All the above-mentioned metaheuristics manipulate a single solution, except for ACO and EC. Although constructive and population-based metaheuristics can be used without recurring to local search, very often their performance can be greatly improved if they are extended to include it. This is the case for both ACO and EC, while GRASP is defined from the very beginning to include local search.

One further important dimension for the classification of metaheuristics concerns the use of memory. Metaheuristics that exploit memory to direct future search are TS, GLS, and ACO. TS either explicitly memorizes previously encountered solutions or memorizes components of previously seen solutions; GLS stores penalties associated with solution components to modify the solutions' evaluation function; and ACO uses pheromones to maintain a memory of past experiences.

It is interesting to note that, for all metaheuristics, there is no general termination criterion. In practice, a number of rules of thumb are used: the maximum CPU time elapsed, the maximum number of solutions generated, the percentage deviation from a lower/upper bound from the optimum, and the maximum number of iterations without improvement in solution quality are examples of such rules. In some cases, metaheuristic-dependent rules of thumb can be defined. An example is TS which can be stopped if the set of solutions in the neighborhood is empty; or SA, where the termination condition is often defined by an annealing schedule.

In conclusion, we see that ACO possesses several characteristics which in their particular combination make it a unique approach: it uses a *population* (colony) of ants which *construct* solutions exploiting a form of *indirect memory* called *artificial pheromones*. The following sections describe in more detail the metaheuristics we mentioned above.

2.4.1   Simulated Annealing

Simulated annealing (Cerný, 1985; Kirkpatrick et al., 1983) is inspired by an analogy between the physical annealing of solids (crystals) and combinatorial optimization problems. In the physical annealing process a solid is first melted and then cooled very slowly, spending a long time at low temperatures, to obtain a perfect lattice structure corresponding to a minimum energy state. SA transfers this process to local search algorithms for combinatorial optimization problems. It does so by associating the set of solutions of the problem attacked with the states of the physical system, the objective function with the physical energy of the solid, and the optimal solutions with the minimum energy states.

SA is a local search strategy which tries to avoid local minima by accepting worse solutions with some probability. In particular, SA starts from some initial solution s and then proceeds as follows: At each step, a solution $s' \in \mathcal{N}(s)$ is generated (often this is done randomly according to a uniform distribution). If s' improves on s, it is accepted; if s' is worse than s, then s' is accepted with a probability which depends on the difference in objective function value $f(s) - f(s')$, and on a parameter T, called temperature. T is lowered (as is also done in the physical annealing process) during the run of the algorithm, reducing in this way the probability of accepting solutions worse than the current one. The probability p_{accept} to accept a solution s' is often defined according to the Metropolis distribution (Metropolis, Rosenbluth, Rosenbluth, Teller, & Teller, 1953):

$$p_{accept}(s, s', T) = \begin{cases} 1, & \text{if } f(s') < f(s); \\ \exp\left(\dfrac{f(s) - f(s')}{T}\right), & \text{otherwise.} \end{cases} \qquad (2.8)$$

Figure 2.2 gives a general algorithmic outline for SA. To implement an SA algorithm, the following parameters and functions have to be specified:

- The function GenerateInitialSolution, that generates an initial solution

- The function InitializeAnnealingParameters that initializes several parameters used in the *annealing schedule*; the parameters comprise

  - an initial temperature T_0

  - the number of iterations to be performed at each temperature (inner loop criterion in figure 2.2)

  - a termination condition (outer loop criterion in figure 2.2)

- The function UpdateTemp that returns a new value for the temperature

- The function GenerateNeighbor that chooses a new solution s' in the neighborhood of the current solution s

- The function AcceptSolution that implements equation (2.8); it decides whether to accept or not the solution returned by GenerateNeighbor

procedure SimulatedAnnealing
  $s \leftarrow$ GenerateInitialSolution
  InitializeAnnealingParameters
  $s_{best} \leftarrow s$
  $n \leftarrow 0$
  while (outer-loop termination condition not met) **do**
    while (inner-loop termination condition not met) **do**
      $s' \leftarrow$ GenerateNeighbor(s)
      $s \leftarrow$ AcceptSolution(T_n, s, s')
      if $(f(s) < f(s_{best}))$ **then**
        $s_{best} \leftarrow s$
      end-if
    end-while
    UpdateTemp(n); $n \leftarrow n + 1$
  end-while
  return s_{best}
end-procedure

Figure 2.2
High-level pseudo-code for simulated annealing (SA).

SA has been applied to a wide variety of problems with mixed success (Aarts, Korst, & van Laarhoven, 1997). It is of special appeal to mathematicians due to the fact that under certain conditions the convergence of the algorithm to an optimal solution can be proved (Geman & Geman, 1984; Hajek, 1988; Lundy & Mees, 1986; Romeo & Sangiovanni-Vincentelli, 1991). Yet, to guarantee convergence to the optimal solution, an impractically slow annealing schedule has to be used and theoretically an infinite number of states has to be visited by the algorithm.

2.4.2   Tabu Search

Tabu search (TS) (Glover, 1989, 1990; Glover & Laguna, 1997) relies on the systematic use of *memory* to guide the search process. It is common to distinguish between *short-term memory*, which restricts the neighborhood $\mathcal{N}(s)$ of the current solution s to a subset $\mathcal{N}'(s) \subseteq \mathcal{N}(s)$, and *long-term memory*, which may extend $\mathcal{N}(s)$ through the inclusion of additional solutions (Glover & Laguna, 1997).

TS uses a local search that, at every step, makes the best possible move from s to a neighbor solution s' even if the new solution is worse than the current one; in this latter case, the move that least worsens the objective function is chosen. To prevent local search from immediately returning to a previously visited solution and, more generally, to avoid cycling, TS can explicitly memorize recently visited solutions and forbid moving back to them. More commonly, TS forbids reversing the effect of recently applied moves by declaring *tabu* those solution attributes that change in the local search. The tabu status of solution attributes is then maintained for a number *tt* of iterations; the parameter *tt* is called the *tabu tenure* or the *tabu list length*. Unfortunately, this may forbid moves toward attractive, unvisited solutions. To avoid such an undesirable situation, an *aspiration criterion* is used to override the tabu status of certain moves. Most commonly, the aspiration criterion drops the tabu status of moves leading to a better solution than the best solution visited so far.

The use of a short-term memory in the search process is probably the most widely applied feature of TS. TS algorithms that only rely on the use of short-term memory are called *simple tabu search* algorithms in Glover (1989). To increase the efficiency of simple TS, long-term memory strategies can be used to intensify or diversify the search. Intensification strategies are intended to explore more carefully promising regions of the search space either by recovering elite solutions (i.e., the best solutions obtained so far) or attributes of these solutions. Diversification refers to the exploration of new search space regions through the introduction of new attribute combinations. Many long-term memory strategies in the context of TS are based on the memorization of the frequency of solution attributes. For a detailed discussion of techniques exploiting long-term memory, see Glover & Laguna (1997).

An algorithmic outline of a simple TS algorithm is given in figure 2.3. The functions needed to define it are the following:

- The function GenerateInitialSolution, which generates an initial solution

- The function InitializeMemoryStructures, which initializes all the memory structures used during the run of the TS algorithm

- The function GenerateAdmissibleSolutions, which is used to determine the subset of neighbor solutions which are not tabu or are tabu but satisfy the aspiration criterion

- The function SelectBestSolution, which returns the best admissible move

- The function UpdateMemoryStructures, which updates the memory structures

To date, TS appears to be one of the most successful and most widely used metaheuristics, achieving excellent results for a wide variety of problems (Glover & Laguna, 1997). Yet this efficiency is often due to a significant fine-tuning effort of a large collection of parameters and different implementation choices (Hertz, Taillard, & de Werra, 1997). However, there have been several proposals such as reactive TS, which try to make TS more robust with respect to parameter settings (Battiti & Tecchiolli, 1994). Interestingly, some theoretical proofs about the behavior of TS exist.

procedure SimpleTabuSearch
  $s \leftarrow$ GenerateInitialSolution
  InitializeMemoryStructures
  $s_{best} \leftarrow s$
  while (termination condition not met) **do**
    $\mathcal{A} \leftarrow$ GenerateAdmissibleSolutions(s)
    $s \leftarrow$ SelectBestSolution($\mathcal{A}$)
    UpdateMemoryStructures
    if $(f(s) < f(s_{best}))$ **then**
      $s_{best} \leftarrow s$
    end-if
  end-while
  return s_{best}
end-procedure

Figure 2.3
High-level pseudo-code for a simple tabu search (TS).

Faigle & Kern (1992) presented a convergence proof for probabilistic TS; Hanafi and Glover proved that several deterministic variants of TS implicitly enumerate the search space and, hence, are also guaranteed to find the optimal solution in finite time (Hanafi, 2000; Glover & Hanafi, 2002).

2.4.3   Guided Local Search

One alternative possibility to escape from local optima is to modify the evaluation function while searching. Guided local search (Voudouris, 1997; Voudouris & Tsang, 1995) is a metaheuristic that makes use of this idea. It uses an augmented cost function $h(s)$, $h : s \mapsto \mathbb{R}$, which consists of the original objective function $f(\cdot)$ plus additional penalty terms pn_i associated with each solution feature i. The augmented cost function is defined as $h(s) = f(s) + \omega \cdot \sum_{i=1}^{n} pn_i \cdot I_i(s)$, where the parameter ω determines the influence of the penalties on the augmented cost function, n is the number of solution features, pn_i is the penalty cost associated with solution feature i, and $I_i(s)$ is an indicator function that takes the value 1 if the solution feature i is present in the solution s and 0 otherwise. A solution feature, for example, in the TSP is an arc and the indicator function tells if a specific arc is used or not.

GLS uses the augmented cost function for choosing local search moves until it gets trapped in a local optimum $\hat{s}$ with respect to $h(\cdot)$. At this point, a utility value $u_i = I_i(\hat{s}) \cdot c_i/(1 + pn_i)$ is computed for each feature, where c_i is the cost of feature i. Features with high costs will have a high utility. The utility values are scaled by pn_i

procedure GuidedLocalSearch
   $s \leftarrow$ GenerateInitialSolution
   InitializePenalties
   $s_{best} \leftarrow s$
   while (termination condition not met) **do**
      $h \leftarrow$ ComputeAugmentedObjectiveFunction
      $\hat{s} \leftarrow$ LocalSearch$(\hat{s}, h)$
      UpdatePenalties$(\hat{s})$
   end-while
   return s_{best}
end-procedure

Figure 2.4
High-level pseudo-code for guided local search (GLS).

to avoid the same high cost features from getting penalized over and over again and the search trajectory from becoming too biased. Then, the penalties of the features with maximum utility are incremented and the augmented cost function is adapted by using the new penalty values. Last, the local search is continued from $\hat{s}$, which, in general, will no longer be locally optimal with respect to the new augmented cost function.

Note that during the local search all solutions encountered must be evaluated with respect to both the original objective function and the augmented cost functions. In fact, the two provide different types of information: the original objective function $f(\cdot)$ determines the quality of a solution, while the augmented cost function is used for guiding the local search.

An algorithmic outline of GLS is given in figure 2.4. The functions to be defined for the implementation of a GLS algorithm are the following:

- The function GenerateInitialSolution, which generates an initial solution
- The function InitializePenalties, which initializes the penalties of the solution features
- The function ComputeAugmentedObjectiveFunction, which computes the new augmented evaluation function after an update of the penalties
- The function LocalSearch, which applies a local search algorithm using the augmented evaluation function
- The function UpdatePenalties, which, once the local search is stuck in a locally optimal solution, updates the penalty vector

GLS has been derived from earlier approaches which dynamically modified the evaluation function during the search like the *breakout method* (Morris, 1993) and GENET (Davenport, Tsang, Wang, & Zhu, 1994). More generally, GLS has tight connections to other weighting schemes like those used in local search algorithms for the satisfiability problem in propositional logic (SAT) (Selman & Kautz, 1993; Frank, 1996) or adaptations of Lagrangian methods to local search (Shang & Wah, 1998). In general, algorithms that modify the evaluation function at computation time are becoming more widely used.

2.4.4   Iterated Local Search

Iterated local search (Lourenço et al., 2002; Martin, Otto, & Felten, 1991) is a simple and powerful metaheuristic, whose working principle is as follows. Starting from an initial solution s, a local search is applied. Once the local search is stuck, the locally optimal solution $\hat{s}$ is perturbed by a move in a neighborhood different from the one used by the local search. This perturbed solution s' is the new starting solution for the local search that takes it to the new local optimum $\hat{s}'$. Finally, an acceptance criterion decides which of the two locally optimal solutions to select as a starting point for the next perturbation step. The main motivation for ILS is to build a randomized walk in a search space of the local optima with respect to some local search algorithm.

An algorithmic outline of ILS is given in figure 2.5. The four functions needed to specify an ILS algorithm are as follows:

- The function GenerateInitialSolution, which generates an initial solution

- The function LocalSearch, which returns a locally optimal solution $\hat{s}$ when applied to s

- The function Perturbation, which perturbs the current solution s generating an intermediate solution s'

- The function AcceptanceCriterion, which decides from which solution the search is continued at the next perturbation step

Additionally, the functions Perturbation and AcceptanceCriterion may also exploit the search history to bias their decisions (Lourenço et al., 2002).

The general idea of ILS was rediscovered by many authors, and has been given many different names, such as *iterated descent* (Baum, 1986), *large-step Markov chains* (Martin et al., 1991), *chained local optimization* (Martin & Otto, 1996), and so on. One of the first detailed descriptions of ILS was given in Martin et al. (1991), although earlier descriptions of the basic ideas underlying the approach exist (Baum,

procedure IteratedLocalSearch
    $s \leftarrow$ GenerateInitialSolution
    $\hat{s} \leftarrow$ LocalSearch(s)
    $s_{best} \leftarrow \hat{s}$
    while (termination condition not met) **do**
        $s' \leftarrow$ Perturbation($\hat{s}$)
        $\hat{s}' \leftarrow$ LocalSearch(s')
        if $(f(\hat{s}') < f(s_{best}))$ **then**
            $s_{best} \leftarrow \hat{s}'$
        end-if
        $\hat{s} \leftarrow$ AcceptanceCriterion($\hat{s}, \hat{s}'$)
    end-while
    return s_{best}
end-procedure

Figure 2.5
High-level pseudo-code for iterated local search (ILS).

1986; Baxter, 1981). Some of the first ILS implementations have shown that the approach is very promising and current ILS algorithms are among the best-performing approximation methods for combinatorial optimization problems like the TSP (Applegate, Bixby, Chvátal, & Cook, 1999; Applegate, Cook, & Rohe, 2003; Johnson & McGeoch, 1997; Martin & Otto, 1996) and several scheduling problems (Brucker, Hurink, & Werner, 1996; Balas & Vazacopoulos, 1998; Congram, Potts, & de Velde, 2002).

2.4.5   Greedy Randomized Adaptive Search Procedures

Greedy randomized adaptive search procedures (Feo & Resende, 1989, 1995) randomize greedy construction heuristics to allow the generation of a large number of different starting solutions for applying a local search.

GRASP is an iterative procedure which consists of two phases, a construction phase and a local search phase. In the construction phase a solution is constructed from scratch, adding one solution component at a time. At each step of the construction heuristic, the solution components are ranked according to some greedy function and a number of the best-ranked components are included in a *restricted candidate list*; typical ways of deriving the restricted candidate list are either to take the best $\gamma\%$ of the solution components or to include all solution components that have a greedy value within some $\delta\%$ of the best-rated solution component. Then, one

procedure GRASP
    while (termination condition not met) **do**
        $s \leftarrow$ ConstructGreedyRandomizedSolution
        $\hat{s} \leftarrow$ LocalSearch(s)
        if $f(\hat{s}) < f(s_{best})$ **then**
            $s_{best} \leftarrow \hat{s}$
        end-if
    end-while
    return s_{best}
end-procedure

Figure 2.6
High-level pseudo-code for greedy randomized adaptive search procedures (GRASP).

of the components of the restricted candidate list is chosen randomly, according to a uniform distribution. Once a full candidate solution is constructed, this solution is improved by a local search phase.

A general outline of the GRASP procedure is given in figure 2.6. For the implementation of a GRASP algorithm we need to define two main functions:

- The function ConstructGreedyRandomizedSolution, which generates a solution
- The function LocalSearch, which implements a local search algorithm

The number of available applications of GRASP is large and several extensions of the basic GRASP algorithm we have presented here have been proposed; see Festa & Resende (2002) and Resende & Ribeiro (2002) for an overview. Regarding theoretical results, it should be mentioned that standard implementations of GRASP use restricted candidate lists and therefore may not converge to the optimal solution (Mockus, Eddy, Mockus, Mockus, & Reklaitis, 1997). One way around this problem is to allow choosing the parameter γ randomly according to a uniform distribution so that occasionally all the solution components are eligible (Resende, Pitsoulis, & Pardalos, 2000).

2.4.6   Evolutionary Computation

Evolutionary computation has become a standard term to indicate problem-solving techniques which use design principles inspired from models of the natural evolution of species.

Historically, there are three main algorithmic developments within the field of EC: evolution strategies (Rechenberg, 1973; Schwefel, 1981), evolutionary programming

(Fogel et al., 1966), and genetic algorithms (Holland, 1975; Goldberg, 1989). Common to these approaches is that they are population-based algorithms that use operators inspired by population genetics to explore the search space (the most typical genetic operators are *reproduction, mutation*, and *recombination*). Each individual in the algorithm represents directly or indirectly (through a decoding scheme) a solution to the problem under consideration. The reproduction operator refers to the process of selecting the individuals that will survive and be part of the next generation. This operator typically uses a bias toward good-quality individuals: The better the objective function value of an individual, the higher the probability that the individual will be selected and therefore be part of the next generation. The recombination operator (often also called crossover) combines parts of two or more individuals and generates new individuals, also called *offspring*. The mutation operator is a unary operator that introduces random modifications to one individual.

Differences among the different EC algorithms concern the particular representations chosen for the individuals and the way genetic operators are implemented. For example, genetic algorithms typically use binary or discrete valued variables to represent information in individuals and they favor the use of recombination, while evolution strategies and evolutionary programming often use continuous variables and put more emphasis on the mutation operator. Nevertheless, the differences between the different paradigms are becoming more and more blurred.

A general outline of an EC algorithm is given in figure 2.7, where *pop* denotes the population of individuals. To define an EC algorithm the following functions have to be specified:

- The function InitializePopulation, which generates the initial population

- The function EvaluatePopulation, which computes the fitness values of the individuals

- The function BestOfPopulation, which returns the best individual in the current population

- The function Recombination, which repeatedly combines two or more individuals to form one or more new individuals

- The function Mutation, which, when applied to one individual, introduces a (small) random perturbation

- The function Reproduction, which generates a new population from the current one

EC is a vast field where a large number of applications and a wide variety of algorithmic variants exist. Because an overview of the EC literature would fill an

procedure EvolutionaryComputationAlgorithm
   $pop \leftarrow$ InitializePopulation
   EvaluatePopulation(pop)
   $s_{best} \leftarrow$ BestOfPopulation(pop)
   while (termination condition not met) **do**
      $pop' \leftarrow$ Recombination(pop)
      $pop'' \leftarrow$ Mutation(pop')
      EvaluatePopulation(pop'')
      $s \leftarrow$ BestOfPopulation(pop'')
      if $f(s) < f(s_{best})$ **then**
         $s_{best} \leftarrow s$
      end-if
      $pop \leftarrow$ Reproduction(pop'')
   end-while
   return s_{best}
end-procedure

Figure 2.7
High-level pseudo-code for an evolutionary computation (EC) algorithm.

entire book, we refer to the following for more details on the subject: Fogel et al., 1966; Fogel, 1995; Holland, 1975; Rechenberg, 1973; Schwefel, 1981; Goldberg, 1989; Michalewicz, 1994; Mitchell, 1996.

Still, one particular EC algorithm, called population-based incremental learning (PBIL) (Baluja & Caruana, 1995), is mentioned here because of its similarities to ACO. PBIL maintains a vector of probabilities called the *generating vector*. Starting from this vector, a population of binary strings representing solutions to the problem under consideration is randomly generated: each string in the population has the i-th bit set to 1 with a probability given by the i-th value on the generating vector. Once a population of solutions is created, the generated solutions are evaluated and this evaluation is used to increase (or decrease) the probabilities of each separate component in the generating vector with the hope that good (bad) solutions in future generations will be produced with higher (lower) probability. It is clear that in ACO the pheromone trail values play a role similar to PBIL's generating vector, and pheromone updating has the same goal as updating the probabilities in the generating vector. A main difference between ACO and PBIL consists in the fact that in PBIL all the components of the probability vector are evaluated independently, so that PBIL works well only when the solution is separable in its components.

2.4.7　Scatter Search

The central idea of scatter search (SS), first introduced by Glover (1977), is to keep a small population of *reference solutions*, called a *reference set*, and to combine them to create new solutions.

A basic version of SS proceeds as follows. It starts by creating a reference set. This is done by first generating a large number of solutions using a *diversification generation method*. Then, these solutions are improved by a local search procedure. (Typically, the number of solutions generated in this way is ten times the size of the reference set [Glover, Laguna, & Martí, 2002], while the typical size of a reference set is usually between ten and twenty solutions.) From these improved solutions, the reference set *rs* is built. The solutions to be put in *rs* are selected by taking into account both their solution quality and their diversity. Then, the solutions in *rs* are used to build a set *c_cand* of subsets of solutions. The solutions in each subset, which can be of size 2 in the simplest case, are candidates for combination. Solutions within each subset of *c_cand* are combined; each newly generated solution is improved by local search and possibly replaces one solution in the reference set. The process of subset generation, solution combination, and local search is repeated until the reference set does not change anymore.

A general outline of a basic SS algorithm is given in figure 2.8, where *pop* denotes a population of candidate solutions. To define an SS algorithm, the following functions have to be specified:

- The function GenerateDiverseSolutions, which generates a population of solutions as candidates for building the first reference set. These solutions must be diverse in the sense that they must be spread over the search space

- The function LocalSearch, which implements an improvement algorithm

- The function BestOfPopulation, which returns the best candidate solution in the current population

- The function GenerateReferenceSet, which generates the initial reference set

- The function GenerateSubsets, which generates the set *c_cand*

- The function SelectSubset, which returns one element of *c_cand*

- The function CombineSolutions, which, when applied to one of the subsets in *c_cand*, returns one or more candidate solutions

- The function WorstOfPopulation, which returns the worst candidate solution in the current population

procedure ScatterSearch
    $pop \leftarrow$ GenerateDiverseSolutions
    $pop \leftarrow$ LocalSearch(pop)
    $s_{best} \leftarrow$ BestOfPopulation(pop)
    $rs \leftarrow$ GenerateReferenceSet(pop)
    new_solution $\leftarrow$ true
    while (new_solution $=$ true) **do**
      new_solution $\leftarrow$ false
      $c\_cand \leftarrow$ GenerateSubsets(rs)
      while ($c\_cand \neq \varnothing$) **do**
        $cc \leftarrow$ SelectSubset(c_cand)
        $s \leftarrow$ CombineSolutions(cc)
        $\hat{s} \leftarrow$ LocalSearch(s)
        $s_{worst} \leftarrow$ WorstOfPopulation(rs)
        if $\hat{s} \notin rs$ and $f(\hat{s}) < f(s_{worst})$ **then**
          UpdateReferenceSet($\hat{s}$)
          new_solution $\leftarrow$ true
        end-if
        if ($f(\hat{s}) < f(s_{best})$) **then**
          $s_{best} \leftarrow \hat{s}$
        end-if
        c_cand $\leftarrow$ c_cand\\cc
      end-while
    end-while
    return s_{best}
end-procedure

Figure 2.8
High-level pseudo-code for scatter search (SS).

▪ The function UpdateReferenceSet, which decides whether a candidate solution should replace one of the solutions in the reference set, and updates the reference set accordingly

SS is a population-based algorithm that shares some similarities with EC algorithms (Glover, 1977; Glover et al., 2002; Laguna & Martí, 2003). Solution combination in SS is analogous to *recombination* in EC algorithms; however, in SS solution combination was conceived as a linear combination of solutions that can lead to both convex and nonconvex combinations of solutions in the reference set (Glover, 1977);

nonconvex combination of solutions allows the generation of solutions that are external to the subspace spanned by the original reference set. See Laguna & Martí (2003) for an overview of implementation principles and of current applications.

2.5   Bibliographical Remarks

Combinatorial Optimization

Combinatorial optimization is a widely studied field for which a large number of textbooks and research articles exist. One of the standard references is the book by Papadimitriou & Steiglitz (1982). There also exist a variety of other textbooks which give rather comprehensive overviews of the field. Examples are books by Lawler (1976), by Nemhauser & Wolsey (1988), and the more recent book by Cook, Cunningham, Pulleyblank & Schrijver (1998). For readers interested in digging into the huge literature on combinatorial optimization, a good starting point is the book of annotated bibliographies edited by Dell'Amico, Maffioli, & Martello (1997).

The standard reference on the theory of $\mathcal{NP}$-completeness is the excellent book by Garey & Johnson (1979). A question of particular interest for researchers in metaheuristics concerns the computational complexity of approximation algorithms. A recent detailed overview of the current knowledge on this subject is given in Hochbaum (1997) and in Ausiello, Crescenzi, Gambosi, Kann, Marchetti-Spaccamela, & Protasi (1999). Of particular interest also is the recently developed complexity theory for local search algorithms, introduced in an article by Johnson, Papadimitriou, & Yannakakis (1988).

ACO Metaheuristic

The first algorithm to fall into the framework of the ACO metaheuristic was Ant System (AS) (Dorigo, 1992; Dorigo, Maniezzo, & Colorni, 1991a, 1996). AS was followed by a number of different algorithmic variants that tried to improve its performance. The ACO metaheuristic, first described in the articles by Dorigo & Di Caro (1999a,b) and Dorigo, Di Caro, & Gambardella (1999), is the result of a research effort directed at building a common framework for these algorithmic variants. Most of the available ACO algorithms are presented in chapter 3 (up-to-date information on ACO is maintained on the Web at www.aco-metaheuristic.org). To be mentioned here is also the international workshop series "ANTS: From Ant Colonies to Artificial Ants" on ant algorithms, where a large part of the contributions focus on different aspects of the ACO metaheuristic (see the Web at iridia.ulb. ac.be/~ants/ for up-to-date information on this workshop series). The proceedings of the most recent workshop of this series in 2002 are published in the *Lecture*

Notes in Computer Science series of Springer-Verlag (Dorigo, Di Caro, & Sampels, 2002a).

Other Metaheuristics

The area of metaheuristics has now become a large field with its own conference series, the Metaheuristics International Conference, which has been held biannually since 1995. After each conference, an edited book covering current research issues in the field is published (Hansen & Ribeiro, 2001; Osman & Kelly, 1996; Voss, Martello, Osman, & Roucairol, 1999).

Single-authored books which give an overview of the whole metaheuristics field are few. An inspiring such book is the recent one by Michalewicz & Fogel (2000). Two other books which cover a number of different metaheuristics are those of Sait & Youssef (1999) and Karaboga & Pham (2000). A recent survey paper is that of Blum & Roli (2003).

As far as single metaheuristics are concerned, we gave basic references to the literature in section 2.4. A large collection of references up to 1996 is provided by Osman & Laporte (1996). A book that gives an extensive overview of local search methods is that edited by Aarts & Lenstra (1997), which contains a number of contributions by leading experts. Another book which gives an overview of a number of metaheuristics (including some variants not covered in this chapter) was edited by Reeves (1995). Recent new metaheuristic ideas are collected in a book edited by Corne, Dorigo, & Glover (1999). Currently, the best overview of the field is the *Handbook of Metaheuristics*, edited by Glover & Kochenberger (2002).

2.6 Things to Remember

- Combinatorial optimization problems arise in many practical and theoretical problems. Often, these problems are very hard to solve to optimality. The theory of $\mathcal{NP}$-completeness classifies the problems according to their difficulty. For many combinatorial optimization problems it has been shown that they belong to the class of $\mathcal{NP}$-hard problems, which means that in the worst case the effort needed to find optimal solutions increases exponentially with problem size, unless $\mathcal{P} = \mathcal{NP}$.

- Exact algorithms try to find optimal solutions and additionally prove their optimality. Despite recent successes, for many $\mathcal{NP}$-hard problems the performance of exact algorithms is not satisfactory and their applicability is often limited to rather small instances.

- Approximate algorithms trade optimality for efficiency. Their main advantage is that in practice they often find reasonably good solutions in a very short time.

▪ A metaheuristic is a set of algorithmic concepts that can be used to define heuristic methods applicable to a wide set of different problems. In other words, a metaheuristic can be seen as a general algorithmic framework which can be applied to different optimization problems with relatively few modifications to make them adapted to a specific problem.

▪ The ACO metaheuristic was inspired by the foraging behavior of real ants. It has a very wide applicability: it can be applied to any combinatorial optimization problem for which a solution construction procedure can be conceived. The ACO metaheuristic is characterized as being a distributed, stochastic search method based on the indirect communication of a colony of (artificial) ants, mediated by (artificial) pheromone trails. The pheromone trails in ACO serve as a distributed numerical information used by the ants to probabilistically construct solutions to the problem under consideration. The ants modify the pheromone trails during the algorithm's execution to reflect their search experience.

▪ The ACO metaheuristic is based on a generic problem representation and the definition of the ants' behavior. Given this formulation, the ants in ACO build solutions to the problem being solved by moving concurrently and asynchronously on an appropriately defined construction graph. The ACO metaheuristic defines the way the solution construction, the pheromone update, and possible *daemon actions*—actions which cannot be performed by a single ant because they require access to nonlocal information—interact in the solution process.

▪ The application of ACO is particularly interesting for (1) $\mathcal{NP}$-hard problems, which cannot be efficiently solved by more traditional algorithms; (2) dynamic shortest-path problems in which some properties of the problem's graph representation change over time concurrently with the optimization process; and (3) problems in which the computational architecture is spatially distributed. The versatility of the ACO metaheuristic has been shown using several example applications.

▪ The ACO metaheuristic is one out of a number of metaheuristics which have been proposed in the literature. Other metaheuristics, including simulated annealing, tabu search, guided local search, iterated local search, greedy randomized adaptive search procedures, and evolutionary computation, have been discussed in this chapter. Several characteristics make ACO a unique approach: it is a constructive, population-based metaheuristic which exploits an indirect form of memory of previous performance. This combination of characteristics is not found in any of the other metaheuristics.

2.7   Thought and Computer Exercises

Exercise 2.1   We have exemplified the application of the ACO metaheuristic to a number of different combinatorial optimization problems. For each of these problems, do the following (for this exercise and the next, consider only the static example problems introduced in this chapter):

1. Define the set of candidate solutions and the set of feasible solutions.

2. Define a greedy construction heuristic. (Answering the following questions may be of some help: What are appropriate solution components? How do you measure the objective function contribution of adding a solution component? Is it always possible to construct feasible candidate solutions? How many different solutions can be generated with the constructive heuristic?)

3. Define a local search algorithm. (Answering the following questions may be of some help: How can local changes be defined? How many solution components are involved in each local search step? How do you choose which neighboring solution to move to? Does the local search always maintain feasibility of solutions?)

Exercise 2.2   Implement the construction heuristics and the local search algorithms defined in the first exercise in your favorite programming language.

Evaluate the performance of the resulting algorithms using test instances. Test instances that have already been used by other researchers are available, for example, at ORLIB mscmga.ms.ic.ac.uk/info.html. Another possibility is to look at www.metaheuristics.org.

How strongly do the local search algorithms improve the solution quality if they are applied to the solutions generated by the construction heuristics?

Exercise 2.3   Develop a description of how to apply the ACO metaheuristic to the combinatorial optimization problems you are familiar with. To do so, answer the following questions: What are the solution components? Are there different ways of defining the solution components? If yes, in which aspects do the definitions differ? How is the construction graph defined? How are the pheromone trails and the heuristic information defined? Are there different ways of defining the heuristic information? How are the constraints treated? How do you implement the ants' behavior and, in particular, how do you construct solutions?

Exercise 2.4   We have introduced three criteria to classify metaheuristics. One is the use of solution construction versus the use of local search; another is the use, or not, of a population of solutions; and the last the use, or not, of a memory within the search process. Additional criteria concern whether the evaluation function is modified during the search or not, whether an algorithm uses several neighborhoods or only a single one, and whether the metaheuristics are inspired by some process occurring in nature. Recapitulate the classification of section 2.4, for the metaheuristics discussed in this chapter. Extend this classification to also include the three additional criteria given above.

Exercise 2.5   There are a number of additional metaheuristics available, some of which are described in *New Ideas in Optimization* (Corne et al., 1999). Develop short descriptions of these metaheuristics in a format similar to that used in this chapter. To do so, first consider the general principles underlying the metaheuristics, develop a general algorithmic outline for the metaheuristic, and describe the functions that need to be defined to implement the metaheuristic. Finally, consider the range of available applications of that metaheuristic and find out about the theoretical knowledge on the convergence behavior of these metaheuristics.

3 Ant Colony Optimization Algorithms for the Traveling Salesman Problem

But you're sixty years old. They can't expect you to keep traveling every week.
—Linda in act 1, scene 1 of *Death of a Salesman*, Arthur Miller, 1949

The traveling salesman problem is an extensively studied problem in the literature and for a long time has attracted a considerable amount of research effort. The TSP also plays an important role in ACO research: the first ACO algorithm, called Ant System, as well as many of the ACO algorithms proposed subsequently, was first tested on the TSP.

There are several reasons for the choice of the TSP as the problem to explain the working of ACO algorithms: it is an important $\mathcal{NP}$-hard optimization problem that arises in several applications; it is a problem to which ACO algorithms are easily applied; it is easily understandable, so that the algorithm behavior is not obscured by too many technicalities; and it is a standard test bed for new algorithmic ideas—a good performance on the TSP is often taken as a proof of their usefulness. Additionally, the history of ACO shows that very often the most efficient ACO algorithms for the TSP were also found to be among the most efficient ones for a wide variety of other problems.

This chapter is therefore dedicated to a detailed explanation of the main members of the ACO family through examples of their application to the TSP: algorithms are described in detail, and a guide to their implementation in a C-like programming language is provided.

3.1 The Traveling Salesman Problem

Intuitively, the TSP is the problem of a salesman who, starting from his hometown, wants to find a shortest tour that takes him through a given set of customer cities and then back home, visiting each customer city exactly once. More formally, the TSP can be represented by a complete weighted graph $G = (N, A)$ with N being the set of nodes representing the cities, and A being the set of arcs. (Note that if the graph is not complete, one can always add arcs to obtain a new, complete graph G' with exactly the same optimal solutions as G; this can be achieved by assigning to the additional arcs weights that are large enough to guarantee that they will not be used in any optimal solution.) Each arc $(i, j) \in A$ is assigned a value (length) d_{ij}, which is the distance between cities i and j, with $i, j \in N$. In the general case of the asymmetric TSP, the distance between a pair of nodes i, j is dependent on the direction of traversing the arc, that is, there is at least one arc (i, j) for which $d_{ij} \neq d_{ji}$. In the symmetric TSP, $d_{ij} = d_{ji}$ holds for all the arcs in A. The goal in the TSP is to find a minimum length Hamiltonian circuit of the graph, where a Hamiltonian circuit is a closed path visiting each of the $n = |N|$ nodes of G exactly once. Thus, an optimal solution to the TSP is a permutation π of the node indices $\{1, 2, \ldots, n\}$ such that the length $f(\pi)$ is minimal, where $f(\pi)$ is given by

$$f(\pi) = \sum_{i=1}^{n-1} d_{\pi(i)\pi(i+1)} + d_{\pi(n)\pi(1)}. \tag{3.1}$$

In the remainder of this chapter we try to highlight differences in performance among ACO algorithms by running computational experiments on instances available from the TSPLIB benchmark library (Reinelt, 1991), which is accessible on the Web at www.iwr.uni-heidelberg.de/groups/comopt/software/TSPLIB95/. TSPLIB instances have been used in a number of influential studies of the TSP (Grötschel & Holland, 1991; Reinelt, 1994; Johnson & McGeoch, 2002) and, in part, they stem from practical applications of the TSP such as drilling holes for printed circuit boards (Reinelt, 1994) or the positioning of X-ray devices (Bland & Shallcross, 1989). Most of the TSPLIB instances are geometric TSP instances, that is, they are defined by the coordinates of a set of points and the distance between these points is computed according to some metric. Figure 3.1 gives two examples of such instances. We refer the reader to the TSPLIB website for a detailed description of how the distances are generated. In any case, independently of which metric is used, in all TSPLIB instances the distances are rounded to integers. The main reason for this choice is of a historical nature: in early computers integer computations were much quicker to perform than computations using floating numbers.

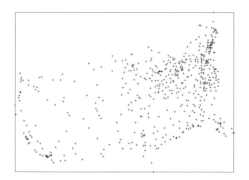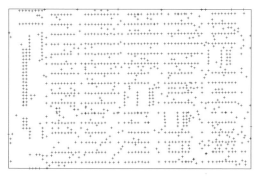

Figure 3.1
Examples of TSP: The figure on the left shows the TSP instance `att532`, which comprises 532 cities in the United States. The figure on the the right shows instance `pcb1173`, which represents the location of 1173 holes to be drilled on a printed circuit board. Each point gives the localization of cities or holes to be drilled, respectively. Both instances are taken from TSPLIB.

3.2   ACO Algorithms for the TSP

ACO can be applied to the TSP in a straightforward way, as described in section 2.3.1; the construction graph $G = (C, L)$, where the set L fully connects the components C, is identical to the problem graph, that is, $C = N$ and $L = A$; the set of states of the problem corresponds to the set of all possible partial tours; and the constraints Ω enforce that the ants construct only feasible tours that correspond to permutations of the city indices. This is always possible, because the construction graph is a complete graph and any closed path that visits all the nodes without repeating any node corresponds to a feasible tour.

In all available ACO algorithms for the TSP, the pheromone trails are associated with arcs and therefore τ_{ij} refers to the desirability of visiting city j directly after city i. The heuristic information is chosen as $\eta_{ij} = 1/d_{ij}$, that is, the heuristic desirability of going from city i directly to city j is inversely proportional to the distance between the two cities. In case $d_{ij} = 0$ for some arc (i, j), the corresponding η_{ij} is set to a very small value. As we discuss later, for implementation purposes pheromone trails are collected into a pheromone matrix whose elements are the τ_{ij}'s. This can be done analogously for the heuristic information.

Tours are constructed by applying the following simple constructive procedure to each ant: (1) choose, according to some criterion, a start city at which the ant is positioned; (2) use pheromone and heuristic values to probabilistically construct a tour by iteratively adding cities that the ant has not visited yet (see figure 3.2), until all cities have been visited; and (3) go back to the initial city. After all ants have completed their tour, they may deposit pheromone on the tours they have followed. We will see that, in some cases, before adding pheromone, the tours constructed by

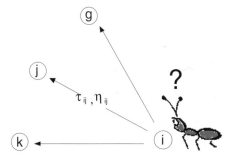

Figure 3.2
An ant arriving in city i chooses the next city to move to as a function of the pheromone values τ_{ij} and of the heuristic values η_{ij} on the arcs connecting city i to the cities j the ant has not visited yet.

```
procedure ACOMetaheuristicStatic
  Set parameters, initialize pheromone trails
  while (termination condition not met) do
    ConstructAntsSolutions
    ApplyLocalSearch        % optional
    UpdatePheromones
  end
end
```

Figure 3.3
Algorithmic skeleton for ACO algorithms applied to "static" combinatorial optimization problems. The application of a local search algorithm is a typical example of a possible daemon action in ACO algorithms.

the ants may be improved by the application of a local search procedure. This high-level description applies to most of the published ACO algorithms for the TSP, one notable exception being Ant Colony System (described in chapter 3, section 3.4.1), in which pheromone evaporation is interleaved with tour construction. In fact, when applied to the TSP and to virtually any other *static* combinatorial optimization problem (see chapter 2, section 2.2), most ACO algorithms employ a more specific algorithmic scheme than the general one of the ACO metaheuristic given in figure 2.1. This algorithm's scheme is shown in figure 3.3; after initializing the parameters and the pheromone trails, these ACO algorithms iterate through a main loop, in which first all of the ants' tours are constructed, then an optional phase takes place in which the ants' tours are improved by the application of some local search algorithm, and finally the pheromone trails are updated. This last step involves pheromone evaporation and the update of the pheromone trails by the ants to reflect their search experience. In figure 3.3 the DaemonActions procedure of figure 2.1 is replaced by the ApplyLocalSearch procedure, and by a routine (not shown in the figure and most often integrated in the UpdatePheromones procedure to facilitate implementation) that helps selecting the ants that should be allowed to deposit pheromone.

As already mentioned, the first ACO algorithm, Ant System (Dorigo, 1992; Dorigo et al., 1991a, 1996), was introduced using the TSP as an example application. AS achieved encouraging initial results, but was found to be inferior to state-of-the-art algorithms for the TSP. The importance of AS therefore mainly lies in the inspiration it provided for a number of extensions that significantly improved performance and are currently among the most successful ACO algorithms. In fact, most of these extensions are direct extensions of AS in the sense that they keep the same solution construction procedure as well as the same pheromone evaporation procedure. These extensions include elitist AS, rank-based AS, and $\mathcal{MAX}$–$\mathcal{MIN}$ AS. The main dif-

Table 3.1
ACO algorithms according to chronological order of appearance

ACO algorithm	TSP	Main references
Ant System (AS)	yes	Dorigo (1992); Dorigo, Maniezzo, & Colorni (1991a,b, 1996)
Elitist AS	yes	Dorigo (1992); Dorigo, Maniezzo, & Colorni (1991a,b, 1996)
Ant-Q	yes	Gambardella & Dorigo (1995); Dorigo & Gambardella (1996)
Ant Colony System	yes	Dorigo & Gambardella (1997a,b)
$\mathcal{MAX}$–$\mathcal{MIN}$ AS	yes	Stützle & Hoos (1996, 2000); Stützle (1999)
Rank-based AS	yes	Bullnheimer, Hartl, & Strauss (1997, 1999c)
ANTS	no	Maniezzo (1999)
Hyper-cube AS	no	Blum, Roli, & Dorigo (2001); Blum & Dorigo (2004)

In the column TSP we indicate whether this ACO algorithm has already been applied to the traveling salesman problem.

ferences between AS and these extensions are the way the pheromone update is performed, as well as some additional details in the management of the pheromone trails. A few other ACO algorithms that more substantially modify the features of AS were also proposed in the literature. These extensions, presented in section 3.4, include Ant-Q and its successor Ant Colony System (ACS), the ANTS algorithm, which exploits ideas taken from lower bounding techniques in mathematical programming, and the hyper-cube framework for ACO. We note that not all available ACO algorithms have been applied to the TSP: exceptions are Maniezzo's ANTS (see section 3.4.2) and ACO implementations based on the hyper-cube framework (see section 3.4.3).

As a final introductory remark, let us note that we do not present the available ACO algorithms in chronological order of their first publication but rather in the order of increasing complexity in the modifications they introduce with respect to AS. The chronological order of the first references and of the main publications on the available ACO algorithms is indicated in table 3.1. Striking is the relatively large gap between 1991–92 and 1995–96. In fact, the seminal publication on AS in *IEEE Transactions on Systems, Man, and Cybernetics*, although submitted in 1991, appeared only in 1996; starting from that publication the interest in ACO has grown very quickly.

3.3   Ant System and Its Direct Successors

In this section we present AS and those ACO algorithms that are largely similar to AS. We do not consider the use of the optional local search phase; the addition of local search to ACO algorithms is the topic of section 3.7.

3.3.1   Ant System

Initially, three different versions of AS were proposed (Dorigo et al., 1991a; Colorni, Dorigo, & Maniezzo, 1992a; Dorigo, 1992). These were called *ant-density*, *ant-quantity*, and *ant-cycle*. Whereas in the ant-density and ant-quantity versions the ants updated the pheromone directly after a move from one city to an adjacent city, in the ant-cycle version the pheromone update was only done after all the ants had constructed the tours and the amount of pheromone deposited by each ant was set to be a function of the tour quality. Nowadays, when referring to AS, one actually refers to ant-cycle since the two other variants were abandoned because of their inferior performance.

The two main phases of the AS algorithm constitute the ants' solution construction and the pheromone update. In AS a good heuristic to initialize the pheromone trails is to set them to a value slightly higher than the expected amount of pheromone deposited by the ants in one iteration; a rough estimate of this value can be obtained by setting, $\forall (i, j)$, $\tau_{ij} = \tau_0 = m / C^{nn}$, where m is the number of ants, and C^{nn} is the length of a tour generated by the nearest-neighbor heuristic (in fact, any other reasonable tour construction procedure would work fine). The reason for this choice is that if the initial pheromone values τ_0's are too low, then the search is quickly biased by the first tours generated by the ants, which in general leads toward the exploration of inferior zones of the search space. On the other side, if the initial pheromone values are too high, then many iterations are lost waiting until pheromone evaporation reduces enough pheromone values, so that pheromone added by ants can start to bias the search.

Tour Construction

In AS, m (artificial) ants concurrently build a tour of the TSP. Initially, ants are put on randomly chosen cities. At each construction step, ant k applies a probabilistic action choice rule, called *random proportional* rule, to decide which city to visit next. In particular, the probability with which ant k, currently at city i, chooses to go to city j is

$$p_{ij}^k = \frac{[\tau_{ij}]^\alpha [\eta_{ij}]^\beta}{\sum_{l \in \mathcal{N}_i^k} [\tau_{il}]^\alpha [\eta_{il}]^\beta}, \quad \text{if } j \in \mathcal{N}_i^k, \tag{3.2}$$

where $\eta_{ij} = 1/d_{ij}$ is a heuristic value that is available a priori, α and β are two parameters which determine the relative influence of the pheromone trail and the heuristic information, and $\mathcal{N}_i^k$ is the feasible neighborhood of ant k when being at city i, that is, the set of cities that ant k has not visited yet (the probability of choosing a

city outside $\mathcal{N}_i^k$ is 0). By this probabilistic rule, the probability of choosing a particular arc (i, j) increases with the value of the associated pheromone trail τ_{ij} and of the
heuristic information value η_{ij}. The role of the parameters α and β is the following. If
$\alpha = 0$, the closest cities are more likely to be selected: this corresponds to a classic
stochastic greedy algorithm (with multiple starting points since ants are initially randomly distributed over the cities). If $\beta = 0$, only pheromone amplification is at work,
that is, only pheromone is used, without any heuristic bias. This generally leads to
rather poor results and, in particular, for values of $\alpha > 1$ it leads to the rapid emergence of a *stagnation* situation, that is, a situation in which all the ants follow the
same path and construct the same tour, which, in general, is strongly suboptimal
(Dorigo, 1992; Dorigo et al., 1996). Good parameter values for the algorithms presented in this section are summarized in box 3.1.

Box 3.1
Parameter Settings for ACO Algorithms without Local Search

Our experimental study of the various ACO algorithms for the TSP has identified parameter settings that result in good performance. For the parameters that are common to almost all the ACO
algorithms, good settings (if no local search is applied) are given in the following table.

ACO algorithm	α	β	ρ	m	τ_0
AS	1	2 to 5	0.5	n	m/C^{nn}
EAS	1	2 to 5	0.5	n	$(e + m)/\rho C^{nn}$
AS_{rank}	1	2 to 5	0.1	n	$0.5r(r - 1)/\rho C^{nn}$
$\mathcal{MMAS}$	1	2 to 5	0.02	n	$1/\rho C^{nn}$
ACS	—	2 to 5	0.1	10	$1/n C^{nn}$

Here, n is the number of cities in a TSP instance. All variants of AS also require some additional
parameters. Good values for these parameters are:

EAS: The parameter e is set to $e = n$.

AS_{rank}: The number of ants that deposit pheromones is $w = 6$.

$\mathcal{MMAS}$: The pheromone trail limits are $\tau_{max} = 1/\rho C^{bs}$ and $\tau_{min} = \tau_{max}(1 - \sqrt[n]{0.05})/((avg - 1) \cdot \sqrt[n]{0.05})$, *where avg is the average number of different choices available to an ant at each step while*
constructing a solution (for a justification of these values see Stützle & Hoos (2000). When applied
to small TSP instances with up to 200 cities, good results are obtained by using always the iteration-
best pheromone update rule, while on larger instances it becomes increasingly important to alternate between the iteration-best and the best-so-far pheromone update rules.

ACS: In the local pheromone trail update rule: $\xi = 0.1$. In the pseudorandom proportional action
choice rule: $q_0 = 0.9$.

It should be clear that in individual instances, different settings may result in much better performance. However, these parameters were found to yield reasonable performance over a significant set of TSP instances.

Each ant k maintains a memory $\mathcal{M}^k$ which contains the cities already visited, in the order they were visited. This memory is used to define the feasible neighborhood $\mathcal{N}_i^k$ in the construction rule given by equation (3.2). In addition, the memory $\mathcal{M}^k$ allows ant k both to compute the length of the tour T^k it generated and to retrace the path to deposit pheromone.

Concerning solution construction, there are two different ways of implementing it: parallel and sequential solution construction. In the parallel implementation, at each construction step all the ants move from their current city to the next one, while in the sequential implementation an ant builds a complete tour before the next one starts to build another one. In the AS case, both choices for the implementation of the tour construction are equivalent in the sense that they do not significantly influence the algorithm's behavior. As we will see, this is not the case for other ACO algorithms such as ACS.

Update of Pheromone Trails
After all the ants have constructed their tours, the pheromone trails are updated. This is done by first lowering the pheromone value on *all* arcs by a constant factor, and then adding pheromone on the arcs the ants have crossed in their tours. Pheromone evaporation is implemented by

$$\tau_{ij} \leftarrow (1 - \rho)\tau_{ij}, \quad \forall (i, j) \in L, \tag{3.3}$$

where $0 < \rho \leq 1$ is the pheromone evaporation rate. The parameter ρ is used to avoid unlimited accumulation of the pheromone trails and it enables the algorithm to "forget" bad decisions previously taken. In fact, if an arc is not chosen by the ants, its associated pheromone value decreases exponentially in the number of iterations. After evaporation, all ants deposit pheromone on the arcs they have crossed in their tour:

$$\tau_{ij} \leftarrow \tau_{ij} + \sum_{k=1}^{m} \Delta\tau_{ij}^k, \quad \forall (i, j) \in L, \tag{3.4}$$

where $\Delta\tau_{ij}^k$ is the amount of pheromone ant k deposits on the arcs it has visited. It is defined as follows:

$$\Delta\tau_{ij}^k = \begin{cases} 1/C^k, & \text{if arc } (i, j) \text{ belongs to } T^k; \\ 0, & \text{otherwise}; \end{cases} \tag{3.5}$$

where C^k, the length of the tour T^k built by the k-th ant, is computed as the sum of the lengths of the arcs belonging to T^k. By means of equation (3.5), the better an ant's tour is, the more pheromone the arcs belonging to this tour receive. In general, arcs that are used by many ants and which are part of short tours, receive more pheromone and are therefore more likely to be chosen by ants in future iterations of the algorithm.

As we said, the relative performance of AS when compared to other metaheuristics tends to decrease dramatically as the size of the test-instance increases. Therefore, a substantial amount of research on ACO has focused on how to improve AS.

3.3.2   Elitist Ant System

A first improvement on the initial AS, called the *elitist strategy* for Ant System (EAS), was introduced in Dorigo (1992) and Dorigo et al., (1991a, 1996). The idea is to provide strong additional reinforcement to the arcs belonging to the best tour found since the start of the algorithm; this tour is denoted as T^{bs} (*best-so-far* tour) in the following. Note that this additional feedback to the best-so-far tour (which can be viewed as additional pheromone deposited by an additional ant called *best-so-far* ant) is another example of a *daemon action* of the ACO metaheuristic.

Update of Pheromone Trails
The additional reinforcement of tour T^{bs} is achieved by adding a quantity e/C^{bs} to its arcs, where e is a parameter that defines the weight given to the best-so-far tour T^{bs}, and C^{bs} is its length. Thus, equation (3.4) for the pheromone deposit becomes

$$\tau_{ij} \leftarrow \tau_{ij} + \sum_{k=1}^{m} \Delta\tau_{ij}^{k} + e\Delta\tau_{ij}^{bs}, \tag{3.6}$$

where $\Delta\tau_{ij}^{k}$ is defined as in equation (3.5) and $\Delta\tau_{ij}^{bs}$ is defined as follows:

$$\Delta\tau_{ij}^{bs} = \begin{cases} 1/C^{bs}, & \text{if arc } (i, j) \text{ belongs to } T^{bs}; \\ 0, & \text{otherwise.} \end{cases} \tag{3.7}$$

Note that in EAS, as well as in all other algorithms presented in section 3.3, pheromone evaporation is implemented as in AS.

Computational results presented in Dorigo (1992) and Dorigo et al. (1991a, 1996) suggest that the use of the elitist strategy with an appropriate value for parameter e allows AS to both find better tours and find them in a lower number of iterations.

3.3.3   Rank-Based Ant System

Another improvement over AS is the *rank-based* version of AS (AS_{rank}), proposed by Bullnheimer et al. (1999c). In AS_{rank} each ant deposits an amount of pheromone that decreases with its rank. Additionally, as in EAS, the best-so-far ant always deposits the largest amount of pheromone in each iteration.

Update of Pheromone Trails
Before updating the pheromone trails, the ants are sorted by increasing tour length and the quantity of pheromone an ant deposits is weighted according to the rank r of the ant. Ties can be solved randomly (in our implementation they are solved by lexicographic ordering on the ant name k). In each iteration only the $(w-1)$ best-ranked ants and the ant that produced the best-so-far tour (this ant does not necessarily belong to the set of ants of the current algorithm iteration) are allowed to deposit pheromone. The best-so-far tour gives the strongest feedback, with weight w (i.e., its contribution $1/C^{bs}$ is multiplied by w); the r-th best ant of the current iteration contributes to pheromone updating with the value $1/C^r$ multiplied by a weight given by $\max\{0, w-r\}$. Thus, the AS_{rank} pheromone update rule is

$$\tau_{ij} \leftarrow \tau_{ij} + \sum_{r=1}^{w-1}(w-r)\Delta\tau_{ij}^r + w\Delta\tau_{ij}^{bs}, \tag{3.8}$$

where $\Delta\tau_{ij}^r = 1/C^r$ and $\Delta\tau_{ij}^{bs} = 1/C^{bs}$. The results of an experimental evaluation by Bullnheimer et al. (1999c) suggest that AS_{rank} performs slightly better than EAS and significantly better than AS.

3.3.4 $\mathcal{MAX}$–$\mathcal{MIN}$ Ant System

$\mathcal{MAX}$–$\mathcal{MIN}$ Ant System ($\mathcal{MMAS}$) (Stützle & Hoos, 1997, 2000; Stützle, 1999) introduces four main modifications with respect to AS. First, it strongly exploits the best tours found: only either the iteration-best ant, that is, the ant that produced the best tour in the current iteration, or the best-so-far ant is allowed to deposit pheromone. Unfortunately, such a strategy may lead to a stagnation situation in which all the ants follow the same tour, because of the excessive growth of pheromone trails on arcs of a good, although suboptimal, tour. To counteract this effect, a second modification introduced by $\mathcal{MMAS}$ is that it limits the possible range of pheromone trail values to the interval $[\tau_{min}, \tau_{max}]$. Third, the pheromone trails are initialized to the upper pheromone trail limit, which, together with a small pheromone evaporation rate, increases the exploration of tours at the start of the search. Finally, in $\mathcal{MMAS}$, pheromone trails are reinitialized each time the system approaches stagnation or when no improved tour has been generated for a certain number of consecutive iterations.

Update of Pheromone Trails

After all ants have constructed a tour, pheromones are updated by applying evaporation as in AS [equation (3.3)], followed by the deposit of new pheromone as
follows:

$$\tau_{ij} \leftarrow \tau_{ij} + \Delta\tau_{ij}^{best}, \qquad\qquad\qquad (3.9)$$

where $\Delta\tau_{ij}^{best} = 1/C^{best}$. The ant which is allowed to add pheromone may be either
the best-so-far, in which case $\Delta\tau_{ij}^{best} = 1/C^{bs}$, or the iteration-best, in which case
$\Delta\tau_{ij}^{best} = 1/C^{ib}$, where C^{ib} is the length of the iteration-best tour. In general, in
$\mathcal{MMAS}$ implementations both the iteration-best and the best-so-far update rules are
used, in an alternate way. Obviously, the choice of the relative frequency with which
the two pheromone update rules are applied has an influence on how greedy the
search is: When pheromone updates are always performed by the best-so-far ant, the
search focuses very quickly around T^{bs}, whereas when it is the iteration-best ant that
updates pheromones, then the number of arcs that receive pheromone is larger and
the search is less directed.

   Experimental results indicate that for small TSP instances it may be best to use
only iteration-best pheromone updates, while for large TSPs with several hundreds of
cities the best performance is obtained by giving an increasingly stronger emphasis to
the best-so-far tour. This can be achieved, for example, by gradually increasing the
frequency with which the best-so-far tour T^{bs} is chosen for the trail update (Stützle,
1999).

Pheromone Trail Limits

In $\mathcal{MMAS}$, lower and upper limits τ_{min} and τ_{max} on the possible pheromone values
on any arc are imposed in order to avoid search stagnation. In particular, the imposed pheromone trail limits have the effect of limiting the probability p_{ij} of selecting
a city j when an ant is in city i to the interval $[p_{min}, p_{max}]$, with $0 < p_{min} \leq p_{ij} \leq
p_{max} \leq 1$. Only when an ant k has just one single possible choice for the next city,
that is $|\mathcal{N}_i^k| = 1$, we have $p_{min} = p_{max} = 1$.

   It is easy to show that, in the long run, the upper pheromone trail limit on any arc
is bounded by $1/\rho C^*$, where C^* is the length of the optimal tour (see proposition 4.1
in chapter 4). Based on this result, $\mathcal{MMAS}$ uses an estimate of this value, $1/\rho C^{bs}$, to
define τ_{max}: each time a new best-so-far tour is found, the value of τ_{max} is updated.
The lower pheromone trail limit is set to $\tau_{min} = \tau_{max}/a$, where a is a parameter
(Stützle, 1999; Stützle & Hoos, 2000). Experimental results (Stützle, 1999) suggest
that, in order to avoid stagnation, the lower pheromone trail limits play a more

important role than τ_{max}. On the other hand, τ_{max} remains useful for setting the pheromone values during the occasional trail reinitializations.

Pheromone Trail Initialization and Reinitialization

At the start of the algorithm, the initial pheromone trails are set to an estimate of the upper pheromone trail limit. This way of initializing the pheromone trails, in combination with a small pheromone evaporation parameter, causes a slow increase in the relative difference in the pheromone trail levels, so that the initial search phase of $\mathcal{MMAS}$ is very explorative.

As a further means of increasing the exploration of paths that have only a small probability of being chosen, in $\mathcal{MMAS}$ pheromone trails are occasionally re-initialized. Pheromone trail reinitialization is typically triggered when the algorithm approaches the stagnation behavior (as measured by some statistics on the pheromone trails) or if for a given number of algorithm iterations no improved tour is found.

$\mathcal{MMAS}$ is one of the most studied ACO algorithms and it has been extended in many ways. In one of these extensions, the pheromone update rule occasionally uses the best tour found since the most recent reinitialization of the pheromone trails instead of the best-so-far tour (Stützle, 1999; Stützle & Hoos, 2000). Another variant (Stützle, 1999; Stützle & Hoos, 1999) exploits the same pseudorandom proportional action choice rule as introduced by ACS [see equation (3.10) below], an ACO algorithm that is presented in section 3.4.1.

3.4   Extensions of Ant System

The ACO algorithms we have introduced so far achieve significantly better performance than AS by introducing minor changes in the overall AS algorithmic structure. In this section we discuss two additional ACO algorithms that, although strongly inspired by AS, achieve performance improvements through the introduction of new mechanisms based on ideas not included in the original AS. Additionally, we present the hyper-cube framework for ACO algorithms.

3.4.1   Ant Colony System

ACS (Dorigo & Gambardella, 1997a,b) differs from AS in three main points. First, it exploits the search experience accumulated by the ants more strongly than AS does through the use of a more aggressive action choice rule. Second, pheromone evaporation and pheromone deposit take place only on the arcs belonging to the best-so-far tour. Third, each time an ant uses an arc (i, j) to move from city i to city

j, it removes some pheromone from the arc to increase the exploration of alternative paths. In the following, we present these innovations in more detail.

Tour Construction
In ACS, when located at city i, ant k moves to a city j chosen according to the so-called *pseudorandom proportional* rule, given by

$$j = \begin{cases} \text{argmax}_{l \in \mathcal{N}_i^k}\{\tau_{il}[\eta_{il}]^\beta\}, & \text{if } q \le q_0; \\ J, & \text{otherwise;} \end{cases} \tag{3.10}$$

where q is a random variable uniformly distributed in $[0, 1]$, q_0 $(0 \le q_0 \le 1)$ is a parameter, and J is a random variable selected according to the probability distribution given by equation (3.2) (with $\alpha = 1$).

In other words, with probability q_0 the ant makes the best possible move as indicated by the learned pheromone trails and the heuristic information (in this case, the ant is exploiting the learned knowledge), while with probability $(1 - q_0)$ it performs a biased exploration of the arcs. Tuning the parameter q_0 allows modulation of the degree of exploration and the choice of whether to concentrate the search of the system around the best-so-far solution or to explore other tours.

Global Pheromone Trail Update
In ACS only one ant (the best-so-far ant) is allowed to add pheromone after each iteration. Thus, the update in ACS is implemented by the following equation:

$$\tau_{ij} \leftarrow (1 - \rho)\tau_{ij} + \rho\Delta\tau_{ij}^{bs}, \quad \forall(i, j) \in T^{bs}, \tag{3.11}$$

where $\Delta\tau_{ij}^{bs} = 1/C^{bs}$. It is important to note that in ACS the pheromone trail update, both evaporation and new pheromone deposit, only applies to the arcs of T^{bs}, not to all the arcs as in AS. This is important, because in this way the computational complexity of the pheromone update at each iteration is reduced from $\mathcal{O}(n^2)$ to $\mathcal{O}(n)$, where n is the size of the instance being solved. As usual, the parameter ρ represents pheromone evaporation; unlike AS's equations (3.3) and (3.4), in equation (3.11) the deposited pheromone is discounted by a factor ρ; this results in the new pheromone trail being a weighted average between the old pheromone value and the amount of pheromone deposited.

In initial experiments, the use of the iteration-best tour was also considered for the pheromone updates. Although for small TSP instances the differences in the final tour quality obtained by updating the pheromones using the best-so-far or the iteration-best tour was found to be minimal, for instances with more than 100 cities the use of the best-so-far tour gave far better results.

Local Pheromone Trail Update

In addition to the global pheromone trail updating rule, in ACS the ants use a local pheromone update rule that they apply immediately after having crossed an arc (i, j) during the tour construction:

$$\tau_{ij} \leftarrow (1 - \xi)\tau_{ij} + \xi\tau_0, \tag{3.12}$$

where ξ, $0 < \xi < 1$, and τ_0 are two parameters. The value of τ_0 is set to be the same as the initial value for the pheromone trails. Experimentally, a good value for ξ was found to be 0.1, while a good value for τ_0 was found to be $1/nC^{nn}$, where n is the number of cities in the TSP instance and C^{nn} is the length of a nearest-neighbor tour. The effect of the local updating rule is that each time an ant uses an arc (i, j) its pheromone trail τ_{ij} is reduced, so that the arc becomes less desirable for the following ants. In other words, this allows an increase in the exploration of arcs that have not been visited yet and, in practice, has the effect that the algorithm does not show a stagnation behavior (i.e., ants do not converge to the generation of a common path) (Dorigo & Gambardella, 1997b). It is important to note that, while for the previously discussed AS variants it does not matter whether the ants construct the tours in parallel or sequentially, this makes a difference in ACS because of the local pheromone update rule. In most ACS implementations the choice has been to let all the ants move in parallel, although there is, at the moment, no experimental evidence in favor of one choice or the other.

Some Additional Remarks

ACS is based on Ant-Q, an earlier algorithm proposed by Gambardella & Dorigo (1995) (see also Dorigo & Gambardella, 1996). In practice, the only difference between ACS and Ant-Q is in the definition of the term τ_0, which in Ant-Q is set to $\tau_0 = \gamma \max_{j \in \mathcal{N}_i^k}\{\tau_{ij}\}$, where γ is a parameter and the maximum is taken over the set of pheromone trails on the arcs connecting the city i on which ant k is positioned to all the cities the ant has not visited yet (i.e., those in the neighborhood $\mathcal{N}_i^k$).

This particular choice for τ_0 was motivated by an analogy with a similar formula used in Q-learning (Watkins & Dayan, 1992), a well-known reinforcement learning algorithm (Sutton & Barto, 1998). Because it was found that setting τ_0 to a small constant value resulted in a simpler algorithm with approximately the same performance, Ant-Q was abandoned.

There also exists an interesting relationship between $\mathcal{MMAS}$ and ACS: they both use pheromone trail limits, although these are explicit in $\mathcal{MMAS}$ and implicit in ACS. In fact, in ACS implementations the pheromone trails can never drop below τ_0 because both pheromone update rules [equations (3.11) and (3.12)] always add an amount of pheromone greater than or equal to τ_0, and the initial pheromone trail value is set to the value τ_0. On the other hand, as discussed in section 4.3.5.2 of chapter 4, it can easily be verified that the pheromone trails can never have a value higher than $1/C^{bs}$. Therefore, in ACS it is implicitly guaranteed that $\forall (i, j) : \tau_0 \leq \tau_{ij} \leq 1/C^{bs}$.

Finally, it should be mentioned that ACS was the first ACO algorithm to use *candidate lists* to restrict the number of available choices to be considered at each construction step. In general, candidate lists contain a number of the best rated choices according to some heuristic criterion. In the TSP case, a candidate list contains for each city i those cities j that are at a small distance. There are several ways to define which cities enter the candidate lists. ACS first sorts the neighbors of a city i according to nondecreasing distances and then inserts a fixed number *cand* of closest cities into i's candidate list. In this case, the candidate lists can be built before solving a TSP instance and they remain fixed during the whole solution process. When located at i, an ant chooses the next city among those of i's candidate list that are not visited yet. Only if all the cities of the candidate list are already marked as visited, is one of the remaining cities chosen. In the TSP case, experimental results have shown that the use of candidate lists improves the solution quality reached by the ACO algorithms. Additionally, it leads to a significant speedup in the solution process (Gambardella & Dorigo, 1996).

3.4.2   Approximate Nondeterministic Tree Search

Approximate nondeterministic tree search (ANTS) (Maniezzo, 1999) is an ACO algorithm that exploits ideas from mathematical programming. In particular, ANTS computes lower bounds on the completion of a partial solution to define the heuristic information that is used by each ant during the solution construction. The name ANTS derives from the fact that the proposed algorithm can be interpreted as an approximate nondeterministic tree search since it can be extended in a straightforward way to a branch & bound (Bertsekas, 1995a) procedure. In fact, in Maniezzo (1999) the ANTS algorithm is extended to an exact algorithm; we refer the interested reader to the original reference for details; here we only present the ACO part of the algorithm.

Apart from the use of lower bounds, ANTS also introduces two additional modifications with respect to AS: the use of a novel action choice rule and a modified pheromone trail update rule.

Use of Lower Bounds

In ANTS, lower bounds on the completion cost of a partial solution are used to compute heuristic information on the attractiveness of adding an arc (i, j). This is achieved by tentatively adding the arc to the current partial solution and by estimating the cost of a complete tour containing this arc by means of a lower bound. This estimate is then used to compute the value η_{ij} that influences the probabilistic decisions taken by the ant during the solution construction: the lower the estimate the more attractive the addition of a specific arc.

The use of lower bounds to compute the heuristic information has the advantage in that otherwise feasible moves can be discarded if they lead to partial solutions whose estimated costs are larger than the best-so-far solution. A disadvantage is that the lower bound has to be computed at each single construction step of an ant and therefore a significant computational overhead might be incurred. To avoid this as much as possible, it is important that the lower bound is computed efficiently.

Solution Construction

The rule used by ANTS to compute the probabilities during the ants' solution construction has a different form than that used in most other ACO algorithms. In ANTS, an ant k that is situated at city i chooses the next city j with a probability given by

$$p_{ij}^k = \frac{\zeta \tau_{ij} + (1 - \zeta)\eta_{ij}}{\sum_{l \in \mathcal{N}_i^k} \zeta \tau_{il} + (1 - \zeta)\eta_{il}}, \quad \text{if } j \in \mathcal{N}_i^k, \tag{3.13}$$

where ζ is a parameter, $0 \le \zeta \le 1$, and $\mathcal{N}_i^k$ is, as before, the feasible neighborhood (as usual, the probability of choosing an arc not belonging to $\mathcal{N}_i^k$ is 0).

An advantage of equation (3.13) is that, when compared to equation (3.2), only one parameter is used. Additionally, simpler operations that are faster to compute, like sums instead of multiplications for combining the pheromone trail and the heuristic information, are applied.

Pheromone Trail Update

Another particularity of ANTS is that it has no explicit pheromone evaporation. Pheromone updates are implemented as follows:

$$\tau_{ij} \leftarrow \tau_{ij} + \sum_{k=1}^{m} \Delta \tau_{ij}^k. \tag{3.14}$$

In the above equation (3.14), $\Delta \tau_{ij}^k$ is given by

$$\Delta \tau_{ij}^k = \begin{cases} \vartheta \left(1 - \dfrac{C^k - LB}{L_{\text{avg}} - LB}\right), & \text{if arc } (i, j) \text{ belongs to } T^k; \\ 0, & \text{otherwise;} \end{cases} \tag{3.15}$$

where ϑ is a parameter, LB is the value of a lower bound on the optimal solution value computed at the start of the algorithm and we have $LB \leq C^*$, where C^* is the length of the optimal tour, and L_{avg} is the moving average of the last l tours generated by the ants, that is, it is the average length of the l most recent tours generated by the algorithm (with l being a parameter of the algorithm). If an ant's solution is worse than the current moving average, the pheromone trail of the arcs used by the ant is decreased; if the ant's solution is better, the pheromone trail is increased. The additional effect of using equation (3.15) is a dynamic scaling of the objective function differences which may be advantageous if in later stages of the search the absolute difference between the ant's solution qualities becomes smaller and, consequently, C^k moves closer to L_{avg}. (Note that once a solution with objective function value equal to LB is found, the algorithm can be stopped, because this means that an optimal solution is found.)

As said before, ANTS has not been applied to the TSP so far, although some limited experiments were performed for the asymmetric TSP (Maniezzo, 2000). The first and main publication of ANTS concerns the quadratic assignment problem (Maniezzo, 1999), for which it obtained very good results (see chapter 5, section 5.2.1).

3.4.3   Hyper-Cube Framework for ACO

The hyper-cube framework for ACO was introduced by Blum, Roli, & Dorigo (2001) to automatically rescale the pheromone values in order for them to lie always in the interval $[0, 1]$. This choice was inspired by the mathematical programming formulation of many combinatorial optimization problems, in which solutions can be represented by binary vectors. In such a formulation, the decision variables, which can assume the values $\{0, 1\}$, typically correspond to the solution components as they are used by the ants for solution construction. A solution to a problem then corresponds to one corner of the n-dimensional hyper-cube, where n is the number of decision variables. One particular way of generating lower bounds for the problem under consideration is to relax the problem, allowing each decision variable to take values in the interval $[0, 1]$. In this case, the set of feasible solutions $\mathcal{S}_{rx}$ consists of all vectors $\vec{v} \in \mathbb{R}^n$ that are convex combinations of binary vectors $\vec{x} \in \mathbb{B}^n$:

$$\vec{v} \in \mathcal{S}_{rx} \Leftrightarrow \vec{v} = \sum_{\vec{x}_i \in \mathbb{B}^n} \gamma_i \cdot \vec{x}_i, \qquad \gamma_i \in [0, 1], \sum \gamma_i = 1.$$

The relationship with ACO becomes clear once we normalize the pheromone values to lie in the interval $[0, 1]$. In this case, the pheromone vector $\vec{\tau} = (\tau_1, \ldots, \tau_n)$

corresponds to a point in $\tilde{S}$; in case $\vec{\tau}$ is a binary vector, it corresponds to a solution of the problem.

When applied to the TSP, a decision variable x_{ij} can be associated with each arc (i, j). This decision variable is set to $x_{ij} = 1$ when the arc (i, j) is used, and to $x_{ij} = 0$ otherwise. In this case, a pheromone value is associated with each decision variable. In fact, the reader may have noticed that this representation corresponds to the standard way of attacking TSPs with ACO algorithms, as presented before.

Pheromone Trail Update Rules
In the hyper-cube framework the pheromone trails are forced to stay in the interval $[0, 1]$. This is achieved by adapting the standard pheromone update rule of ACO algorithms. Let us explain the necessary change considering the pheromone update rule of AS [equations (3.3) and (3.4)]. The modified rule is given by

$$\tau_{ij} \leftarrow (1 - \rho)\tau_{ij} + \rho \sum_{k=1}^{m} \Delta\tau_{ij}^{k}, \tag{3.16}$$

where, to compute the rightmost term, instead of equation (3.7) we use

$$\Delta\tau_{ij}^{k} = \begin{cases} \dfrac{1/C^k}{\sum_{h=1}^{m}(1/C^h)}, & \text{if arc } (i, j) \text{ is used by ant } k; \\ 0, & \text{otherwise.} \end{cases} \tag{3.17}$$

This pheromone trail update rule guarantees that the pheromone trails remain smaller than 1; the update rule is also illustrated in figure 3.4: The new pheromone vector can be interpreted as a shift of the old pheromone vector toward the vector given by the weighted average of the solutions used in the pheromone update.

3.5   Parallel Implementations

The very nature of ACO algorithms lends them to be parallelized in the data or population domains. In particular, many parallel models used in other population-based algorithms can be easily adapted to ACO. Most parallelization strategies can be classified into *fine-grained* and *coarse-grained* strategies. Characteristic of fine-grained parallelization is that very few individuals are assigned to single processors and that frequent information exchange among the processors takes place. In coarse-grained approaches, on the contrary, larger subpopulations or even full populations are assigned to single processors and information exchange is rather rare. See, for example, Cantú-Paz (2000) for an overview.

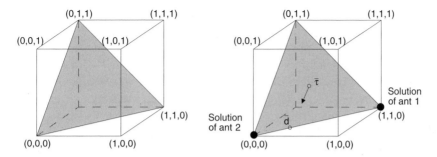

Figure 3.4
Left: Assume that the set of feasible solutions consists of the three vectors $(0, 0, 0), (1, 1, 0)$, and $(0, 1, 1)$. Then, the pheromone vector $\vec{\tau}$ moves over the gray shaded area. Right: The two solutions $(0, 0, 0)$ and $(1, 1, 0)$ have been generated by the ants and are used for the pheromone trail update: $\vec{\tau}$ will be shifted toward $\vec{d}$. Note that $\vec{d}$ is the weighted average of the two solutions, so that it belongs to the segment connecting them [in the example $(0, 0, 0)$ is considered of higher quality than $(1, 1, 0)$, and therefore $\vec{\tau}$ is closer to $(0, 0, 0)$ than to $(1, 1, 0)$].

Fine-grained parallelization schemes have been investigated with parallel versions of AS for the TSP on the Connection Machine CM-2 adopting the approach of attributing a single processing unit to each ant (Bolondi & Bondanza, 1993). Experimental results showed that communication overhead can be a major problem with this approach, since ants end up spending most of their time communicating the modifications they made to pheromone trails. Similar negative results have been reported by Bullnheimer, Kotsis, and Strauss (1998).

As shown by several researches (Bolondi & Bondanza, 1993; Bullnheimer et al., 1998; Krüger, Merkle, & Middendorf, 1998; Middendorf, Reischle, & Schmeck, 2002; Stützle, 1998b), coarse-grained parallelization schemes are much more promising for ACO. In this case, p colonies run in parallel on p processors.

Stützle (1998b) has considered the extreme case in which there is no communication among the colonies. It is equivalent to the parallel independent run of many ACO algorithms, and is the easiest way to parallelize randomized algorithms. The computational results presented in Stützle (1998b) show that this approach is very effective.

A number of other researchers have considered the case in which information among the colonies is exchanged at certain intervals. For example, Bullnheimer et al. (1998) proposed the *partially asynchronous parallel implementation* (PAPI). In PAPI, pheromone information was exchanged among the colonies every fixed number of iterations and a high speedup was experimentally observed. Krüger et al. (1998) investigated the type of information that should be exchanged among the colonies and how this information should be used to update the colonies' pheromone trail information. Their results showed that it is better to exchange the best solutions found so far and to use them in the pheromone update rather than to exchange complete pheromone matrices. Middendorf et al. (2002), extending the original work of Michel & Middendorf (1998), investigated different ways of exchanging solutions among m ant colonies. They let colonies exchange information every fixed number of iterations. The information exchanged is (1) the best-so-far solution that is shared among all colonies, and (2) either the locally best-so-far solutions or the w iteration-best ants, or a combination of the two, that are sent to neighbor colonies, where the neighborhood was organized as a directed ring. Their main observation was that the best results were obtained by limiting the information exchange to the locally best solutions.

Some preliminary work on the parallel implementation of an ACO algorithm on a shared memory architecture using OpenMP (Chandra, Dagum, Kohr, Maydan, McDonald, & Menon, 2000) is presented in Delisle, Krajecki, Gravel, & Gagné (2001).

3.6   Experimental Evaluation

In order to establish a meaningful comparison of the different versions of ACO discussed in the previous sections, we have reimplemented all of them using the TSP as an application problem, with the exception of ANTS, for which no application to the TSP has been reported in the literature. The resulting software package is available for download at www.aco-metaheuristic.org/aco-code/. We used this software package to study the dependence of the ACO algorithms' behavior on particular configurations or parameters. All the experiments were performed either on a 700 MHz Pentium III double-processor machine with 512 MB of RAM or on a 1.2 GHz Athlon MP double-processor machine with 1 GB of RAM; both machines were running SUSE Linux 7.3. These experiments should be understood as giving an indication of the general behavior of the available ACO algorithms when applied to $\mathcal{NP}$-hard combinatorial optimization problems and as an illustration of what happens when ACO algorithms are combined with local search algorithms. On the contrary, the experiments are not meant to present results competitive with current state-of-the-art algorithms for the TSP. In fact, current state-of-the-art algorithms for the TSP exploit complex data structures and local search routines that have not been implemented for ACO. Nevertheless, the results of our study are interesting because most of our findings remain true when ACO is applied to other $\mathcal{NP}$-hard problems.

3.6.1 The Behavior of ACO Algorithms

Artificial ants iteratively sample tours through a loop that includes a tour construction biased by the artificial pheromone trails and the heuristic information. The main mechanism at work in ACO algorithms that triggers the discovery of good tours is the positive feedback given through the pheromone update by the ants: the shorter the ant's tour, the higher the amount of pheromone the ant deposits on the arcs of its tour. This in turn leads to the fact that these arcs have a higher probability of being selected in the subsequent iterations of the algorithm. The emergence of arcs with high pheromone values is further reinforced by the pheromone trail evaporation that avoids an unlimited accumulation of pheromones and quickly decreases the pheromone level on arcs that only very rarely, or never, receive additional pheromone.

This behavior is illustrated in figure 3.5, where AS is applied to the 14-city TSP instance `burma14` from TSPLIB. The figure gives a visual representation of the pheromone matrix: pheromone trail levels are translated into gray scale, where black

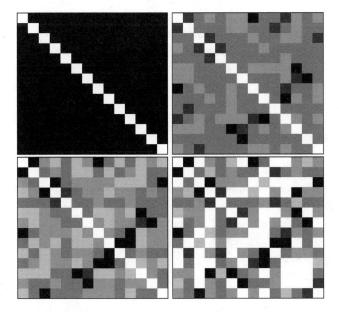

Figure 3.5
A visual representation of the pheromone matrix. The pheromone values on the arcs, stored in the pheromone matrix, are translated into gray-scale values; the darker an entry, the higher the associated pheromone trail value. The plots, from upper left to lower right, show the pheromone value for AS applied to TSPLIB instance `burma14` with 14 cities after 0, 5, 10, and 100 iterations. Note the symmetry with respect to the main diagonal, which is due to the fact that `burma14` is a symmetric TSP instance.

represents the highest pheromone trails and white the lowest ones. The four plots give snapshots of the pheromone matrix after 0, 5, 10, and 100 iterations (from upper left to lower right). At the beginning, all the matrix's cells are black except for those on the diagonal which are always white because they are initialized to zero and never updated. After five iterations, the differences between the pheromone trails are still not very manifest; this is due to the fact that pheromone evaporation and pheromone update could be applied only five times and therefore large differences between the pheromone trails could not be established yet. Also, after five iterations the pheromone trails are still rather high, which is due to the large initial pheromone values. As the algorithm continues to iterate, the differences between the pheromone values become stronger and finally a situation is reached in which only few connections have a large amount of pheromone associated with them (and therefore a large probability of being chosen) and several connections have pheromone values close to zero, making a selection of these connections very unlikely.

With good parameter settings, the long-term effect of the pheromone trails is to progressively reduce the size of the explored search space so that the search concentrates on a small number of promising arcs. Yet, this behavior may become undesirable, if the concentration is so strong that it results in an early stagnation of the search (remember that search stagnation is defined as the situation in which all the ants follow the same path and construct the same solution). In such an undesirable situation the system has ceased to explore new possibilities and no better tour is likely to be found anymore.

Several measures may be used to describe the amount of exploration an ACO algorithm still performs and to detect stagnation situations. One of the simplest possibilities is to compute the standard deviation σ_L of the length of the tours the ants construct after every iteration—if σ_L is zero, this is an indication that all the ants follow the same path (although σ_L can go to zero also in the very unlikely case in which the ants follow different tours of the same length).

Because the standard deviation depends on the absolute values of the tour lengths, a better choice is the use of the variation coefficient, defined as the quotient between the standard deviation of the tour lengths and the average tour length, which is independent of the scale.

The distance between tours gives a better indication of the amount of exploration the ants perform. In the TSP case, a way of measuring the distance $dist(T, T')$ between two tours T and T' is to count the number of arcs contained in one tour but not in the other. A decrease in the average distance between the ants' tours indicates that preferred paths are appearing, and if the average distance becomes zero, then the system has entered search stagnation. A disadvantage of this measure is that it is computationally expensive: there are $\mathcal{O}(n^2)$ possible pairs to be compared and each single comparison has a complexity of $\mathcal{O}(n)$.

While these measures only use the final tours constructed by the ants, the λ-branching factor, $0 < \lambda < 1$, introduced in Dorigo & Gambardella (1997b), measures the distribution of the pheromone trail values more directly. Its definition is based on the following notion: If for a given city i the concentration of pheromone trail on almost all the incident arcs becomes very small but is large for a few others, the freedom of choice for extending partial tours from that city is very limited. Consequently, if this situation arises simultaneously for all the nodes of the graph, the part of the search space that is effectively searched by the ants becomes relatively small. The λ-branching factor for a city i is defined as follows: If τ_{max}^i is the maximal and τ_{min}^i the minimal pheromone trail value on arcs incident to node i, the λ-branching factor is given by the number of arcs incident to i that have a pheromone trail value $\tau_{ij} \geq \tau_{min}^i + \lambda(\tau_{max}^i - \tau_{min}^i)$. The value of λ ranges over the interval $[0,1]$, while the values of the λ-branching factors range over the interval $[2, n-1]$, where n is the number of nodes in the construction graph (which, in the TSP case, is the same as the number of cities). The average λ-branching factor $\bar{\lambda}$ is the average of the λ-branching factors of all nodes and gives an indication of the size of the search space effectively being explored by the ants. If, for example, $\bar{\lambda}$ is very close to 3, on average only three arcs for each node have a high probability of being chosen. Note that in the TSP the minimal $\bar{\lambda}$ is 2, because for each city there must be at least two arcs used by the ants to reach and to leave the city while building their solutions.

A disadvantage of the λ-branching factor is that its values depend on the setting of the parameter λ. Another possibility for a measure of stagnation would be to use the average $\bar{\mathcal{E}} = \sum_{i=1}^{n} \mathcal{E}_i/n$ of the entropies $\mathcal{E}_i$ of the selection probabilities at each node:

$$\mathcal{E}_i = -\sum_{j=1}^{l} p_{ij} \log p_{ij}, \tag{3.18}$$

where p_{ij} is the probability of choosing arc (i, j) when being in node i, and l, $1 \leq l \leq n-1$, is the number of possible choices. Still another way to measure stagnation is given by the following formula:

$$\frac{\sum_{\tau_{ij} \in T} \min\{\tau_{max} - \tau_{ij}, \tau_{ij} - \tau_{min}\}}{n^2}, \tag{3.19}$$

whose value tends to 0 as the algorithm moves toward stagnation.

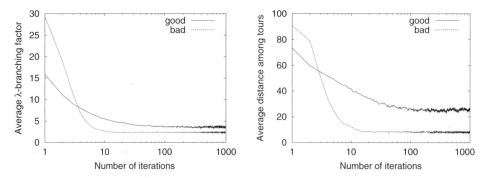

Figure 3.6
Bad behavior because of early stagnation: The plots give (left) the average λ-branching factor, $\lambda = 0.05$, and (right) the average distance among the tours generated by AS on the symmetric TSPLIB instance kroA100. "Good" system behavior is observed setting parameters to $\alpha = 1$, $\beta = 2$, $m = n$. "Bad" system behavior is observed setting parameters $\alpha = 5$, $\beta = 0$, $m = n$.

Behavior of AS

In this section we show the typical behavior of the average λ-branching factor and of the average distance among tours when AS has parameter settings that result in either good or bad algorithm performance. The parameter settings are denoted by *good* and *bad* in figure 3.6, and the values used are $\alpha = 1$, $\beta = 2$, $m = n$ and to $\alpha = 5$, $\beta = 0$, $m = n$ respectively. Figure 3.6 shows that for bad parameter settings the λ-branching factor converges to its minimum value much faster than for good parameter settings (λ is set to 0.05). A similar situation occurs when observing the average distance between tours. In fact, the experimental results of Dorigo et al. (1996) suggest that AS enters stagnation behavior if α is set to a large value, and does not find high-quality tours if α is chosen to be much smaller than 1. Dorigo et al. (1996) tested values of $\alpha \in \{0, 0.5, 1, 2, 5\}$. An example of bad system behavior that occurs if the amount of exploration is too large is shown in figure 3.7. Here, *good* refers to the same parameter setting as above and *bad* to the setting $\alpha = 1$, $\beta = 0$, and $m = n$. For both stagnation measures, average λ-branching factor and average distance between tours, the algorithm using the bad parameter setting is not able to focus the search on the most promising parts of the search space.

The overall result suggests that for AS good parameter settings are those that find a reasonable balance between a too narrow focus of the search process, which in the worst case may lead to stagnation behavior, and a too weak guidance of the search, which can cause excessive exploration.

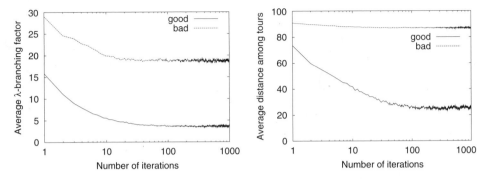

Figure 3.7
Bad behavior because of excessive exploration: The plots give (left) the average λ-branching factor, $\lambda = 0.05$, and (right) the average distance among the tours generated by AS on the symmetric TSPLIB instance kroA100. "Good" system behavior is observed setting parameters to $\alpha = 1, \beta = 2, m = n$. "Bad" system behavior is observed setting parameters $\alpha = 1, \beta = 0, m = n$.

Behavior of Extensions of AS

One particularity of AS extensions is that they direct the ants' search in a more aggressive way. This is mainly achieved by a stronger emphasis given to the best tours found during each iteration (e.g., in $\mathcal{MMAS}$) or the best-so-far tour (e.g., in ACS). We would expect that this stronger focus of the search is reflected by statistical measures of the amount of exploration. Figure 3.8 indicates the development of the λ-branching factor and the average distance between tours as observed in AS, EAS, AS_{rank}, $\mathcal{MMAS}$, and ACS. For this comparison we used the same parameter settings as in box 3.1, except for the value of β which was set to 2 for all algorithms.

The various ACO algorithms show, in part, strongly different behaviors, which gives an indication that there are substantial differences in their ways of directing the search. While ACS shows a low λ-branching factor and small average distances between the tours throughout the algorithm's entire run, for the others a transition from a more explorative search phase, characterized by a rather high average λ-branching factor, to an exploitation phase, characterized by a very low average λ-branching factor, can be observed. While this transition happens very soon in AS and AS_{rank}, it occurs only later in $\mathcal{MMAS}$. On the other hand, AS_{rank} is the only algorithm that enters stagnation when run for a sufficiently high number of iterations. This observation also suggests that AS_{rank} could profit from occasional pheromone trail reinitializations, as was proposed for $\mathcal{MMAS}$ (Stützle & Hoos, 2000).

It is interesting to note that, although $\mathcal{MMAS}$ also converges to the minimum average λ-branching factor, which suggests stagnation behavior, the average distance

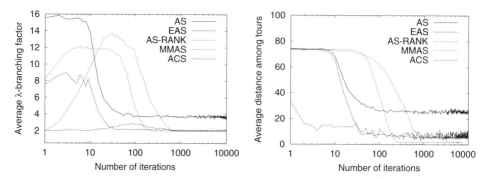

Figure 3.8
Comparing AS extensions: The plots give (left) the average λ-branching factor, $\lambda = 0.05$, and (right) the average distance among the tours for several ACO algorithms on the symmetric TSPLIB instance kroA100. Parameters were set as in box 3.1, except for β which was set to $\beta = 2$ for all the algorithms.

between the tours it generates remains significantly higher than zero. The reason for this apparently contradictory result is that $\mathcal{MMAS}$ uses pheromone trail limits τ_{max} and τ_{min}. So, even when the pheromone trails on the arcs of a tour reach the value τ_{max} and all others have the value τ_{min}, new tours will still be explored.

A common characteristic of all of the AS extensions is that their search is focused on a specific region of the search space. An indication of this is given by the lower λ-branching factor and the lower average distance between the tours of these extensions when compared to AS. Because of this, AS extensions need to be endowed with features intended to counteract search stagnation.

It should be noted that the behavior of the various ACO algorithms also depends strongly on the parameter settings. For example, it is easy to force $\mathcal{MMAS}$ to converge much faster to good tours by making the search more aggressive through the use of only the best-so-far update or by a higher evaporation rate. Nevertheless, the behavior we show in figure 3.8 is typical for reasonable parameter settings (see box 3.1).

In the following, we discuss the behavior of $\mathcal{MMAS}$ and ACS in more detail. This choice is dictated by the fact that these two algorithms are the most used and often the best-performing of ACO algorithms.

Behavior of $\mathcal{MMAS}$

Of the ACO algorithms considered in this chapter, $\mathcal{MMAS}$ has the longest explorative search phase. This is mainly due to the fact that pheromone trails are initialized to the initial estimate of τ_{max}, and that the evaporation rate is set to a low value (a value that gives good results for long runs of the algorithm was found to be $\rho = 0.02$). In fact, because of the low evaporation rate, it takes time before significant differences among the pheromone trails start to appear.

When this happens, $\mathcal{MMAS}$ behavior changes from explorative search to a phase of exploitation of the experience accumulated in the form of pheromone trails. In this phase, the pheromone on the arcs corresponding to the best-found tour rises up to the maximum value τ_{max}, while on all the other arcs it decreases down to the minimum value τ_{min}. This is reflected by an average λ-branching factor of 2.0. Nevertheless, the exploration of tours is still possible, because the constraint on the minimum value of pheromone trails has the effect of giving to each arc a minimum probability $p_{min} > 0$ of being chosen. In practice, during this exploitation phase $\mathcal{MMAS}$ constructs tours that are similar to either the best-so-far or the iteration-best tour, depending on the algorithm implementation.

Behavior of ACS

ACS uses a very aggressive search that focuses from the very beginning around the best-so-far tour T^{bs}. In other words, it generates tours that differ only in a relatively small number of arcs from the best-so-far tour T^{bs}. This is achieved by choosing a large value for q_0 in the pseudorandom proportional action choice rule [see equation (3.10)], which leads to tours that have many arcs in common with the best-so-far tour. An interesting aspect of ACS is that while arcs are traversed by ants, their associated pheromone is diminished, making them less attractive, and therefore favoring the exploration of still unvisited arcs. Local updating has the effect of lowering the pheromone on visited arcs so that they will be chosen with a lower probability by the other ants in their remaining steps for completing a tour. As a consequence, the ants never converge to a common tour, as is also shown in figure 3.8.

3.6.2 Comparison of Ant System with Its Extensions

There remains the final question about the solution quality returned by the various ACO algorithms. In figure 3.9 we compare the development of the average solution quality measured in twenty-five trials for instance d198 (left side) and in five trials for instance rat783 (right side) of several ACO algorithms as a function of the computation time, which is indicated in seconds on the x-axis. We found experimentally that all extensions of AS achieve much better final solutions than AS, and in all cases the worst final solution returned by the AS extensions is better than the average final solution quality returned by AS.

In particular, it can be observed that ACS is the most aggressive of the ACO algorithms and returns the best solution quality for very short computation times.

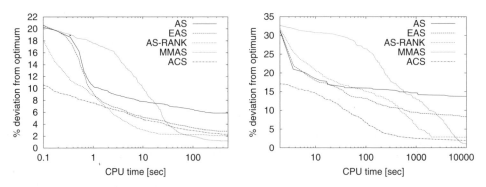

Figure 3.9
Comparing AS extensions: The plots give the development of the average percentage deviation from the optimum as a function of the computation time in seconds for AS, EAS, AS_{rank}, $\mathcal{MMAS}$, and ACS for the symmetric TSPLIB instances d198 (left), and rat783 (right). Parameters were set as indicated in box 3.1, except for β, which was set to $\beta = 5$ for all algorithms.

Differently, $\mathcal{MMAS}$ initially produces rather poor solutions and in the initial phases it is outperformed even by AS. Nevertheless, its final solution quality, for these two instances, is the best among the compared ACO algorithms.

These results are consistent with the findings of the various published research papers on AS extensions: in all these publications it was found that the respective extensions improved significantly over AS performance. Comparisons among the several AS extensions indicate that the best performing variants are $\mathcal{MMAS}$ and ACS, closely followed by AS_{rank}.

3.7   ACO plus Local Search

The vast literature on metaheuristics tells us that a promising approach to obtaining high-quality solutions is to couple a local search algorithm with a mechanism to generate initial solutions. As an example, it is well known that, for the TSP, iterated local search algorithms are currently among the best-performing algorithms. They iteratively apply local search to initial solutions that are generated by introducing modification to some locally optimal solutions (see chapter 2, section 2.4.4, for a detailed description of iterated local search).

ACO's definition includes the possibility of using local search (see figure 3.3); once ants have completed their solution construction, the solutions can be taken to their local optimum by the application of a local search routine. Then pheromones are updated on the arcs of the locally optimized solutions. Such a coupling of solution

construction with local search is a promising approach. In fact, because ACO's solution construction uses a different neighborhood than local search, the probability
that local search improves a solution constructed by an ant is quite high. On the
other hand, local search alone suffers from the problem of finding good starting solutions; these solutions are provided by the artificial ants.

In the following, we study how the performance of one of the ACO algorithms
presented before, $\mathcal{MMAS}$, is improved when coupled with a local search. To do
so, we implemented three of the most used types of local search for the TSP: 2-opt,
2.5-opt, and 3-opt. 2-opt was explained in box 2.4 (with the name 2-exchange), while
2.5-opt and 3-opt are explained in box 3.2. All three implementations exploit three
standard speedup techniques: the use of nearest-neighbor lists of limited length (here
20), the use of a fixed radius nearest-neighbor search, and the use of *don't look bits*.
These techniques together make the computation time increase subquadratically with
the instance size. See Bentley (1992) and Johnson & McGeoch (1997) for details on
these speedup techniques.

3.7.1   How to Add Local Search to ACO Algorithms?

There exist a large number of possible choices when combining local search with
ACO algorithms. Some of these possibilities relate to the fundamental question of
how effective and how efficient the local search should be. In fact, in most local
search procedures, the better the solution quality returned, the higher the computation time required. This translates into the question whether for a given computation
time it is better to frequently apply a quick local search algorithm that only slightly
improves the solution quality of the initial solutions, or whether a slow but more
effective local search should be used less frequently.

Other issues are related to particular parameter settings and to which solutions the
local search should be applied. For example, the number of ants to be used, the
necessity to use heuristic information or not, and which ants should be allowed to
improve their solutions by a local search, are all questions of particular interest
when an ACO algorithm is coupled with a local search routine. An interesting question is whether the implementation choices done and the parameter values chosen in
the case of ACO algorithms are still the best once local search is added. In general,
there may be significant differences regarding particular parameter settings. For
example, for $\mathcal{MMAS}$ it was found that when applied without local search, a good
strategy is to frequently use the iteration-best ant to update pheromone trails. Yet,
when combined with local search a stronger emphasis of the best-so-far update
seemed to improve performance (Stützle, 1999).

Box 3.2
2.5-opt and 3-opt

The 3-opt neighborhood consists of those tours that can be obtained from a tour s by replacing at most three of its arcs. The removal of three arcs results in three partial tours that can be recombined into a full tour in eight different ways. However, only four of these eight ways involve the introduction of three new arcs, the other four reduce to 2-opt moves (see box 2.4 for details on the 2-opt neighborhood). (Note that in a 3-opt local search procedure 2-opt moves are also examined.) The figure below gives one particular example of a 3-opt exchange move.

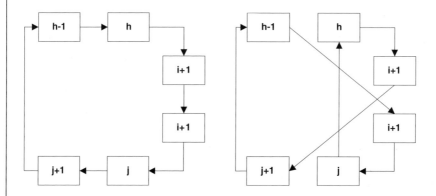

2.5-opt is a local search algorithm that includes a strongly restricted version of a 3-opt move on top of a 2-opt local search. When checking for an improving 2-opt move, it is also checked whether inserting the city between a city i and its successor, as illustrated in the figure below, results in an improved tour.

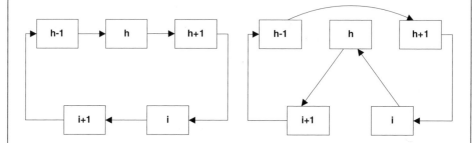

2.5-opt leads only to a small, constant overhead in computation time over that required by a 2-opt local search but, as experimental results show (Bentley, 1992), it leads to significantly better tours. However, the tour quality returned by 2.5-opt is still significantly worse than that of 3-opt. Implementations of the above-mentioned local search procedures not using any speedup techniques result in the following time complexities for a single neighborhood search: $\mathcal{O}(n^2)$ for 2-opt and 2.5-opt, and $\mathcal{O}(n^3)$ for 3-opt.

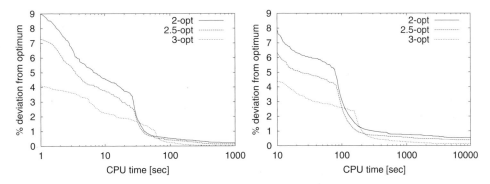

Figure 3.10
Comparing local search procedures: The plots give the average percentage deviation from the optimal tour as a function of the CPU time in seconds for $\mathcal{MMAS}$ using a 2-opt, a 2.5-opt, and a 3-opt local search procedure for the symmetric TSPLIB instances `pcb1173` (left) and `pr2392` (right). Parameters are set to the values given in box 3.3.

In the following, we give some exemplary results, focusing our attention on $\mathcal{MMAS}$ and ACS. In particular, we examine the influence that the strength of the local search, the number of ants, and the use of heuristic information have on the algorithms' performance.

Strength of the Local Search
We combined $\mathcal{MMAS}$ with 2-opt, 2.5-opt, and 3-opt local search procedures. While the solution quality returned by these local search algorithms increases from 2-opt to 3-opt, the same is true for the necessary computation time to identify local optima (Reinelt, 1994; Johnson & McGeoch, 2002).

Figure 3.10 plots the solution quality as a function of the CPU time. For the largest amount of computation time, $\mathcal{MMAS}$ combined with 3-opt gives the best average solution quality. The fact that for a short interval of time $\mathcal{MMAS}$ combined with 2-opt or 2.5-opt gives slightly better results than $\mathcal{MMAS}$ combined with 3-opt can be explained as follows. First, remember (see section 3.6.2 and figure 3.9) that $\mathcal{MMAS}$ moves from an initial explorative phase to an exploitation phase by increasing over time the relative frequency with which the best-so-far pheromone update is applied with respect to the iteration-best pheromone update. Second, we have seen (see box 3.2) that 3-opt has a higher time complexity than 2-opt and 2.5-opt. This means that an iteration of the $\mathcal{MMAS}$ with 3-opt algorithm requires more CPU time than an iteration of $\mathcal{MMAS}$ with 2-opt or 2.5-opt. Therefore, the explanation for the observed temporary better behavior of $\mathcal{MMAS}$ with 2-opt or 2.5-opt is that there is a

Box 3.3
Parameter Settings for ACO Algorithms with Local Search

The only ACO algorithms that have been applied with local search to the TSP are ACS and $\mathcal{MMAS}$. Good settings, obtained experimentally (see, e.g., Stützle & Hoos [2000] for $\mathcal{MMAS}$ and Dorigo & Gambardella [1997b] for ACS), for the parameters common to both algorithms are indicated below.

ACO algorithm	α	β	ρ	m	τ_0
$\mathcal{MMAS}$	1	2	0.2	25	$1/\rho C^{nn}$
ACS	—	2	0.1	10	$1/n C^{nn}$

The remaining parameters are:

$\mathcal{MMAS}$: τ_{max} is set, as in box 3.1, to $\tau_{max} = 1/(\rho C^{bs})$, while $\tau_{min} = 1/(2n)$. For the pheromone deposit, the schedule for the frequency with which the best-so-far pheromone update is applied is

$$f_{bs} = \begin{cases} \infty & \text{if } i \le 25 \\ 5 & \text{if } 26 \le i \le 75 \\ 3 & \text{if } 76 \le i \le 125 \\ 2 & \text{if } 126 \le i \le 250 \\ 1 & \text{otherwise} \end{cases} \tag{3.20}$$

where f_{bs} is the number of algorithm iterations between two updates performed by the best-so-far ant (in the other iterations it is the iteration-best ant that makes the update) and i is the iteration counter of the algorithm.

ACS: We have $\xi = 0.1$ and $q_0 = 0.98$.

Common to both algorithms is also that after each iteration all the tours constructed by the ants are improved by the local search. Additionally, in $\mathcal{MMAS}$ occasional pheromone trail reinitializations are applied. This is done when the average λ-branching factor becomes smaller than 2.00001 and if for more than 250 iterations no improved tour has been found.

 Note that on individual instances different settings may result in much better performance.

period of time in which while $\mathcal{MMAS}$ with 3-opt is still in the explorative phase, $\mathcal{MMAS}$ with 2-opt and $\mathcal{MMAS}$ with 2.5-opt are already in the exploitation phase.

In any case, once the final tour quality obtained by the different variants is taken into account, the computational results clearly suggest that the use of more effective local searches improves the solution quality of $\mathcal{MMAS}$.

Number of Ants

In a second series of experiments we investigated the role of the number of ants m on the final performance of $\mathcal{MMAS}$. We ran $\mathcal{MMAS}$ using parameter settings of $m \in \{1, 2, 5, 10, 25, 50, 100\}$ leaving all other choices the same. The result was that on small problem instances with up to 500 cities, the number of ants did not matter very

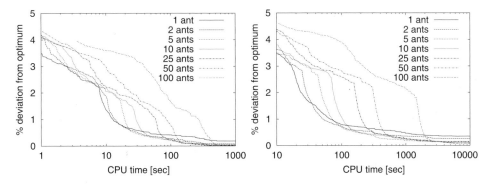

Figure 3.11
Varying the number of ants used: The plots give the average percentage deviation from the optimal tour as a function of the CPU time in seconds for $\mathcal{MMAS}$ with 3-opt using a number of ants varying from 1 ant to 100 ants on the symmetric TSPLIB instances pcb1173 (left) and pr2392 (right). Parameters are set to the values given in box 3.3.

much with respect to the best final performance. In fact, the best trade-off between solution quality and computation time seems to be obtained when using a small number of ants—between two and ten. Yet, on the larger instances, the usefulness of having a population of ants became more apparent. For instances with more than 500 cities the worst computational results were always obtained when using only one ant and the second worst results when using two ants (see figure 3.11).

Heuristic Information
It is well known that when ACO algorithms are applied to the TSP without local search, the heuristic information is essential for the generation of high-quality tours. In fact, in the initial phases of the search, the pheromones, being set to initial random values, do not guide the artificial ants, which end up constructing (and reinforcing) tours of very bad quality. The main role of the heuristic information is to avoid this, by biasing ants so that they build reasonably good tours from the very beginning. Once local search is added to the ACO implementation, the randomly generated initial tours become good enough. It is therefore reasonable to expect that heuristic information is no longer necessary.

Experiments with $\mathcal{MMAS}$ and ACS on the TSP confirmed this conjecture: when used with local search, even without using heuristic information, very high-quality tours were obtained. For example, figure 3.12 plots the average percentage deviation from the optimal tour as a function of CPU time obtained with $\mathcal{MMAS}$ and ACS with local search on the symmetric TSPLIB instance pcb1173. The figure shows that $\mathcal{MMAS}$ without heuristic information converged in most cases somewhat

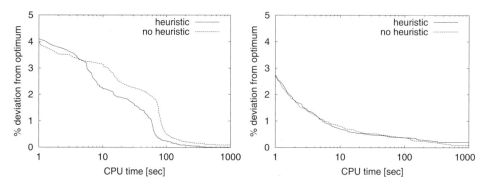

Figure 3.12
The role of heuristic information when using local search: The plots give the average percentage deviation from the optimal tour as a function of the CPU time in seconds for $\mathcal{MMAS}$ (left) and ACS (right) with local search on the symmetric TSPLIB instance `pcb1173`, with and without the use of heuristic information during tour construction.

slower to tours that were slightly worse than those obtained using heuristic information, while in most cases ACS's final tour length with heuristic information was slightly worse than without.

One might argue that the question whether heuristic information is used or not is just a matter of parameter settings (not using heuristic information is simply achieved by setting $\beta = 0$). Yet, the importance of our computational results is somewhat more far-reaching. While in the TSP the distance between cities is an obvious and computationally inexpensive heuristic to use, in other problems it may be much more difficult to find, or expensive to compute, meaningful heuristic information which helps to improve performance. Fortunately, if no such obvious heuristic information exists, our computational results suggest that using an ACO algorithm incorporating local search may be enough to achieve good results.

Lamarckian versus Darwinian Pheromone Updates
Let us reconsider the choice of the tour that is used to deposit pheromones after a local search: Each ant produces a tour, say s_1, which is then transformed into another tour, say s_2, by the local search. Then the pheromones are updated. As our goal is to maximize the quality of the final tour s_2, pheromone updates must be proportional to the quality of s_2, not s_1. Once this is accepted, there are still two ways of updating the pheromones:

- We reinforce the pheromones corresponding to the final tour s_2, or
- we reinforce the pheromones corresponding to the intermediate tour s_1.

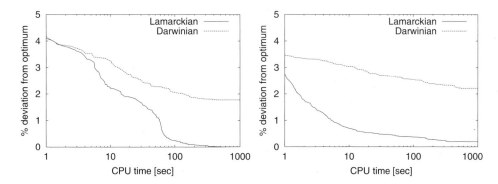

Figure 3.13
Lamarckian versus Darwinian pheromone updates: The plots give the average percentage deviation from the optimal tour as a function of the CPU time in seconds for $\mathcal{MM}$AS (left) and ACS (right) using Lamarckian and Darwinian pheromone updates on the symmetric TSPLIB instance `pcb1173`.

By analogy with similar procedures in the area of genetic algorithms (Whitley, Gordon, & Mathias, 1994), we call the first alternative the Lamarckian approach, and the second the Darwinian approach.

The main argument supporting the Lamarckian approach is that it is reasonable to think that, if the search of the ACO algorithm can be biased by the better tour s_2, then it would be stupid to use the worse tour s_1. In fact, in published ACO implementations, only the Lamarckian alternative has been used. On the other hand, the main argument in favor of the Darwinian approach is the view that what ACO algorithms with local search really do is to learn a way to generate good initial solutions for the local search, where "good" means that the initial solutions allow local search to reach good local optima.

In figure 3.13, we report some results we obtained with $\mathcal{MM}$AS and ACS on one of our test instances. As can be observed, for the TSP case, the Lamarckian approach outperforms by far the Darwinian approach. Analogous tests on other TSP instances and other problems, like the quadratic assignment problem, confirmed this observation and we conjecture that for most combinatorial optimization problems the Lamarckian approach is preferable.

3.8 Implementing ACO Algorithms

This section describes in detail the steps that have to be taken to implement an ACO algorithm for the TSP. Because the basic considerations for the implementation of different ACO algorithm variants are very similar, we mainly focus on AS and indicate, where appropriate, the necessary changes for implementing other ACO algorithms.

A first implementation of an ACO algorithm can be quite straightforward. In fact, if a greedy construction procedure like a nearest-neighbor heuristic is available, one can use as a construction graph the same graph used by the construction procedure, and then it is only necessary to (1) add pheromone trail variables to the construction graph and (2) define the set of artificial ants to be used for constructing solutions in such a way that they implement, according to equation (3.2), a randomized version of the construction procedure. It must be noted, however, that in order to have an efficient implementation, often additional data structures are required, like arrays to store information which, although redundant, make the processing much faster. In the following, we describe the steps to be taken to obtain an efficient implementation of AS. We will give a *pseudo-code* description of a possible implementation in a C-like notation. This description is general enough to allow a reader with some previous experience in procedural or object-oriented programming (a university-level first-year programming course should suffice) to implement an efficient version of any of the ACO algorithms presented in this chapter. Additionally, a C code of several ACO algorithms is available online at www.aco-metaheuristic.org/aco-code/.

3.8.1   Data Structures

As a first step, the basic data structures have to be defined. These must allow storing the data about the TSP instance and the pheromone trails, and representing artificial ants.

Figure 3.14 gives a general outline of the main data structures that are used for the implementation of an ACO algorithm, which includes the data for the problem representation and the data for the representation of the ants, as explained below.

Problem Representation

Intercity Distances.   Often a symmetric TSP instance is given as the coordinates of a number of n points. In this case, one possibility would be to store the x and y coordinates of the cities in two arrays and then compute on the fly the distance between the cities as needed. However, this leads to a significant computational overhead: obviously, it is more reasonable to precompute all intercity distances and to store them in a symmetric *distance matrix* with n^2 entries. In fact, although for symmetric TSPs we only need to store $n(n-1)/2$ distinct distances, it is more efficient to use an n^2 matrix to avoid performing additional operations to check whether, when accessing a generic distance $d(i,j)$, entry (i,j) or entry (j,i) of the matrix should be used.

% Representation of problem data

integer *dist[n][n]*   % distance matrix
integer *nn_list[n][nn]*   % matrix with nearest neighbor lists of depth *nn*
real   *pheromone[n][n]* % pheromone matrix
real   *choice_info[n][n]* % combined pheromone and heuristic information

% Representation of ants

structure *single_ant*
begin
   integer *tour_length* % the ant's tour length
   integer *tour[n + 1]* % ant's memory storing (partial) tours
   integer *visited[n]*   % visited cities
end

single_ant *ant[m]*   % structure of type single_ant

Figure 3.14
Main data structures for the implementation of an ACO algorithm for the TSP.

Note that for very large instances it may be necessary to compute distances on the fly, if it is not possible (or too expensive) to keep the full distance matrix in the main memory. Fortunately, in these cases there exist some intermediate possibilities, such as storing the distances between a city and the cities of its nearest-neighbor list, that greatly reduce the necessary amount of computation. It is also important to know that, for historical reasons, in almost all the TSP literature, the distances are stored as integers. In fact, in old computers integer operations used to be much faster than operations on real numbers, so that by setting distances to be integers, much more efficient code could be obtained.

Nearest-Neighbor Lists.   In addition to the distance matrix, it is convenient to store for each city a list of its nearest neighbors. Let d_i be the list of the distances from a city i to all cities j, with $j = 1, \ldots n$ and $i \neq j$ (we assume here that the value d_{ii} is assigned a value larger than d_{max}, where d_{max} is the maximum distance between any two cities). The nearest-neighbor list of a city i is obtained by sorting the list d_i according to nondecreasing distances, obtaining a sorted list d_i'; ties can be broken randomly. The position r of a city j in city i's nearest-neighbor list $nn\_list[i]$ is the index of the distance d_{ij} in the sorted list d_i', that is, $nn\_list[i][r]$ gives the identifier (index) of the r-th nearest city to city i (i.e., $nn\_list[i][r] = j$). Nearest-neighbor lists for all cities can be constructed in $\mathcal{O}(n^2 \log n)$ (in fact, you have to repeat a sorting algorithm over $n - 1$ cities for each city).

An enormous speedup is obtained for the solution construction in ACO algorithms, if the nearest-neighbor list is cut off after a constant number nn of nearest neighbors, where typically nn is a small value ranging between 15 and 40. In this case, an ant located in city i chooses the next city among the nn nearest neighbors of i; in case the ant has already visited all the nearest neighbors, then it makes its selection among the remaining cities. This reduces the complexity of making the choice of the next city to $\mathcal{O}(1)$, unless the ant has already visited all the cities in $nn\_list[i]$. However, it should be noted that the use of truncated nearest-neighbor lists can make it impossible to find the optimal solution.

Pheromone Trails. In addition to the instance-related information, we also have to store for each connection (i, j) a number τ_{ij} corresponding to the pheromone trail associated with that connection. In fact, for symmetric TSPs this requires storing $n(n - 1)/2$ distinct pheromone values, because we assume that $\tau_{ij} = \tau_{ji}, \ \forall (i, j)$. Again, as was the case for the distance matrix, it is more convenient to use some redundancy and to store the pheromones in a symmetric n^2 matrix.

Combining Pheromone and Heuristic Information. When constructing a tour, an ant located on city i chooses the next city j with a probability which is proportional to the value of $[\tau_{ij}]^\alpha [\eta_{ij}]^\beta$. Because these very same values need to be computed by each of the m ants, computation times may be significantly reduced by using an additional matrix *choice_info*, where each entry *choice_info*$[i][j]$ stores the value $[\tau_{ij}]^\alpha [\eta_{ij}]^\beta$. Again, in the case of a symmetric TSP instance, only $n(n - 1)/2$ values have to be computed, but it is convenient to store these values in a redundant way as in the case of the pheromone and the distance matrices. Additionally, one may store the η_{ij}^β values in a further matrix *heuristic* (not implemented in the code associated with the book) to avoid recomputing these values after each iteration, because the heuristic information stays the same throughout the whole run of the algorithm (some tests have shown that the speedup obtained when no local search is used is approximately 10%, while no significant differences are observed when local search is used). Finally, if some distances are zero, which is in fact the case for some of the benchmark instances in the TSPLIB, then one may set them to a very small positive value to avoid division by zero.

Speeding Up the Pheromone Update. Further optimization can be introduced by restricting the computation of the numbers in the *choice_info* matrix to the connections between a city and the cities of its nearest-neighbor list. In fact, this technique, which is exploited in the implementation of the various ACO algorithms in the accompanying code, strongly reduces the computation time when ACO algorithms are applied to large TSP instances with several hundreds or thousands of cities.

Representing Ants

An ant is a simple computational agent which constructs a solution to the problem at hand, and may deposit an amount of pheromone $\Delta\tau$ on the arcs it has traversed. To do so, an ant must be able to (1) store the partial solution it has constructed so far, (2) determine the feasible neighborhood at each city, and (3) compute and store the objective function value of the solutions it generates.

The first requirement can easily be satisfied by storing the partial tour in a sufficiently large array. For the TSP we represent tours by arrays of length $n + 1$, where at position $n + 1$ the first city is repeated. This choice makes easier some of the other procedures like the computation of the tour length.

The knowledge of the partial tour at each step is sufficient to allow the ant to determine whether a city j is in its feasible neighborhood: it is enough to scan the partial tour for the occurrence of city j. If city j has not been visited yet, then it is member of the feasible neighborhood; otherwise it is not. Unfortunately, this simple way of determining the feasible neighborhood involves an operation of worst-case complexity $\mathcal{O}(n)$ for each city i, resulting in a high computational overhead. The simplest way around this problem is to associate with each ant an additional array *visited* whose values are set to *visited*[j] = 1 if city i has already been visited by the ant, and to *visited*[j] = 0 otherwise. This array is updated by the ant while it builds a solution.

Finally, the computation of the tour length, stored by the ant in the *tour_length* variable, can easily be done by summing the length of the n arcs in the ant's tour.

Hence, an ant may be represented by a structure that comprises one variable *tour_length* to store the ant's objective function value, one $(n + 1)$-dimensional array *tour* to store the ant's tour, and one n-dimensional array *visited* to store the visited cities (note that in figure 3.14, the array *visited*, part of the data structure *single_ant*, is declared of type **integer**; however, to save memory, it could be declared of type **Boolean**).

Overall Memory Requirement

For representing all the necessary data for the problem we need four matrices of dimension $n \times n$ for representing the distance matrix, the pheromone matrix, the heuristic information matrix, and the *choice_info* matrix, and a matrix of size $n \times nn$ for the nearest-neighbor lists. Additionally, for each of the ants we need two arrays of size $(n + 1)$ and n to store, respectively, the tour and the visited cities, as well as an integer for storing the tour's length. Finally, we need a variable for representing each of the m ants. Since the number of ants is typically either a small constant (this is the case for ACS and Ant-Q or for most ACO algorithms with local search) or on the order of n (this is the case for AS variants without use of local search), the overall memory requirement is $\mathcal{O}(n^2)$. In addition to these main data structures, it is also necessary to store intermediate results, such as the best solution found so far, and statistical information about the algorithm performance; nevertheless, these additional data require only a very minor amount of memory when compared to the data for representing the colony of ants and the problem.

To derive a more exact estimate of the memory requirements, we can assume that representing an integer value on a computer takes 4 bytes and representing a "real" number takes 8 bytes. Additionally, we assume the number of ants to be $m = n$, and we do not consider the heuristic information matrix. The estimate is obtained as follows (see figure 3.14): $24n^2$ bytes are necessary for the problem data ($4n^2$ for the distance matrix, $4n^2$ for the matrix nn_list, $8n^2$ for the pheromone matrix, and $8n^2$ for the matrix $choice\_info$), while $8n^2$ bytes are needed for the representation of the ants (there are $m = n$ ants and each ant requires two integer arrays of length n). The overall memory requirement can therefore be assumed to be roughly $32n^2$ bytes, which is a slight underestimate of the real memory consumption. Memory requirements increase strongly with problem size, because the memory requirement is quadratic in n. However, the memory requirements are reasonable when considering the memory available in current computers (using the above $32n^2$ estimate, instances of up to 4000 cities can be tackled by ACO algorithms with a computer with 512 MB of RAM), and the fact that the problem instances of most combinatorial optimization problems to which ACO has been applied (see chapter 5 for an overview) are typically much smaller than those of the TSP, so that memory consumption is rarely an issue.

3.8.2   The Algorithm

The main tasks to be considered in an ACO algorithm are the solution construction, the management of the pheromone trails, and the additional techniques such as local search. In addition, the data structures and parameters need to be initialized and some statistics about the run need to be maintained. In figure 3.15 we give a high-level view of the algorithm, while in the following we give some details on how to implement the different procedures of AS in an efficient way.

Data Initialization

In the data initialization, (1) the instance has to be read; (2) the distance matrix has to be computed; (3) the nearest-neighbor lists for all cities have to be computed; (4)

```
procedure ACOforTSP
    InitializeData
    while (not terminate) do
        ConstructSolutions
        LocalSearch
        UpdateStatistics
        UpdatePheromoneTrails
    end-while
end-procedure
```

Figure 3.15
High-level view of an ACO algorithm for the TSP.

```
procedure InitializeData
    ReadInstance
    ComputeDistances
    ComputeNearestNeighborLists
    ComputeChoiceInformation
    InitializeAnts
    InitializeParameters
    InitializeStatistics
end-procedure
```

Figure 3.16
Procedure to initialize the algorithm.

the pheromone matrix and the *choice_info* matrix have to be initialized; (5) the ants
have to be initialized; (6) the algorithm's parameters must be initialized; and (7) some
variables that keep track of statistical information, such as the used CPU time, the
number of iterations, or the best solution found so far, have to be initialized. A possible organization of these tasks into several data initialization procedures is indicated in figure 3.16.

Termination Condition
The program stops if at least one termination condition applies. Possible termination
conditions are: (1) the algorithm has found a solution within a predefined distance
from a lower bound on the optimal solution quality; (2) a maximum number of tour
constructions or a maximum number of algorithm iterations has been reached; (3) a
maximum CPU time has been spent; or (4) the algorithm shows stagnation behavior.

```
      procedure ConstructSolutions
1         for k = 1 to m do
2            for i = 1 to n do
3               ant[k].visited[i] ← false
4            end-for
5         end-for
6         step ← 1
7         for k = 1 to m do
8            r ← random{1,...,n}
9            ant[k].tour[step] ← r
10           ant[k].visited[r] ← true
11        end-for
12        while (step < n) do
13           step ← step + 1
14           for k = 1 to m do
15              ASDecisionRule(k, step)
16           end-for
17        end-while
18        for k = 1 to m do
19           ant[k].tour[n + 1] ← ant[k].tour[1]
20           ant[k].tour_length ← ComputeTourLength(k)
21        end-for
      end-procedure
```

Figure 3.17
Pseudo-code of the solution construction procedure for AS and its variants.

Solution Construction

The tour construction is managed by the procedure ConstructSolutions, shown in figure 3.17. The solution construction requires the following phases.

1. First, the ants' memory must be emptied. This is done in lines 1 to 5 of procedure ConstructSolutions by marking all cities as unvisited, that is, by setting all the entries of the array *ants.visited* to *false* for all the ants.

2. Second, each ant has to be assigned an initial city. One possibility is to assign each ant a random initial city. This is accomplished in lines 6 to 11 of the procedure. The function random returns a random number chosen according to a uniform distribution over the set $\{1, \ldots, n\}$.

3. Next, each ant constructs a complete tour. At each construction step (see the procedure in figure 3.17) the ants apply the AS action choice rule [equation (3.2)]. The procedure ASDecisionRule implements the action choice rule and takes as parameters the ant identifier and the current construction step index; this is discussed below in more detail.

4. Finally, in lines 18 to 21, the ants move back to the initial city and the tour length of each ant's tour is computed. Remember that, for the sake of simplicity, in the tour representation we repeat the identifier of the first city at position $n + 1$; this is done in line 19.

As stated above, the solution construction of all of the ants is synchronized in such a way that the ants build solutions in parallel. The same behavior can be obtained, for all AS variants, by ants that construct solutions sequentially, because the ants do not change the pheromone trails at construction time (this is not the case for ACS, in which case the sequential and parallel implementations give different results).

While phases (1), (2), and (4) are very straightforward to code, the implementation of the action choice rule requires some care to avoid large computation times. In the action choice rule an ant located at city i probabilistically chooses to move to an unvisited city j based on the pheromone trails τ_{ij}^{α} and the heuristic information η_{ij}^{β} [see equation (3.2)].

Here we give pseudo-codes for the action choice rule with and without consideration of candidate lists. The pseudo-code for the first variant ASDecisionRule is given in figure 3.18. The procedure works as follows: first, the current city c of ant k is determined (line 1). The probabilistic choice of the next city then works analogously to the *roulette wheel* selection procedure of evolutionary computation (Goldberg, 1989): each value *choice_info*$[c][j]$ of a city j that ant k has not visited yet determines a slice on a circular roulette wheel, the size of the slice being proportional to the weight of the associated choice (lines 2–10). Next, the wheel is spun and the city to which the marker points is chosen as the next city for ant k (lines 11–17). This is implemented by

1. summing the weight of the various choices in the variable *sum_probabilities*,

2. drawing a uniformly distributed random number r from the interval $[0, sum\_probabilities]$,

3. going through the feasible choices until the sum is greater or equal to r.

Finally, the ant is moved to the chosen city, which is marked as visited (lines 18 and 19).

procedure ASDecisionRule(k, i)
    input   k   % ant identifier
    input   i   % counter for construction step

```
1    c ← ant[k].tour[i − 1]
2    sum_probabilities = 0.0
3    for j = 1 to n do
4       if ant[k].visited[ j] then
5          selection_probability[ j] ← 0.0
6       else
7          selection_probability[ j] ← choice_info[c][ j]
8          sum_probabilities ← sum_probabilities + selection_probability[ j]
9       end-if
10   end-for
11   r ← random[0, sum_probabilities]
12   j ← 1
13   p ← selection_probability[ j]
14   while (p < r) do
15      j ← j + 1
16      p ← p + selection_probability[ j]
17   end-while
18   ant[k].tour[i] ← j
19   ant[k].visited[ j] ← true
     end-procedure
```

Figure 3.18
AS without candidate lists: pseudo-code for the action choice rule.

These construction steps are repeated until the ants have completed a tour. Since each ant has to visit exactly n cities, all the ants complete the solution construction after the same number of construction steps.

When exploiting candidate lists, the procedure ASDecisionRule needs to be adapted, resulting in the procedure NeighborListASDecisionRule, given in figure 3.19. A first change is that when choosing the next city, one needs to identify the appropriate city index from the candidate list of the current city c. This results in changes of lines 3 to 10 of figure 3.18: the maximum value of index j is changed from n to nn in line 3 and the test performed in line 4 is applied to the j-th nearest neighbor given by $nn\_list[c][j]$. A second change is necessary to deal with the situation in which all the cities in the candidate list have already been visited by ant k. In this case, the variable $sum\_probabilities$ keeps its initial value 0.0 and one city out of those not in

procedure NeighborListASDecisionRule(k, i)

    input   k   % ant identifier

    input   i   % counter for construction step

1     $c \leftarrow ant[k].tour[i-1]$

2     $sum\_probabilities \leftarrow 0.0$

3   **for** $j = 1$ **to** nn **do**

4   **if** $ant[k].visited[nn\_list[c][j]]$ **then**

5         $selection\_probability[j] \leftarrow 0.0$

6   **else**

7         $selection\_probability[j] \leftarrow choice\_info[c][nn\_list[c][j]]$

8         $sum\_probabilities \leftarrow sum\_probabilities + selection\_probability[j]$

9   **end-if**

10   **end-for**

11   **if** $(sum\_probabilities = 0.0)$ **then**

12   ChooseBestNext(k, i)

13   **else**

14      $r \leftarrow$ random$[0, sum\_probabilities]$

15      $j \leftarrow 1$

16      $p \leftarrow selection\_probability[j]$

17   **while** $(p < r)$ **do**

18        $j \leftarrow j + 1$

19        $p \leftarrow p + selection\_probability[j]$

20   **end-while**

21      $ant[k].tour[i] \leftarrow nn\_list[c][j]$

22      $ant[k].visited[nn\_list[c][j]] \leftarrow true$

23   **end-if**

    end-procedure

Figure 3.19
AS with candidate lists: pseudo-code for the action choice rule.

procedure ChooseBestNext(k, i)
  input   k   % ant identifier
  input   i   % counter for construction step
  $v \leftarrow 0.0$
  $c \leftarrow ant[k].tour[i - 1]$
  for $j = 1$ **to** n **do**
    if not $ant[k].visited[j]$ **then**
      if $choice\_info[c][j] > v$ **then**
        $nc \leftarrow j$ % city with maximal $\tau^{\alpha}\eta^{\beta}$
        $v \leftarrow choice\_info[c][j]$
      end-if
    end-if
  end-for
  $ant[k].tour[i] \leftarrow nc$
  $ant[k].visited[nc] \leftarrow true$
end-procedure

Figure 3.20
AS: pseudo-code for the procedure ChooseBestNext.

the candidate list is chosen: the procedure ChooseBestNext (see pseudo-code in figure 3.20) is used to identify the city with maximum value of $[\tau_{ij}]^{\alpha}[\eta_{ij}]^{\beta}$ as the next to move to.

It is clear that by using candidate lists the computation time necessary for the ants to construct solutions can be significantly reduced, because the ants choose from among a much smaller set of cities. Yet, the computation time is reduced only if the procedure ChooseBestNext does not need to be applied too often. Fortunately, as also suggested by the computational results presented in Gambardella & Dorigo (1996), this seems not to be the case.

Local Search

Once the solutions are constructed, they may be improved by a local search procedure. While a simple 2-opt local search can be implemented in a few lines, the implementation of an efficient variant is somewhat more involved. This is already true to some extent for the implementation of the 3-opt local search, and even more for the Lin-Kernighan heuristic. Since the details of the local search are not important for understanding how ACO algorithms can be coded efficiently, we refer to the accompanying code (available at www.aco-metaheuristic.org/aco-code/) for more information on the local search implementation.

procedure ASPheromoneUpdate
  Evaporate
  for $k = 1$ **to** m **do**
    DepositPheromone(k)
  end-for
  ComputeChoiceInformation
end-procedure

Figure 3.21
AS: management of the pheromone updates.

procedure Evaporate
  for $i = 1$ **to** n **do**
    for $j = i$ **to** n **do**
      $pheromone[i][j] \leftarrow (1 - \rho) \cdot pheromone[i][j]$
      $pheromone[j][i] \leftarrow pheromone[i][j]$ % pheromones are symmetric
    end-for
  end-for
end-procedure

Figure 3.22
AS: implementation of the pheromone evaporation procedure.

Pheromone Update

The last step in an iteration of AS is the pheromone update. This is implemented by the procedure ASPheromoneUpdate (figure 3.21), which comprises two pheromone update procedures: pheromone evaporation and pheromone deposit. The first one, Evaporate (figure 3.22), decreases the value of the pheromone trails on all the arcs (i, j) by a constant factor ρ. The second one, DepositPheromone (figure 3.23), adds pheromone to the arcs belonging to the tours constructed by the ants. Additionally, the procedure ComputeChoiceInformation computes the matrix *choice_info* to be used in the next algorithm iteration. Note that in both procedures care is taken to guarantee that the pheromone trail matrix is kept symmetric, because of the symmetric TSP instances.

When attacking large TSP instances, profiling the code showed that the pheromone evaporation and the computation of the *choice_info* matrix can require a considerable amount of computation time. On the other hand, when using candidate lists in the solution construction, only a small part of the entries of the pheromone

procedure DepositPheromone(k)
  input   k   % ant identifier
  $\Delta\tau \leftarrow 1/ant[k].tour\_length$
  for $i = 1$ **to** n **do**
    $j \leftarrow ant[k].tour[i]$
    $l \leftarrow ant[k].tour[i + 1]$
    $pheromone[j][l] \leftarrow pheromone[j][l] + \Delta\tau$
    $pheromone[l][j] \leftarrow pheromone[j][l]$
  end-for
end-procedure

Figure 3.23
AS: implementation of the pheromone deposit procedure.

matrix are ever required. Therefore, the exploitation of candidate lists speeds up also the pheromone update. In fact, the use of candidate lists with a constant number of nearest neighbors reduces the complexity of these two procedures to $\mathcal{O}(n)$, although with a large constant hidden in the $\mathcal{O}(\cdot)$ notation.

Concerning pheromone depositing, we note that, differently from AS, the best-performing ACO algorithms typically allow only one or, at most, very few ants to deposit pheromone. In this case, the complexity of the pheromone deposit is of order $\mathcal{O}(n)$. Therefore, only for AS and EAS is the complexity of the pheromone trail deposit procedure $\mathcal{O}(n^2)$ if the number of ants m is set to be proportional to n, as suggested in the original papers (Dorigo et al., 1991a,b, 1996; Bauer et al., 2000).

Note that this type of speedup technique for the pheromone trail update is not necessary for ACS, because in ACS only the pheromone trails of arcs that are crossed by some ant have to be changed and the number of ants in each iteration is a low constant.

Statistical Information about ACO Algorithm Behavior
The last step in the implementation of AS is to store statistical data on algorithm behavior (examples are the best-found solution since the start of the algorithm run, or the iteration number at which the best solution was found). Details about these procedures are available at www.aco-metaheuristic.org/aco-code/.

3.8.3   Changes for Implementing Other ACO Algorithms

When implementing AS variants, most of the above-described procedures remain unchanged. Some of the necessary adaptations are described in the following:

procedure ACSLocalPheromoneUpdate(k, i)

   input   k   % ant identifier

   input   i   % counter for construction step

   $h \leftarrow ant[k].tour[i - 1]$

   $j \leftarrow ant[k].tour[i]$

   $pheromone[h][j] \leftarrow (1 - \xi)pheromone[h][j] + \xi\tau_0$

   $pheromone[j][h] \leftarrow pheromone[h][j]$

   $choice\_info[h][j] \leftarrow pheromone[h][j] \cdot \exp(1/dist[h][j], \beta)$

   $choice\_info[j][h] \leftarrow choice\_info[h][j]$

end-procedure

Figure 3.24
Implementation of the local pheromone update in ACS.

- When depositing pheromone, the solution may be given some weight, as is the case in EAS and AS$_{rank}$. This can be accomplished by simply adding a weight factor as an additional argument of the procedure DepositPheromone.

- $\mathcal{MMAS}$ has to keep track of the pheromone trail limits. The best way to do so is to integrate this into the procedure ASPheromoneUpdate.

- Finally, the search control of some of the AS variants may need minor changes. Examples are occasional pheromone trail reinitializations or the schedule for the frequency of the best-so-far update according to equation (3.20) in $\mathcal{MMAS}$.

Unlike AS variants, the implementation of ACS requires more significant changes, as listed in the following:

- The implementation of the pseudorandom proportional action choice rule [see equation (3.10)] requires the generation of a random number q uniformly distributed in the interval $[0, 1]$ and the application of the procedure ChooseBestNext if $q < q_0$, or of the procedure ASDecisionRule otherwise.

- The local pheromone update [equation (3.12)] can be managed by the procedure ACSLocalPheromoneUpdate (see figure 3.24) that is always invoked immediately after an ant moves to a new city.

- The implementation of the global pheromone trail update [equation (3.11)] is similar to the procedure for the local pheromone update except that pheromone trails are modified only on arcs belonging to the best-so-far tour.

- Note that the integration of the computation of new values for the matrix *choice_ info* into the local and the global pheromone trail update procedures avoids having to modify this matrix in any other part of the algorithm except for the initialization.

3.9   Bibliographical Remarks

The Traveling Salesman Problem

The TSP is one of the oldest and most studied combinatorial optimization problems. The first references to the TSP and closely related problems date back to the 19th century (see the overview paper on the history of combinatorial optimization by Schrjiver [2002] and the webpage Solving TSPs accessible at www.math.princeton. edu/tsp/ for more details). The TSP has been studied intensively in both operations research and computer science since the '50s. Therefore, it is not surprising that a large number of different algorithmic techniques were either applied to the TSP or developed because of the challenge posed by this problem. Up to the early '80s these approaches comprised mainly construction heuristics (Clarke & Wright, 1964; Christofides, 1976; Golden & Stewart, 1985; Bentley, 1992), iterative improvement algorithms (Flood, 1956; Croes, 1958; Lin, 1965; Lin & Kernighan, 1973), and exact methods like branch & bound or branch & cut (Dantzig, Fulkerson, & Johnson, 1954; Grötschel, 1981; Padberg & Grötschel, 1985; Grötschel & Holland, 1991; Applegate et al., 1995). An in-depth overview of these early approaches is given in Lawler et al. (1985). Extensive experimental evaluations of construction heuristics and iterative improvement algorithms may be found in Bentley (1992), Reinelt (1994), and Johnson & McGeoch (1997, 2002).

Since the beginning of the '80s, more and more metaheuristics have been tested on the TSP. In fact, the TSP was the first problem to which simulated annealing, one of the first metaheuristic approaches, was applied (Cerný, 1985; Kirkpatrick, Gelatt, & Vecchi, 1983). Following SA, virtually any metaheuristic used the TSP as a test problem. These include tabu search (Knox, 1994; Zachariasen & Dam, 1996), guided local search (Voudouris & Tsang, 1999), evolutionary algorithm (Merz & Freisleben, 1997; Walters, 1998), ACO algorithms (see this chapter and Stützle & Dorigo, 1999b), and iterated local search (ILS) (Baum, 1986; Martin et al., 1991; Johnson & McGeoch, 1997; Applegate et al., 2003).

The state of the art (until 1997) for solving symmetric TSPs with heuristics is summarized in the overview article by Johnson & McGeoch (1997). This article contains a discussion of the relative performance of different metaheuristic approaches to the TSP and concludes that ILS algorithms using fine-tuned implementations of the Lin-Kernighan heuristic (Lin & Kernighan, 1973) are the most successful. The most recent effort in collecting the state of the art for TSP solving by heuristic methods was undertaken by the "8th DIMACS Implementation Challenge on the TSP"; details of this benchmark challenge can be found at www.research.att.com/~dsj/chtsp/ and the results as of February 2002, including construction heuristics, iterative improvement algorithms, metaheuristics, and more TSP-specific approaches, are summarized in a paper by Johnson & McGeoch (2002). The conclusion of this recent undertaking is that, when running time is not much of a concern, the best-performing algorithms appear to be the tour-merging approach (a TSP-specific heuristic) of Applegate et al. (1999) and the iterated version of Helsgaun's Lin-Kernighan variant (Helsgaun, 2000). In this context, it is interesting to note that the iterated version of Helsgaun's implementation of the Lin-Kernighan heuristic uses a constructive approach (as does ant colony optimization) to generate the initial tours for the local searches, where the best-so-far solution strongly biases the tour construction.

Finally, let us note that the results obtained with exact algorithms for the TSP are quite impressive. As of spring 2002, the largest instance provably solved to optimality comprises 15112 cities. Solving such a large instance required a network of 110 processors and took a total time estimated to be equivalent to 22.6 CPU-years on a 500 MHz, EV6 Alpha processor (more details on optimization algorithms for the TSP, the most recent results, and the source code of these algorithms are available at www.math.princeton.edu/tsp/). Although these results show the enormous progress that has been made by exact methods, they also divert attention from the fact that these results on the TSP are not really representative of the performance of exact algorithms on many other combinatorial optimization problems. There are in fact a large number of problems that become intractable for exact algorithms, even for rather small instances.

ACO Algorithms

The currently available results obtained by ACO algorithms applied to the TSP are not competitive with the above-mentioned approaches. By adding more sophisticated local search algorithms like the implementation of the Lin-Kernighan heuristic available at www.math.princeton.edu/tsp/ or Helsgaun's variant of the Lin-Kernighan heuristic, ACO's computational results on the TSP can certainly be strongly improved, but it is an open question whether the results of the best algorithms available can be reached. Nevertheless, as already stated in the introduction, the main importance of the TSP for the ACO research field is that it is a problem on which the behavior of ACO algorithms can be studied without obscuring the algorithm behavior by many technicalities. In fact, the best-performing variants of ACO algorithms on the TSP often reach world-class performance on many other problems (see chapter 5 for several such applications).

In addition to the ACO algorithms discussed in this chapter, recently a new variant called best-worst Ant System (BWAS) was proposed (Cordón et al., 2000;

Cordón, de Viana, & Herrera, 2002). It introduces three main variations with respect to AS. First, while using, in a way similar to $\mathcal{MM}$AS and ACS, an aggressive update rule in which only the best-so-far ant is allowed to deposit pheromone, it also exploits the worst ant of the current iteration to subtract pheromone on the arcs it does not have in common with the best-so-far solution. Second, BWAS relies strongly on search diversification through the frequent reinitialization of the pheromone trails. Third, as an additional means for diversifying the search, it introduces pheromone mutation, a concept borrowed from evolutionary computation. The influence of these three features was systematically analyzed in Cordón et al. (2002). The currently available results, however, are not fully conclusive, so that it is not possible to judge BWAS's performance with respect to the currently best-performing ACO algorithms for the TSP: $\mathcal{MM}$AS and ACS.

ACO algorithms have also been tested on the asymmetric TSP (Dorigo & Gambardella, 1997b; Stützle & Hoos, 2000; Stützle & Dorigo, 1999b), where the distance between a pair of nodes i, j is dependent on the direction of traversing the arc. ACO algorithms for the symmetric TSP can be extended very easily to the asymmetric case, by taking into account that in general $\tau_{ij} \neq \tau_{ji}$, because the direction in which the arcs are traversed has to be taken into account. Experimental results suggest that ACO algorithms can find optimal solutions to ATSP instances with up to a few hundred nodes. In the ATSP case, at the time the research was done, results competitive with those obtained by other metaheuristics were obtained. However, recent results on algorithmic approaches to the ATSP (see Johnson, Gutin, McGeoch, Yeo, Zhang, & Zverovitch, 2002) suggest that current ACO algorithms do not reach state-of-the-art performance for the ATSP.

It is also worth mentioning that there are at least two ant algorithms not fitting into the ACO metaheuristic framework that have been applied to combinatorial optimization problems. These are *Hybrid Ant System* (HAS) by Gambardella, Taillard, & Dorigo (1999b) and *Fast Ant System* (FANT) by Taillard (1998). HAS does not use pheromone trails to construct solutions but to guide a solution modification process similar to perturbation moves as used in ILS. FANT differs from ACO algorithms mainly in the pheromone management process and in the avoidance of explicit evaporation of pheromones, which are decreased by occasional reinitializations. Both HAS and FANT were applied to the quadratic assignment problem and were found to yield good performance. However, adaptations of both to the TSP resulted in significantly worse performance than, for example, $\mathcal{MM}$AS (Stützle & Linke, 2002).

The combination of ACO algorithms with local search was considered for the first time by Maniezzo et al. (1994) for the application of AS to the quadratic assignment problem. When applied to the TSP, local search was combined for the first time with ACS by Dorigo & Gambardella (1997b) and with $\mathcal{MMAS}$ by Stützle & Hoos (1996). For a detailed experimental investigation of the influence of parameter settings on the final performance of ACO algorithms for the TSP, see those papers.

3.10   Things to Remember

- The TSP was the first combinatorial optimization problem to be attacked by ACO. The first ACO algorithm, called Ant System, achieved good performance on small TSP instances but showed poor scaling behavior to large instances. The main role of AS was that of a "proof-of-concept" that stimulated research on better-performing ACO algorithms as well as applications to different types of problems.

- Nowadays, a large number of different ACO algorithms are available. All of these algorithms include a strong exploitation of the best solutions found during the search and the most successful ones add explicit features to avoid premature stagnation of the search. The main differences between the various AS extensions consist of the techniques used to control the search process. Experimental results show that for the TSP, but also for other problems, these variants achieve a much better performance than AS.

- When applying ACO algorithms to the TSP, the best performance is obtained when the ACO algorithm uses a local optimizer to improve the solutions constructed by the ants. As we will see in chapter 5, this is typical for the application of ACO to $\mathcal{NP}$-hard optimization problems.

- When using local search, it is typically sufficient to apply a small constant number of ants to achieve high performance, and experimental results suggest that in this case the role played by the heuristic information becomes much less important.

- The implementation of ACO algorithms is often rather straightforward, as shown in this chapter via the example of the implementation of AS for the TSP. Nevertheless, care should be taken to make the code as efficient as possible.

- The implementation code for most of the ACO algorithms presented in this chapter is available at www.aco-metaheuristic.org/aco-code/.

3.11   Computer Exercises

Exercise 3.1   In all ACO algorithms for the TSP the amount of pheromone deposited by an ant is proportional to the ant's tour length. Modify the code in such a way that the amount of pheromone deposited is a constant and run tests with the various ACO algorithms.

For which ACO algorithms would you expect that this change does not influence the performance very strongly? Why?

Exercise 3.2   Use a profiler to identify how much computation time is taken by the different procedures (solution construction, pheromone evaporation, local search, etc.) of the ACO algorithms. Identify the computationally most expensive parts.

Exercise 3.3   There exist some ACO algorithms that were proposed in the literature but that have never been applied to the symmetric TSP. These ACO algorithms include the ANTS algorithm (Maniezzo, 1999) and the hyper-cube framework for ACO (Blum et al., 2001; Blum & Dorigo, 2004). Extend the implementation of the ACO algorithms that is available at www.aco-metaheuristic.org/aco-code/ to include these two ACO algorithms.

Hint: For ANTS care has to be taken that the computation of the lower bounds is as efficient as possible, because this is done at each construction step of each ant.

Exercise 3.4   ACO algorithms have mainly been tested on Euclidean TSP instances available from TSPLIB. Many TSP algorithms are experimentally tested on random distance matrix instances, where each entry in the distance matrix is a random number sampled from some interval. Download a set of such instances from the webpage of the 8th DIMACS Implementation Challenge on the TSP (www.research.att.com/~dsj/chtsp/) and test the ACO algorithms on these types of instances.

Exercise 3.5   The implementations described in this chapter were designed for attacking symmetric TSP problems. Adapt the available code to solve ATSP instances.

Exercise 3.6   Compare the results obtained with the ACO algorithms to those obtained with the approaches described in the review paper on heuristics for the asymmetric TSP by Johnson et al. (2002).

Exercise 3.7   The solution construction procedure used in all ACO algorithms is a randomized form of the nearest-neighbor heuristic, in which at each step the closest, still unvisited, city to the current city is chosen and becomes the current city. However, a large number of other solution construction procedures exist (e.g., see Bentley, 1992; Reinelt, 1994; Johnson & McGeoch, 2002). Promising results have been reported among others for the *savings heuristic*, the *greedy heuristic*, and the *insertion heuristic*.

Adapt the ACO algorithms' code so that these construction heuristics can be used in place of the nearest-neighbor heuristic.

Exercise 3.8   Combine the available ACO algorithms with implementations of the Lin-Kernighan heuristic. You may adapt the publicly available Lin-Kernighan codes of the Concorde distribution (available at www.math.princeton.edu/tsp/concorde.html) or Keld Helsgaun's Lin-Kernighan variant (available at www.dat.ruc.dk/~keld/research/LKH/) and use these to improve the solutions generated by the ants (do not forget to ask the authors of the original code for permission to modify/adapt it).

Exercise 3.9   Extend the available code for the TSP to the sequential ordering problem (see chapter 2, section 2.3.2, for a definition of the problem). For a description of an ACO approach to the SOP, see chapter 5, section 5.1.1.

4 Ant Colony Optimization Theory

In theory, there is no difference between theory and practice. But in practice, there is a difference!
—Author unknown

The brief history of the ant colony optimization metaheuristic is mainly a history of experimental research. Trial and error guided all early researchers and still guides most of the ongoing research efforts. This is the typical situation for virtually all existing metaheuristics: it is only after experimental work has shown the practical interest of a novel metaheuristic that researchers try to deepen their understanding of the metaheuristic's functioning not only through more and more sophisticated experiments but also by means of an effort to build a theory. Typically, the first theoretical problem considered is the one concerning convergence: will the metaheuristic find the optimal solution if given enough resources? Other questions that are often investigated are the speed of convergence, principled ways of setting the metaheuristic's parameters, relations to existing approaches, identification of problem characteristics that make the metaheuristic more likely to be successful, understanding the importance of the different metaheuristic components, and so on. In this chapter we address those problems for which we have an answer at the time of writing. In particular, we discuss the convergence of some types of ACO algorithms to the optimal solution and the relationship between ACO and other well-known techniques such as stochastic gradient ascent.

4.1 Theoretical Considerations on ACO

When trying to prove theoretical properties for the ACO metaheuristic, the researcher faces a first major problem: ACO's very general definition. Although generality is a desirable property—it allows putting in the same framework ant-based algorithms applied to discrete optimization problems that range from static problems such as the traveling salesman problem to time-varying problems such as routing in telecommunications networks—it makes theoretical analysis much more complicated, if not impossible. A rapid look at the ACO metaheuristic description in figure 2.1 of chapter 2, should convince the reader that ACO as such is not amenable to a theoretical analysis of the type necessary to prove, for example, convergence. Even the simplified version of ACO shown in figure 3.3 of chapter 3, which can be applied only to static combinatorial optimization problems, is too loosely defined to allow for theoretical work. It is for this reason that the convergence proofs presented in the forthcoming sections do not apply to the metaheuristic itself, but to particular ACO algorithms, such as the $\mathcal{MAX}$–$\mathcal{MIN}$ Ant System or the Ant Colony System (see sections 3.3.4 and 3.4.1 of chapter 3).

The first theoretical aspect of ACO that we consider in this chapter is the *convergence problem*: Does the algorithm considered eventually find the optimal solution? This is an interesting question, because ACO algorithms are stochastic search procedures in which the bias due to the pheromone trails could prevent them from ever reaching the optimum. It is important to note that, when considering a stochastic optimization algorithm, there are at least two possible types of convergence: *convergence in value* and *convergence in solution*. Informally, and making the hypothesis that in case of problems with more than one optimal solution we are interested in convergence toward any of them, when studying convergence in value we are interested in evaluating the probability that the algorithm will generate an optimal solution at least once. On the contrary, when studying convergence in solution we are interested in evaluating the probability that the algorithm reaches a state which keeps generating the same optimal solution. In the following, we discuss both types of convergence for some subsets of ACO algorithms. Note, however, that although in general convergence in solution is a stronger and more desirable result to prove than convergence in value, in optimization we are interested in finding the optimal solution once (after it has been found the problem is solved and the algorithm can be stopped), so that convergence in value is all that we need.

In the following, we define two ACO algorithms called $\mathrm{ACO}_{bs,\tau_{min}}$ and $\mathrm{ACO}_{bs,\tau_{min}(\theta)}$, and we prove convergence results for both of them: convergence in value for ACO algorithms in $\mathrm{ACO}_{bs,\tau_{min}}$ and convergence in solution for ACO algorithms in $\mathrm{ACO}_{bs,\tau_{min}(\theta)}$. Here, θ is the iteration counter of the ACO algorithm and $\tau_{min}(\theta)$ indicates that the τ_{min} parameter may change during a run of the algorithm. We then show that these proofs continue to hold when typical elements of ACO, such as local search and heuristic information, are introduced. Finally, we discuss the meaning of these results and we show that the proof of convergence in value applies directly to two of the experimentally most successful ACO algorithms: $\mathcal{MMAS}$ and ACS.

Unfortunately, no results are currently available on the *speed of convergence* of any ACO algorithm. Therefore, although we can prove convergence, we currently have no other way to measure algorithmic performance than to run extensive experimental tests.

Another theoretical aspect that is investigated in this chapter is the formal relationship between ACO and other approaches. In particular, following Dorigo et al. (2002c) and Zlochin, Birattari, Meuleau, & Dorigo (2001), we put ACO in the more general framework of model-based search, so that it is possible to better understand the relations between ACO, stochastic gradient ascent, and the more recent cross-entropy method (De Bonet, Isbell, & Viola, 1997; Rubinstein, 2001).

4.2   The Problem and the Algorithm

In this section we briefly summarize the problem description and the algorithms that we have encountered in chapters 2 and 3. As in chapter 2, we consider an instance of a minimization problem $(\mathcal{S}, f, \Omega)$, where $\mathcal{S}$ is the *set of (candidate) solutions*, f is the *objective function*, which assigns an objective function (cost) value $f(s)$ to each candidate solution $s \in \mathcal{S}$, and Ω is a *set of constraints*, which defines the set of *feasible* candidate solutions. The goal is to find an optimal solution s^*, that is, a feasible candidate solution of minimum cost.

An important difference with chapter 2, however, is that here we consider only static problems for which topology and costs remain fixed in time; in fact, the convergence proofs we present in the following are meaningless in the case of time-varying problems where an algorithm must be able to follow the dynamics inherent to the problem.

The instance $(\mathcal{S}, f, \Omega)$ is mapped on a problem that can be characterized by the following list of items:

- A finite set $C = \{c_1, c_2, \ldots, c_{N_C}\}$ of *components*.

- A finite set $\mathcal{X}$ of *states* of the problem, defined in terms of all possible sequences $x = \langle c_i, c_j, \ldots, c_h, \ldots \rangle$ over the elements of C. The length of a sequence x, that is, the number of components in the sequence, is expressed by $|x|$. The maximum possible length of a sequence is bounded by a positive constant $n < +\infty$.

- The set of (candidate) solutions $\mathcal{S}$ is a subset of $\mathcal{X}$ (i.e., $\mathcal{S} \subseteq \mathcal{X}$). (In other words, candidate solutions are identified with specific states.)

- A set of feasible states $\tilde{\mathcal{X}}$, with $\tilde{\mathcal{X}} \subseteq \mathcal{X}$, defined via a problem-dependent test that verifies that it is not impossible to complete a sequence $x \in \tilde{\mathcal{X}}$ into a solution satisfying the constraints Ω.

- A non-empty set $\mathcal{S}^*$ of optimal solutions, with $\mathcal{S}^* \subseteq \tilde{\mathcal{X}}$ and $\mathcal{S}^* \subseteq \mathcal{S}$.

Additionally, as was discussed in chapter 2, section 2.2.1, a cost $g(s)$ is associated with each candidate solution $s \in \mathcal{S}$. In the following, we set $g(s) \equiv f(s) \; \forall s \in \tilde{\mathcal{S}}$, where $\tilde{\mathcal{S}} \subseteq \mathcal{S}$ is the set of feasible candidate solutions defined via the constraints Ω. Note that in the current definitions the time t does not appear, because in this chapter we consider only static (i.e., not time-varying) problems.

As we have seen in previous chapters, given the above formulation, artificial ants build candidate solutions by performing randomized walks on the *construction graph*, that is, the completely connected graph $G_C = (C, L)$, where the nodes are the components C, and the set L fully connects these components. The random walk of the

procedure AntSolutionConstruction

    Select a start node c_1 according to some problem dependent criterion

    $h \leftarrow 1$

    $x_h \leftarrow \langle c_1 \rangle$

    while $(x_h \notin \mathcal{S}$ and $\mathcal{N}_i^k \neq \varnothing)$ **do**

        $j \leftarrow$ SelectNextNode$(x_h, \mathcal{T})$

        $x_h \leftarrow x_h \otimes j$

    end-while

    if $x_h \in \mathcal{S}$ **then**

        return x_h

        else abort

    end-if

end-procedure

Figure 4.1
High-level pseudo-code for the procedure AntSolutionConstruction applied by ant k. The operator $\otimes$ denotes the addition of a component j to the partial solution x_h. The procedure either returns a full solution s, or is aborted. The SelectNextNode$(x_h, \mathcal{T})$ is given by equation (4.1).

artificial ants is biased by *pheromone trails* τ, gathered in a vector $\mathcal{T}$. As in previous chapters, we restrict our attention to the case in which pheromone trails are associated with connections, so that τ_{ij} is the pheromone associated with the connection between components i and j. It is straightforward to extend algorithms and proofs to the other cases.

The algorithm is initialized by setting the pheromone trails to an initial value $\tau_0 > 0$ (remember that τ_0 is a parameter of the algorithm). At each iteration of the algorithm, ants are positioned on nodes chosen according to some problem-dependent criterion. While moving from one node to another of the graph G_C, constraints Ω are used to prevent ants from building infeasible solutions. The solution construction behavior of a generic ant k, called AntSolutionConstruction, is described in figure 4.1.

In this procedure, the function SelectNextNode$(x_h, \mathcal{T})$ returns the next node j chosen according to the following probability:

$$P_{\mathcal{T}}(c_{h+1} = j \,|\, x_h) = \begin{cases} \dfrac{F_{ij}(\tau_{ij})}{\sum_{(i,l) \in \mathcal{N}_i^k} F_{il}(\tau_{il})}, & \text{if } (i,j) \in \mathcal{N}_i^k; \\ 0, & \text{otherwise;} \end{cases} \qquad (4.1)$$

where (i,j) belongs to $\mathcal{N}_i^k$ iff the sequence $x_{h+1} = \langle c_1, c_2, \ldots, c_h, j \rangle$ built by ant k satisfies the constraints Ω (i.e., $x_{h+1} \in \tilde{\mathcal{X}}$) and $F_{ij}(z)$ is some nondecreasing, monotonic function. Note that by writing $F_{ij}(z)$ instead of $F(z)$ we indicate that the function $F(z)$ may be different on each arc. In practice, in all ACO implementations we are aware of, the dependence on the arc is due to the fact that pheromone values are composed with some function of an arc-specific information η_{ij} called "heuristic visibility." As we have seen in chapter 3, most commonly $F_{ij}(z) = z^\alpha \eta_{ij}^\beta$, where $\alpha, \beta > 0$ are parameters [a notable exception is equation (3.13), used in Maniezzo's ANTS].

If it happens during solution construction that $x_h \notin \mathcal{S}$ and $\mathcal{N}_i^k = \varnothing$, that is, the construction process has reached a dead end, the AntSolutionConstruction procedure is aborted and the current state x_h is discarded. (This situation may be prevented by allowing artificial ants to build infeasible solutions as well. In such a case a penalty term reflecting the degree of infeasibility is usually added to the cost function.)

For certain problems, it can be useful to use a more general scheme, where F depends on the pheromone values of several "related" connections, rather than just a single one. Moreover, instead of the *random proportional rule* above, different selection schemes, such as the *pseudorandom proportional rule* [see equation (3.10)], may be considered.

Once all the ants have terminated their AntSolutionConstruction procedure, a pheromone update phase is started in which pheromone trails are modified. Let s^{bs} be the best-so-far solution (i.e., the best feasible solution found since the first iteration of the algorithm) and s^θ be the iteration-best solution (i.e., the best feasible solution obtained in the current iteration θ); $f(s^{bs})$ and $f(s^\theta)$ are the corresponding objective function values. The pheromone update procedure decreases the value of the pheromone trails on *all* connections in L by a small factor ρ, called the evaporation rate, and then increases the value of the pheromone trails on the connections belonging to s^{bs} (in the literature, adding pheromone only to those arcs that belong to the best-so-far solution is known as the *global-best pheromone update* (Dorigo & Gambardella, 1997b), but is more appropriately referred to as *best-so-far update* in the following).

All the different schemes for pheromone update discussed in chapter 3, section 3.3 (i.e., AS's and its extensions' pheromone update rules) can be described using the GenericPheromoneUpdate procedure shown in figure 4.2. Here θ is the index of the current iteration, S_i is the set of solutions generated in the i-th iteration, ρ is the evaporation rate ($0 < \rho \leq 1$), and $q_f(s \mid S_1, \ldots, S_\theta)$ is some "quality function," which is typically required to be nonincreasing with respect to f (i.e., $f(s_1) > f(s_2) \Rightarrow q_f(s_1) \leq q_f(s_2)$), and is defined over the "reference set" $\hat{S}_\theta$, as discussed in the following.

Different ACO algorithms may use different quality functions and reference sets. For example, in AS the quality function is simply $1/f(s)$ and the reference set $\hat{S}_\theta = S_\theta$. In many of the extensions of AS, either the *iteration-best update* or the

procedure GenericPheromoneUpdate
  foreach $(i, j) \in L$ **do**
    $\tau_{ij} \leftarrow (1 - \rho)\tau_{ij}$
  end-foreach
  foreach $s \in \hat{S}_\theta$ **do**
    foreach $(i, j) \in s$ **do**
      $\tau_{ij} \leftarrow \tau_{ij} + q_f(s \,|\, S_1, \ldots, S_\theta)$
    end-foreach
  end-foreach
end-procedure

Figure 4.2
High-level pseudo-code for the procedure GenericPheromoneUpdate. θ is the index of the current iteration, S_i is the sample (i.e., the set of generated solutions) in the i-th iteration, ρ, $0 < \rho \leq 1$, is the evaporation rate, and $q_f(s \,|\, S_1, \ldots, S_\theta)$ is some "quality function," which is typically required to be nonincreasing with respect to f and is defined over the "reference set" $\hat{S}_\theta$, as discussed in the text.

best-so-far update is used: in the first case the reference set is a singleton containing the best solution within S_θ (if there are several iteration-best solutions, one of them is chosen randomly). In the best-so-far update, the reference set contains the best among all the iteration-best solutions (and if there is more than one, the earliest one is chosen). In some cases a combination of the two update methods is used.

In case a good lower bound on the optimal solution cost is available, one may use the following quality function, as done in ANTS (Maniezzo, 1999) [see also equation (3.15)]:

$$q_f(s \,|\, S_1, \ldots, S_\theta) = \tau_0 \left(1 - \frac{f(s) - LB}{f_{\text{avg}} - LB} \right) = \tau_0 \frac{f_{\text{avg}} - f(s)}{f_{\text{avg}} - LB}, \tag{4.2}$$

where f_{avg} is the average of the costs of the last k solutions and LB is a lower bound on the optimal solution cost. With this quality function, the solutions are evaluated by comparing their cost to the average cost of the other recent solutions, rather than by using the absolute cost values. In addition, the quality function is automatically scaled based on the proximity of the average cost to the lower bound, and no explicit pheromone evaporation is performed.

As we have seen (see chapter 3, section 3.4.1), the pheromone update used in ACS differs slightly from the generic update described above. In ACS there is no general pheromone evaporation applied to all connections as in AS and its extensions. On the contrary, the only pheromones that evaporate are those associated with the arcs of the best-so-far solution: the best-so-far update computes a weighted sum between the old pheromone trail and the amount deposited, where the evaporation rate ρ determines the weights of the two values [see equation (3.11)]. Additionally, the pheromones are decreased by the ants during solution construction by means of the local pheromone update rule [see equation (3.12)].

Two additional modifications of the generic update are found in $\mathcal{MAX}$–$\mathcal{MIN}$ Ant System and in the hyper-cube framework for ACO (see chapter 3, sections 3.3.4 and 3.4.3, respectively). $\mathcal{MMAS}$ puts limits on the minimum value of pheromone trails. With this modification, the probability of generating any particular solution is kept above some positive threshold, which helps prevent search stagnation and premature convergence to suboptimal solutions. In the hyper-cube framework for ACO, an automatic scaling of the pheromone values is implemented.

4.3 Convergence Proofs

In this section, we study the convergence properties of some important subsets of ACO algorithms. First, we define the $\text{ACO}_{bs,\tau_{min}}$ algorithm and we prove its convergence in value with probability 1. Then, we define the $\text{ACO}_{bs,\tau_{min}(\theta)}$ algorithm and we prove its convergence in solution. After showing that both proofs continue to hold when local search and heuristic information are added, we discuss the meaning of the proofs and show that the convergence in value proof applies to the experimentally most successful ACO algorithms.

Using the notation of the previous section, $\text{ACO}_{bs,\tau_{min}}$ is defined as follows. First, in the ant solution construction procedure the initial location of each ant is chosen in a problem-specific way (often this is done using a uniform random distribution), and $F_{ij}(\tau_{ij}) \equiv F(\tau_{ij})$ [i.e., we remove the dependence of the function F on the arc (i, j) to which it is applied; for the algorithms presented in chapter 3, this corresponds to removing the dependence on the heuristic η; this dependence is reintroduced in section 4.3.3]. Additionally, to ease the following derivations, we assume $F(\tau_{ij})$ to be of the form used in almost all ACO algorithms: $F(\tau_{ij}) = \tau_{ij}^{\alpha}$, where $0 < \alpha < +\infty$ is a parameter. The probabilistic construction rule of equation (4.1) applied by the ants to build solutions becomes

$$P_{\mathcal{T}}(c_{h+1} = j \mid x_h) = \begin{cases} \dfrac{\tau_{ij}^{\alpha}}{\sum_{(i,l) \in \mathcal{N}_i^k} \tau_{il}^{\alpha}}, & \text{if } (i, j) \in \mathcal{N}_i^k; \\[2ex] 0, & \text{otherwise.} \end{cases} \tag{4.3}$$

Second, the pheromone update procedure is implemented by choosing $\hat{S}_\theta = s^{bs}$ (i.e., the reference set contains only the best-so-far solution) and, additionally, a

procedure $\text{ACO}_{bs,\tau_{min}}$ PheromoneUpdate

  foreach $(i,j) \in L$ **do**

    $\tau_{ij} \leftarrow (1-\rho)\tau_{ij}$

  end-foreach

  if $f(s^\theta) < f(s^{bs})$ **then**

    $s^{bs} \leftarrow s^\theta$

  end-if

  foreach $(i,j) \in s^{bs}$ **do**

    $\tau_{ij} \leftarrow \tau_{ij} + q_f(s^{bs})$

  end-foreach

  foreach (i,j) **do**

    $\tau_{ij} \leftarrow \max\{\tau_{min}, \tau_{ij}\}$

  end-foreach

end-procedure

Figure 4.3
High-level pseudo-code for the $\text{ACO}_{bs,\tau_{min}}$ PheromoneUpdate procedure. s^θ and s^{bs} are the iteration-best and best-so-far solutions respectively, while τ_{min} is a parameter.

lower limit $\tau_{min} > 0$ is put on the value of pheromone trails. In practice, the Generic-PheromoneUpdate procedure of figure 4.2 becomes the $\text{ACO}_{bs,\tau_{min}}$ PheromoneUpdate procedure shown in figure 4.3.

The value τ_{min} is a parameter of $\text{ACO}_{bs,\tau_{min}}$; in the following we assume that $\tau_{min} < q_f(s^*)$. This can be achieved by setting, for example, $\tau_0 = q_f(s')/2$, where s' is a solution used to initialize $\text{ACO}_{bs,\tau_{min}}$.

The choice of the name $\text{ACO}_{bs,\tau_{min}}$ for this algorithm is due to the fact that the best-so-far solution is used to update pheromones and that a lower limit τ_{min} on the range of feasible pheromone trails is introduced. (Note that in the original paper in which $\text{ACO}_{bs,\tau_{min}}$ was introduced [Stützle & Dorigo, 2002], the algorithm was called $\text{ACO}_{gb,\tau_{min}}$.)

4.3.1　Convergence in Value

In this subsection we prove that $\text{ACO}_{bs,\tau_{min}}$ is guaranteed to find an optimal solution with a probability that can be made arbitrarily close to 1 if given enough time (convergence in value). However, as we indicate in section 4.3.2, we cannot prove convergence in solution for $\text{ACO}_{bs,\tau_{min}}$.

Before proving the first theorem, it is convenient to show that, due to pheromone evaporation, the maximum possible pheromone level τ_{max} is asymptotically bounded.

Proposition 4.1 *For any τ_{ij} it holds:*

$$\lim_{\theta \to \infty} \tau_{ij}(\theta) \leq \tau_{max} = \frac{q_f(s^*)}{\rho}. \tag{4.4}$$

Proof The maximum possible amount of pheromone added to any arc (i, j) after any iteration is $q_f(s^*)$. Clearly, at iteration 1 the maximum possible pheromone trail is $(1 - \rho)\tau_0 + q_f(s^*)$, at iteration 2 it is $(1 - \rho)^2\tau_0 + (1 - \rho)q_f(s^*) + q_f(s^*)$, and so on. Hence, due to pheromone evaporation, the pheromone trail at iteration θ is bounded by

$$\tau_{ij}^{max}(\theta) = (1 - \rho)^\theta \tau_0 + \sum_{i=1}^{\theta}(1 - \rho)^{\theta-i} q_f(s^*).$$

As $0 < \rho \leq 1$, this sum converges asymptotically to

$$\tau_{max} = \frac{q_f(s^*)}{\rho}. \qquad \square$$

Proposition 4.2 *Once an optimal solution s^* has been found, it holds that*

$$\forall (i, j) \in s^* : \lim_{\theta \to \infty} \tau_{ij}^*(\theta) = \tau_{max} = \frac{q_f(s^*)}{\rho},$$

where τ_{ij}^* is the pheromone trail value on connections $(i, j) \in s^*$.

Proof Once an optimal solution has been found, remembering that $\forall \theta \geq 1$, $\tau_{ij}^*(\theta) \geq \tau_{min}$ and that the best-so-far update rule is used, we have that $\tau_{ij}^*(\theta)$ monotonically increases. The proof of proposition 4.2 is basically a repetition of the proof of proposition 4.1, restricted to the connections of the optimal solution (τ_0 is replaced by $\tau_{ij}^*(\theta^*)$ in the proof of proposition 4.1, where θ^* is the iteration in which the first optimal solution was found). $\qquad \square$

Proposition 4.1 implies that, for the following proof of theorem 4.1, the only essential point is that $\tau_{min} > 0$, because τ_{max} will anyway be bounded by pheromone evaporation. Proposition 4.2 additionally states that, once an optimal solution has been found, the value of the pheromone trails on all connections of s^* converges to $\tau_{max} = q_f(s^*)/\rho$.

We can now prove the following theorem.

Theorem 4.1 *Let $P^*(\theta)$ be the probability that the algorithm finds an optimal solution at least once within the first θ iterations. Then, for an arbitrarily small $\varepsilon > 0$ and for a sufficiently large θ it holds that*

$P^*(\theta) \geq 1 - \varepsilon$,

and, asymptotically, $\lim_{\theta \to \infty} P^*(\theta) = 1$.

Proof Due to the pheromone trail limits τ_{min} and τ_{max} we can guarantee that any feasible choice in equation (4.3) for any partial solution x_h is made with a probability $p_{min} > 0$. A trivial lower bound for p_{min} is given by

$$p_{min} \geq \hat{p}_{min} = \frac{\tau_{min}^{\alpha}}{(N_C - 1)\tau_{max}^{\alpha} + \tau_{min}^{\alpha}}, \tag{4.5}$$

where N_C is the cardinality of the set C of components. (For the derivation of this bound we consider the following "worst-case" situation: the pheromone trail associated with the desired decision is τ_{min}, while all the other feasible choices—there are at most $N_C - 1$—have an associated pheromone trail of τ_{max}.) Then, any generic solution s', including any optimal solution $s^* \in \mathcal{S}^*$, can be generated with a probability $\hat{p} \geq \hat{p}_{min}^n > 0$, where $n < +\infty$ is the maximum length of a sequence. Because it is sufficient that one ant finds an optimal solution, a lower bound for $P^*(\theta)$ is given by

$$\hat{P}^*(\theta) = 1 - (1 - \hat{p})^{\theta}.$$

By choosing a sufficiently large θ, this probability can be made larger than any value $1 - \varepsilon$. Hence, we have that $\lim_{\theta \to \infty} \hat{P}^*(\theta) = 1$. $\square$

4.3.2   Convergence in Solution

In this subsection we prove convergence in solution for $\mathrm{ACO}_{bs, \tau_{min}(\theta)}$, which differs from $\mathrm{ACO}_{bs, \tau_{min}}$ by allowing a change in value for τ_{min} while solving a problem. That is, we prove that, in the limit, any arbitrary ant of the colony will construct the optimal solution with probability 1. This cannot be proved if we impose, as done in $\mathrm{ACO}_{bs, \tau_{min}}$, a small, positive lower bound on the lower pheromone trail limits because in this case at any iteration θ each ant can construct any solution with a nonzero probability. The key of the proof is therefore to allow the lower pheromone trail limits to decrease over time toward zero, but making this decrement slow enough to guarantee that the optimal solution is eventually found. We call $\mathrm{ACO}_{bs, \tau_{min}(\theta)}$ the modification of $\mathrm{ACO}_{bs, \tau_{min}}$ obtained in this way, where $\tau_{min}(\theta)$ indicates the dependence of the lower pheromone trail limits on the iteration counter.

The proof of convergence in solution is organized in two theorems. First, in theorem 4.2 (in a way analogous to what was done in the proof of theorem 4.1) we prove that it can still be guaranteed that an optimal solution is found with a probability converging to 1 when lower pheromone trail limits of the $ACO_{bs, \tau_{min}(\theta)}$ algorithm decrease toward 0 at not more than logarithmic speed (in other words, we prove that $ACO_{bs, \tau_{min}(\theta)}$ converges in value). Next, in theorem 4.3 we prove, under the same conditions, convergence in solution of $ACO_{bs, \tau_{min}(\theta)}$.

Theorem 4.2   *Let the lower pheromone trail limits in $ACO_{bs, \tau_{min}(\theta)}$ be*

$$\forall \theta \geq 1, \quad \tau_{min}(\theta) = \frac{d}{\ln(\theta + 1)},$$

with d being a constant, and let $P^*(\theta)$ be the probability that the algorithm finds an optimal solution at least once within the first θ iterations. Then it holds that

$$\lim_{\theta \to \infty} P^*(\theta) = 1.$$

Proof   Differently from what was done in the proof of theorem 4.1, here we prove that an upper bound on the probability of *not* constructing the optimal solution is 0 in the limit (i.e., the optimal solution is found in the limit with probability 1). Let the event E_θ denote that iteration θ is the iteration in which an optimal solution is found for the first time. The event $\bigwedge_{\theta=1}^{\infty} \neg E_\theta$ that no optimal solution is ever found, implies that also one arbitrary, but fixed, optimal solution s^* is never found. Therefore, an upper bound to the probability $P(\bigwedge_{\theta=1}^{\infty} \neg E_\theta)$ is given by $P(s^*$ is never traversed), that is:

$$P\left(\bigwedge_{\theta=1}^{\infty} \neg E_\theta \right) \leq P(s^* \text{ is never traversed}). \tag{4.6}$$

Now, in a way similar to what was done for theorem 4.1, we can guarantee that at a generic iteration θ any feasible choice according to equation (4.3) can be made with a probability p_{min} bounded as follows:

$$p_{min} \geq \hat{p}_{min}(\theta) = \frac{\tau_{min}^\alpha(\theta)}{(N_C - 1)\tau_{max}^\alpha + \tau_{min}^\alpha(\theta)}$$

$$\geq \frac{\tau_{min}^\alpha(\theta)}{N_C \tau_{max}^\alpha} = \hat{p}'_{min}(\theta).$$

Then, a lower bound on the probability that a fixed ant k is constructing the optimal solution s^* is given by $\hat{p}(\theta) \geq (\hat{p}'_{min}(\theta))^n$, where $n < +\infty$ is the maximum length of a sequence. This bound is independent of what happened before iteration θ. Therefore, we can give the following upper bound on the right side of equation (4.6):

$$P(s^* \text{ is never traversed}) \leq \prod_{\theta=1}^{\infty} (1 - (\hat{p}'_{min}(\theta))^n)$$

$$= \prod_{\theta=1}^{\infty} \left(1 - \left(\frac{\tau_{min}^{\alpha}(\theta)}{N_C \tau_{max}^{\alpha}} \right)^n \right). \tag{4.7}$$

We now must prove that this product is equal to 0. To do so, we consider its logarithm

$$\sum_{\theta=1}^{\infty} \ln \left[1 - \left(\frac{\tau_{min}^{\alpha}(\theta)}{N_C \tau_{max}^{\alpha}} \right)^n \right],$$

and we show that the resulting series, starting from some finite number l, grows quicker than the harmonic series, so that it diverges to $-\infty$, which implies that the original product is equal to 0. First, remember that $\tau_{min}(\theta) = d/\ln(\theta + 1)$. Then

$$\sum_{\theta=1}^{\infty} \ln \left[1 - \left(\frac{\tau_{min}^{\alpha}(\theta)}{N_C \tau_{max}^{\alpha}} \right)^n \right] = \sum_{\theta=1}^{\infty} \ln \left[1 - \left(\frac{\left(\frac{d}{\ln(\theta + 1)} \right)^{\alpha}}{N_C \tau_{max}^{\alpha}} \right)^n \right]$$

$$= \sum_{\theta=1}^{\infty} \ln \left[1 - \left(\frac{d_1}{(\ln(\theta + 1))^{\alpha}} \right)^n \right]$$

$$\leq -d_1^n \sum_{\theta=1}^{\infty} \left(\frac{1}{\ln(\theta + 1)} \right)^{\alpha n} = -\infty,$$

where $d_1 = d^{\alpha}/N_C \tau_{max}^{\alpha}$.

The inequality holds because for any $x < 1$, $\ln(1 - x) \leq -x$. The equality holds because $\sum_x (\ln x)^{-i}$ is a diverging series. To see the latter, note that for each positive constant $\delta > 0$ and for sufficiently large x, $(\ln x)^i \leq \delta \cdot x$, and therefore $\delta/(\ln x)^i \geq 1/x$. It then suffices to remember that $\sum_x 1/x$ is the harmonic series, which is known to diverge to ∞.

These derivations say that an upper bound for the logarithm of the product given in equation (4.7) and, hence, the logarithm on the right side of equation (4.6), is $-\infty$; therefore, the product given in equation (4.7) and the right side of equation (4.6) have to be 0, that is, the probability of never finding the optimal solution $(P(\bigwedge_{\theta=1}^{\infty} \neg E_\theta))$ is 0. This proves that an optimal solution will be found with probability 1. $\square$

In the limiting case, once the optimal solution has been found, we can estimate an ant's probability of constructing an optimal solution when following the stochastic policy of the algorithm. In fact, it can be proved that any ant will in the limit construct the optimal solution with probability 1—that is, we can prove convergence in solution. Before the proof of this assertion, it is convenient to show that the pheromone trails of connections that do not belong to the optimal solution asymptotically converge to 0.

Proposition 4.3 *Once an optimal solution has been found and for any $\tau_{ij}(\theta)$ such that $(i, j) \notin s^*$ it holds that*

$$\lim_{\theta \to \infty} \tau_{ij}(\theta) = 0.$$

Proof After the optimal solution has been found, connections not belonging to the optimal solution do not receive pheromone anymore. Thus, their value can only decrease. In particular, after one iteration $\tau_{ij}(\theta^* + 1) = \max\{\tau_{min}(\theta), (1 - \rho)\tau_{ij}(\theta^*)\}$, after two iterations $\tau_{ij}(\theta^* + 2) = \max\{\tau_{min}(\theta), (1 - \rho)^2 \tau_{ij}(\theta^*)\}$, and so on ($\theta^*$ is the iteration in which s^* was first found). Additionally, we have that $\lim_{\theta \to \infty} d/\ln(\theta^* + \theta + 1) = 0$ and $\lim_{\theta \to \infty} (1 - \rho)^\theta \tau_{ij}(\theta^* + \theta) = 0$. Therefore, $\lim_{\theta \to \infty} \tau_{ij}(\theta^* + \theta) = 0$. $\qquad\qquad\square$

Theorem 4.3 *Let θ^* be the iteration in which the first optimal solution has been found and $P(s^*, \theta, k)$ be the probability that an arbitrary ant k constructs s^* in the θ-th iteration, with $\theta > \theta^*$. Then it holds that*

$$\lim_{\theta \to \infty} P(s^*, \theta, k) = 1.$$

Proof Let ant k be located on component i and (i, j) be a connection of s^*. A lower bound $\hat{p}_{ij}^*(\theta)$ for the probability $p_{ij}^*(\theta)$ that ant k makes the "correct choice" (i, j) is given by the term

$$\hat{p}_{ij}^*(\theta) = \frac{(\tau_{ij}^*(\theta))^\alpha}{(\tau_{ij}^*(\theta))^\alpha + \sum_{(i,h) \notin s^*} (\tau_{ih}(\theta))^\alpha}.$$

Because of propositions 4.2 and 4.3 we have

$$\hat{p}_{ij}^* = \lim_{\theta \to \infty} \hat{p}_{ij}^*(\theta) = \frac{\lim_{\theta \to \infty} (\tau_{ij}^*(\theta))^\alpha}{\lim_{\theta \to \infty} (\tau_{ij}^*(\theta))^\alpha + \sum_{(i,h) \notin s^*} \lim_{\theta \to \infty} (\tau_{ih}(\theta))^\alpha}$$

$$= \frac{\tau_{max}^\alpha}{\tau_{max}^\alpha + \sum_{(i,h) \notin s^*} 0^\alpha} = 1.$$

Hence, in the limit any fixed ant will construct the optimal solution with probability 1, because at each construction step it takes the correct decision with probability 1.   □

4.3.3   Additional Features of ACO Algorithms

As we have seen in chapter 3, many ACO algorithms include some features that are present neither in $\text{ACO}_{bs,\tau_{min}}$ nor in $\text{ACO}_{bs,\tau_{min}(\theta)}$. The most important are the use of local search algorithms to improve the solutions constructed by the ants and the use of heuristic information in the choice of the next component. Therefore, a natural question is how these features affect the convergence proof for $\text{ACO}_{bs,\tau_{min}}$. Note that here and in the following, although the remarks made about $\text{ACO}_{bs,\tau_{min}}$ in general also apply to $\text{ACO}_{bs,\tau_{min}(\theta)}$, for simplicity we often refer only to $\text{ACO}_{bs,\tau_{min}}$.

Let us first consider the use of local search. Local search tries to improve an ant's solution s by iteratively applying small, local changes to it. Typically, the best solution s' found by the local search is returned and used to update the pheromone trails. It is rather easy to see that the use of local search neither affects the convergence properties of $\text{ACO}_{bs,\tau_{min}}$, nor those of $\text{ACO}_{bs,\tau_{min}(\theta)}$. In fact, the validity of both convergence proofs depends only on the way solutions are constructed and not on the fact that the solutions are taken or not to their local optima by a local search routine.

A priori available information on the problem can be used to derive heuristic information that biases the probabilistic decisions taken by the ants. When incorporating such heuristic information into $\text{ACO}_{bs,\tau_{min}}$, the most common choice is $F_{ij}(\tau_{ij}) = [\tau_{ij}]^{\alpha}[\eta_{ij}]^{\beta}$, as explained in section 4.2. In this case equation (4.3), becomes

$$P_{\mathcal{T}}(c_{h+1} = j \mid x_h) = \begin{cases} \dfrac{[\tau_{ij}]^{\alpha}[\eta_{ij}]^{\beta}}{\sum_{(i,l) \in \mathcal{N}_i^k}[\tau_{il}]^{\alpha}[\eta_{il}]^{\beta}}, & \text{if } (i,j) \in \mathcal{N}_i^k; \\ \\ 0, & \text{otherwise}; \end{cases} \tag{4.8}$$

where η_{ij} measures the heuristic desirability of adding solution component j. In fact, neither theorem 4.1 nor theorems 4.2 and 4.3 are affected by the heuristic information, if we have $0 < \eta_{ij} < +\infty$ for each $(i,j) \in L$ and $\beta < \infty$. In fact, with these assumptions η is limited to some (instance-specific) interval $[\eta_{min}, \eta_{max}]$, with $\eta_{min} > 0$ and $\eta_{max} < +\infty$. Then, the heuristic information has only the effect of changing the lower bounds on the probability p_{min} of making a specific decision [see, e.g., equation (4.5), or the analogous estimates in the proofs of theorems 4.2 and 4.3].

4.3.4   What Does the Proof Really Say?

It is instructive to understand what theorems 4.1 to 4.3 really tell us. First, theorem 4.1 says that, when using a fixed positive lower bound on the pheromone trails, $ACO_{bs, \tau_{min}}$ is guaranteed to find the optimal solution. Theorem 4.2 extends this result by saying that we essentially can keep this property for $ACO_{bs, \tau_{min}(\theta)}$ algorithms, if we decrease the bound τ_{min} to 0 slowly enough. (Unfortunately, theorem 4.2 cannot be proved for the exponentially fast decrement of the pheromone trails obtained by a constant pheromone evaporation rate, which most ACO algorithms use.) However, the proofs do not say anything about the time required to find an optimal solution, which can be astronomically large. A similar limitation applies to other well-known convergence proofs, such as those formulated for simulated annealing by Hajek (1988) and by Romeo & Sangiovanni-Vincentelli (1991). Finally, theorem 4.3 shows that a sufficiently slow decrement of the lower pheromone trail limits leads to the effect that the algorithm converges to a state in which all the ants construct the optimal solution over and over again. In fact, for this latter result it is essential that in the limit the pheromone trails go to 0. If, as is done in $ACO_{bs, \tau_{min}}$, a fixed lower bound τ_{min} is set, it can only be proved that the probability of constructing an optimal solutions is larger than $1 - \hat{\varepsilon}(\tau_{min}, \tau_{max})$, where $\hat{\varepsilon}$ is a function of τ_{min} and τ_{max} (Stützle & Dorigo, 2002).

  Because in practice we are more interested in finding an optimal solution at least once than in generating it over and over again, let us have a closer look at the role played by τ_{min} and τ_{max} in the proof of theorem 4.1: the smaller the ratio τ_{max}/τ_{min}, the larger the lower bound $\hat{p}_{min}$ given in the proof. This is important, because the larger the $\hat{p}_{min}$, the smaller the worst-case estimate of the number of iterations θ needed to assure that an optimal solution is found with a probability larger than $1 - \varepsilon$. In fact, the tightest bound is obtained if all pheromone trails are the same, that is, for the case of uniformly random solution construction; in this case we would have $\hat{p}_{min} = 1/N_C$ (note that this fact is independent of the tightness of the lower bounds used in theorem 4.1). This somewhat counterintuitive result is due to the fact that our proof is based on a worst-case analysis: we need to consider the worst-case situation in which the bias in the solution construction introduced by the pheromone trails is counterproductive and leads to suboptimal solutions; that is, we have to assume that the pheromone trail level associated with the connection an ant needs to pass for constructing an optimal solution is τ_{min}, while on the other connections it is much higher—in the worst case corresponding to τ_{max}. In practice, however, as shown by the results of many published experimental works (see Dorigo & Di Caro, 1999b; Dorigo et al., 1999; Dorigo & Stützle, 2002, as well as chapter 5 of this book, for an overview), this does not happen, and the bias introduced by the pheromone trails does indeed help to speed up convergence to an optimal solution.

4.3.5   Convergence of Some ACO Algorithms

As mentioned, from the point of view of the researcher interested in applications of the algorithm, the interesting part of the discussed convergence proofs is theorem 4.1, which guarantees that $ACO_{bs,\tau_{min}}$ will find an optimal solution if run long enough.

It is therefore interesting that this theorem also applies to ACO algorithms that differ from $ACO_{bs,\tau_{min}}$ in the way the pheromone update procedure is implemented. In general, theorem 4.1 applies to any ACO algorithm for which the probability $P(s)$ of constructing a solution $s \in S$ always remains greater than a small constant $\varepsilon > 0$. In $ACO_{bs,\tau_{min}}$ this is a direct consequence of the fact that $0 < \tau_{min} < \tau_{max} < +\infty$, which was obtained by (1) explicitly setting a minimum value τ_{min} for pheromone trails, (2) limiting the amount of pheromone that the ants may deposit after each iteration, that is, $\forall s, g(s) < z < +\infty$, (3) letting pheromone evaporate over time, that is, by setting $\rho > 0$, and (4) by the particular form of the function $F(\tau_{ij})$ chosen. We call the subset of ACO algorithms that satisfy these conditions $ACO_{\tau_{min}}$. $ACO_{bs,\tau_{min}}$ differs from $ACO_{\tau_{min}}$ in that it additionally imposes the use of the best-so-far update rule. Therefore, $ACO_{bs,\tau_{min}}$ can be seen as a particular case of $ACO_{\tau_{min}}$. By definition, theorem 4.1 holds for any algorithm in $ACO_{\tau_{min}}$. In the following, we show that $\mathcal{MMAS}$ and ACS, two of the experimentally most successful ACO algorithms, belong to $ACO_{\tau_{min}}$.

$\mathcal{MAX}$–$\mathcal{MIN}$ Ant System

It is easy to show that $\mathcal{MMAS}$, described in detail in chapter 3, section 3.3.4, belongs to $ACO_{\tau_{min}}$. In fact, there are only two minor differences between $\mathcal{MMAS}$ and $ACO_{bs,\tau_{min}}$. First, $\mathcal{MMAS}$ uses an explicit value for τ_{max} instead of an implicit one as $ACO_{bs,\tau_{min}}$ does. In fact, this is a very minor difference, because $\mathcal{MMAS}$ uses the upper pheromone trail limit defined by proposition 4.1 as an estimate of τ_{max}. Second, $\mathcal{MMAS}$ uses a somewhat more general pheromone update rule than $ACO_{bs,\tau_{min}}$. Like $ACO_{bs,\tau_{min}}$, $\mathcal{MMAS}$ uses only one solution to select the connections on which to add pheromone, but it allows a choice between the iteration-best solution s^θ and the best-so-far solution s^{bs}. During the run the best-so-far solution is chosen more and more often, until reaching a situation in which pheromone is added only to connections belonging to s^{bs}. It is therefore clear that theorem 4.1 holds for $\mathcal{MMAS}$.

Ant Colony System

Ant Colony System also belongs to $ACO_{\tau_{min}}$, although this is more difficult to see. ACS differs in three main points from $ACO_{bs,\tau_{min}}$. First, it uses the *pseudorandom*

procedure ACSGlobalPheromoneUpdate
   if $f(s^\theta) < f(s^{bs})$ **then**
     $s^{bs} \leftarrow s^\theta$
   end-if
   foreach $(i, j) \in s^{bs}$ **do**
     $\tau_{ij} \leftarrow (1 - \rho)\tau_{ij} + \rho q_f(s^{bs})$
   end-foreach
end-procedure

Figure 4.4
High-level pseudo-code for the ACSGlobalPheromoneUpdate procedure. ρ is the pheromone evaporation.

proportional rule [see equation (3.10), section 3.4.1]: at each construction step an ant either chooses, with probability q_0, the connection with the largest pheromone trail value, or it performs, with probability $(1 - q_0)$, a biased exploration according to equation (4.3). Second, ACS does not apply pheromone evaporation to all connections, but only to those belonging to the best-so-far solution. The update rule used in ACS is given by the ACSGlobalPheromoneUpdate procedure shown in figure 4.4.

Third, during the construction of the solution each ant in ACS uses a local pheromone trail update rule that it applies immediately after having crossed a connection (i, j). Consider the situation in which an ant has built a partial solution $x_h = \langle c_1, c_2, \ldots, c_{h-1}, i \rangle$ and it adds a component j so that the new partial solution is $x_{h+1} = \langle c_1, c_2, \ldots, i, j \rangle$. Then, the pheromone trail on connection (i, j) is updated according to the rule:

$$\tau_{ij} \leftarrow (1 - \xi)\tau_{ij} + \xi\tau_0. \tag{4.9}$$

That is, a fraction ξ of the trail is evaporated and a small amount τ_0 is added. In practice, the effect of the local updating rule is to decrease the pheromone trail on the visited connection, making in this way the connection less desirable for the following ants.

It is convenient to remark that the two pheromone update rules used in ACS are of the form $a_{h+1} = (1 - \psi)a_h + \psi b$ for $h \geq 1$, where a_{h+1} and a_h are $\tau_{ij}(\theta + 1)$ and $\tau_{ij}(\theta)$, and, respectively, $b = q_f(s^{bs}), \tau_0$, and $\psi = \rho, \xi$. Thus we have

$$a_h = (1 - \psi)^h a_0 + b[1 - (1 - \psi)^h], \tag{4.10}$$

which is b in the limit as $h \to \infty$. The sequence decreases for $a_0 > b$ (with maximum a_0) and increases for $a_0 < b$ (with maximum b).

Now the question is: How does the convergence in value result of ACO$_{bs, \tau_{min}}$ transfer to ACS? First, we observe that in ACS the maximum amount of pheromone is limited by $\tau_{max} = \tau_{max}^{ACS} = q_f(s^*)$ (this bound is obtained without considering the local pheromone update). Second, because τ_0 is chosen to be smaller than $q_f(s^{bs})$, no pheromone trail value can fall below τ_0 and therefore τ_0 gives a lower bound on the pheromone trail of any solution component (i, j).

The next step is to show that any feasible solution can be constructed with a non-zero probability. The easiest way to see this is to rewrite the probability of making some fixed choice (i, j) in ACS. Let us assume that connection (i, j) does not have the largest pheromone trail associated. Then the probability of choosing connection (i, j) can be calculated as the product of the probability of making a randomized choice, which is $1 - q_0$, and the probability of choosing connection (i, j) in this randomized choice. A bound for the latter is given by $\hat{p}_{min}$ in equation (4.5). Therefore, a lower bound for the probability of making any specific choice at any construction step is $(1 - q_0) \cdot \hat{p}_{min}$ and theorem 4.1 directly applies to ACS.

It is interesting, however, to note that AS, as well as some of its variants (e.g., EAS, section 3.3.2, and the rank-based AS, section 3.3.3) do not belong to $ACO_{\tau_{min}}$. In fact, in these three algorithms there is no lower bound to the value of pheromone trails and therefore the pheromones can become null much faster than imposed by the bounds of theorem 4.2.

In any case, ACS and $\mathcal{MMAS}$ were shown to perform better than AS and its variants on many standard benchmark problems such as the TSP and the QAP. Therefore, we are in the fortunate case in which ACO algorithms for which convergence can be proven theoretically show a better performance in practice.

4.4   ACO and Model-Based Search

Up to now we have taken the classic view in which the ACO metaheuristic is seen as a class of stochastic search procedures working in the space of the solutions of a combinatorial optimization problem. Under this interpretation, artificial ants are stochastic constructive heuristics that build better and better solutions to a combinatorial optimization problem by using and updating pheromone trails. In other words, our attention has been directed to the stochastic constructive procedure used by the ants and to how the ants use the solutions they build to bias the search of future ants by changing pheromone values.

In this section we change perspective and interpret ACO algorithms as methods for searching in the space of pheromone values with the goal of maximizing the probability of generating good solutions. In other terms, we interpret the construction graph G_C and the associated pheromones $\mathcal{T}$ as a parametric probability distribution used by ACO to generate solutions to the considered problem. And we interpret the set of solutions generated by the artificial ants as a sample used to update the parameters of the probability distribution, that is, the pheromone trails. Adopting this view, it is natural to understand ACO as a member of model-based search algorithms, as explained in the following. This view of ACO allows drawing interesting analogies with methods such as stochastic gradient ascent and cross-entropy.

4.4.1 Model-Based Search

In the field of metaheuristics for combinatorial optimization, following a classification similar to the one found in machine learning (Quinlan, 1993b), two antithetic approaches can be identified: the *instance-based* and the *model-based* approach.

Most of the classic search methods may be considered instance-based, since they generate new candidate solutions using solely the current solution or the current "population" of solutions. Typical representatives of this class are evolutionary computation algorithms (Fogel et al., 1966; Holland, 1975; Rechenberg, 1973; Schwefel, 1981; Goldberg, 1989) or local search and its variants, such as, for example, simulated annealing (Kirkpatrick et al., 1983; Cerný, 1985; Hajek, 1988) and iterated local search (Lourenço et al., 2002) (an exception is tabu search (Glover, 1989), which uses additional information in the form of tabu lists).

On the other hand, in the last decade several new methods, which are classified as *model-based search* (MBS) algorithms in Zlochin et al. (2001), have been proposed, the best-known example being *estimation of distribution algorithms* (Mühlenbein & Paass, 1996; Pelikan, Goldberg, & Lobo, 1999; Larrañaga & Lozano, 2001). In model-based search algorithms, candidate solutions are generated using a parameterized probabilistic model that is updated using the previously seen solutions in such a way that the search will concentrate on the regions containing high-quality solutions. The general approach is described schematically in figure 4.5.

At a very general level, the MBS approach attempts to solve the minimization problem defined in section 4.2 by repeating the following two steps:

- Candidate solutions are constructed using some parameterized probabilistic model, that is, a parameterized probability distribution over the solution space.

- Candidate solutions are evaluated and then used to modify the probabilistic model in a way that is deemed to bias future sampling toward low-cost solutions. Note that the model's structure may be fixed in advance, with solely the model's parameters being updated, or alternatively, the structure of the model may be allowed to change as well.

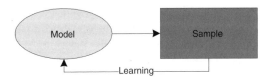

Figure 4.5
Schematic description of the model-based search (MBS) approach.

For any algorithm belonging to this general scheme, two components, corresponding to the two steps above, need to be instantiated:

- A parameterized probabilistic model that allows an efficient generation of the candidate solutions.

- An update rule for the model's parameters or structure, or both.

It is important to note that the term "model" is used here to denote an adaptive stochastic mechanism for generating candidate solutions, and not an approximate description of the environment, as done, for example, in reinforcement learning (Sutton & Barto, 1998). There is, however, a rather close connection between these two usages of the term "model," as the model adaptation in combinatorial optimization may be considered an attempt to model (in the reinforcement learning sense) the structure of the "promising" solutions. For a formal interpretation of ACO in terms of the reinforcement learning literature, see Birattari, Di Caro, & Dorigo (2002a).

It is easy to see that the ACO metaheuristic belongs to the MBS framework. First, the probabilistic model used is the coupling of the *construction graph* with the set of stochastic procedures called *artificial ants*, where the model parameters are the pheromone trails associated with the construction graph. Second, the model update rules are the rules used to update pheromone trails.

As discussed in chapter 3, the pheromone update rules proposed in the literature are of a somewhat heuristic nature and lack a theoretical justification. In the following, we show how *stochastic gradient ascent* (SGA) (Robbins & Monroe, 1951; Mitchell, 1997) and the *cross-entropy* (CE) method (De Bonet et al., 1997; Rubinstein, 2001) can be used to derive mathematically well-founded model update rules. We start with a discussion of the use of SGA and CE within the MBS framework, without being restricted to a particular type of probabilistic model. Then, we cast both SGA and the CE method into the ACO framework, and we show that some existing ACO updates approximate SGA while others can be rederived as a particular implementation of the CE method.

4.4.2 SGA and CE in the MBS Framework

In this section we discuss two systematic approaches to the update of the model's parameters in the MBS framework, namely SGA and the CE method. As in section 4.2, we consider a minimization problem $(\mathcal{S}, f, \Omega)$, where $\mathcal{S}$ is the set of (candidate) solutions, f is the objective function, which assigns an objective function (cost) value $f(s)$ to each candidate solution $s \in \mathcal{S}$, and Ω is a set of constraints, which defines the set of feasible candidate solutions. The goal of the minimization problem is to find an optimal solution s^*, that is, a feasible candidate solution of minimum cost. The set of all optimal solutions is denoted by $\mathcal{S}^*$.

Throughout the remainder of this section we assume that a space $\mathcal{M}$ of possible probabilistic models is given and that it is expressive enough. More formally, this means that we assume that for every possible solution s, the distribution $\delta_s(\cdot)$, defined as $\delta_s(s') = 1$, if $s' = s$, and $\delta_s(s') = 0$ otherwise, belongs to $\mathcal{M}$ [note that this condition may be relaxed by assuming that $\delta_s(\cdot)$ is in the closure of $\mathcal{M}$, that is, that there exists a sequence $P_i \in \mathcal{M}$ for which $\lim_{i \to \infty} P_i = \delta_s(\cdot)$]. This "expressiveness" assumption is needed in order to guarantee that the sampling can concentrate in the proximity of any solution, the optimal solution in particular.

Additionally, we assume that the model structure is fixed, and that the model space $\mathcal{M}$ is parameterized by a vector $\mathcal{T} \in \Phi \subset \mathbb{R}^w$, where Φ is a w-dimensional parameter space. In other words, $\mathcal{M} = \{P_{\mathcal{T}}(\cdot) \mid \mathcal{T} \in \Phi\}$ and for any $s \in \mathcal{S}$ the function $P_{\mathcal{T}}(s)$ is continuously differentiable with respect to $\mathcal{T}$.

The original optimization problem may be replaced by the following equivalent continuous *maximization problem*:

$$\mathcal{T}^* = \underset{\mathcal{T}}{\operatorname{argmax}} \ \mathcal{E}(\mathcal{T}), \tag{4.11}$$

where $\mathcal{E}(\mathcal{T}) = E_{\mathcal{T}}(q_f(s))$, $E_{\mathcal{T}}$ denotes expectation with respect to $P_{\mathcal{T}}$, and $q_f(s)$ is a fixed *quality function*, which is strictly decreasing with respect to f, that is: $q_f(s_1) < q_f(s_2) \Leftrightarrow f(s_1) > f(s_2)$.

It may be easily verified that, under the "expressiveness" assumption we made about the model space, the support of $P_{\mathcal{T}^*}$ (i.e., the set $\{s \mid P_{\mathcal{T}^*}(s) > 0\}$) is necessarily contained in $\mathcal{S}^*$. This implies that solving the problem given by equation (4.11) is equivalent to solving the original combinatorial optimization problem.

Stochastic Gradient Ascent

A possible way of searching for a (possibly local) optimum of the problem given by equation (4.11) is to use the gradient ascent method. In other words, gradient ascent may be used as a heuristic to change $\mathcal{T}$ with the goal of solving equation (4.11). The

procedure GradientAscent
   InitializeAlgorithmParameters
   $\mathcal{T} \leftarrow$ InitializeModelParameters
   while (termination condition not met) **do**
      $\nabla \mathcal{E}(\mathcal{T}) \leftarrow$ CalculateGradient($\mathcal{T}$)
      $\mathcal{T} \leftarrow \mathcal{T} + \alpha \nabla \mathcal{E}(\mathcal{T})$
   end-while
end-procedure

Figure 4.6
High-level pseudo-code for the GradientAscent procedure. The CalculateGradient($\mathcal{T}$) procedure is given by equation (4.12).

gradient ascent procedure, shown in figure 4.6, starts from some initial $\mathcal{T}$ (possibly randomly generated). Then, at each iteration it calculates the gradient $\nabla \mathcal{E}(\mathcal{T})$ and updates $\mathcal{T}$ to become $\mathcal{T} + \alpha \nabla \mathcal{E}(\mathcal{T})$, where α is a step-size parameter (which, in a more sophisticated implementation, could be made a function of the iteration counter θ: $\alpha = \alpha_\theta$).

The gradient can be calculated (bearing in mind that $\nabla \ln f = \nabla f / f$) as follows:

$$\nabla \mathcal{E}(\mathcal{T}) = \nabla E_{\mathcal{T}}(q_f(s)) = \nabla \sum_s q_f(s) P_{\mathcal{T}}(s) = \sum_s q_f(s) \nabla P_{\mathcal{T}}(s) =$$

$$= \sum_s P_{\mathcal{T}}(s) q_f(s) \frac{\nabla P_{\mathcal{T}}(s)}{P_{\mathcal{T}}(s)} = \sum_s P_{\mathcal{T}}(s) q_f(s) \nabla \ln P_{\mathcal{T}}(s) =$$

$$= E_{\mathcal{T}}(q_f(s) \nabla \ln P_{\mathcal{T}}(s)). \tag{4.12}$$

However, the gradient ascent algorithm cannot be implemented in practice, because for its evaluation a summation over the whole search space is needed. A more practical alternative is the use of *stochastic gradient ascent* (Baird & Moore, 1999; Bertsekas, 1995b; Williams, 1992), which replaces the expectation in equation (4.12) by an empirical mean of a sample generated from $P_{\mathcal{T}}$.

The update rule for the stochastic gradient then becomes

$$\mathcal{T}_{\theta+1} = \mathcal{T}_\theta + \alpha \sum_{s \in S_\theta} q_f(s) \nabla \ln P_{\mathcal{T}_\theta}(s), \tag{4.13}$$

where S_θ is the sample at iteration θ.

In order to derive a practical algorithm from the SGA approach, we need a model for which the derivatives of the $\ln P_{\mathcal{T}}(\cdot)$ can be calculated efficiently. In section 4.4.3

we show how this can be done in the context of the iterative construction scheme used in the ACO metaheuristic.

Cross-Entropy Method

The basic ideas behind the CE method for combinatorial optimization were originally proposed in the *mutual-information-maximizing input clustering* (MIMIC) algorithm of De Bonet et al. (1997). They were later further developed by Rubinstein (1999, 2001), who was the first to use the *cross-entropy* name to denote this class of algorithms. For the derivation of the CE method in this section we follow Dorigo et al. (2002) and Zlochin et al. (2001).

Starting from some initial distribution $P_0 \in \mathcal{M}$, the CE method inductively builds a series of distributions $P_\theta \in \mathcal{M}$, in an attempt to increase the probability of generating low-cost solutions after each iteration. A tentative way to achieve this goal is to set $P_{\theta+1}$ equal to $\hat{P}$, where $\hat{P}$ is a value proportional to P_θ as follows:

$$\hat{P} \propto P_\theta q_f, \tag{4.14}$$

where q_f is, again, some quality function, depending on the cost value.

If this were possible, then, for time-independent quality functions, after θ iterations we would obtain $P_\theta \propto P_0 (q_f)^\theta$. Consequently, as $\theta \to \infty$, P_θ would converge to a probability distribution restricted to $\mathcal{S}^*$. Unfortunately, even if the distribution P_θ belongs to the family $\mathcal{M}$, the distribution $\hat{P}$ as defined by equation (4.14) does not necessarily remain in $\mathcal{M}$, hence some sort of projection is needed (see exercise 4.2).

Accordingly, a natural candidate for the projection $P_{\theta+1}$ is the distribution $P \in \mathcal{M}$ that minimizes the *Kullback-Leibler divergence* (Kullback, 1959), which is a commonly used measure of the difference between two distributions:

$$D(\hat{P}\|P) = \sum_s \hat{P}(s) \ln \frac{\hat{P}(s)}{P(s)}, \tag{4.15}$$

or equivalently the *cross-entropy*:

$$-\sum_s \hat{P}(s) \ln P(s). \tag{4.16}$$

Since $\hat{P} \propto P_\theta q_f$, the CE minimization is equivalent to the following maximization problem:

$$P_{\theta+1} = \underset{P \in \mathcal{M}}{\operatorname{argmax}} \sum_s P_\theta(s) q_f(s) \ln P(s). \tag{4.17}$$

It should be noted that, unlike SGA, in the CE method the quality function is only required to be nonincreasing with respect to the cost and may also depend on the iteration index, either deterministically or stochastically. For example, it might depend on the points sampled so far. One common choice is $q_f(s) = I(f(s) < f_\theta)$, where $I(\cdot)$ is an indicator function, and f_θ is, for example, some quantile (e.g., lower 10%) of the cost distribution during the last iteration. (We remind the reader that the indicator function $I(f(s) < f_\theta)$ is such that $I(f(s) < f_\theta) = 1$ if $f(s) < f_\theta$ and 0 otherwise.)

In a way similar to what happened with the gradient of equation (4.12), the maximization problem given by equation (4.17) cannot be solved in practice, because the evaluation of the function $\sum_s P_\theta(s) q_f(s) \ln P(s)$ requires summation over the whole solution space. Once again, a finite sample approximation can be used instead:

$$P_{\theta+1} = \underset{P \in \mathcal{M}}{\operatorname{argmax}} \sum_{s \in S_\theta} q_f(s) \ln P(s), \qquad (4.18)$$

where S_θ is a sample from P_θ.

Note that if the quality function is of the form $I(f(s) < const)$, then equation (4.18) defines a *maximum-likelihood* model, with the sample used for estimation being restricted to the top-quality solutions. With other quality functions, equation (4.18) may be interpreted as defining a weighted maximum-likelihood estimate.

In some relatively simple cases, some of which are discussed in section 4.4.3, the problem [equation (4.18)] can be solved exactly. In general, however, the analytic solution is unavailable. Still, even if the exact solution is not known, some iterative methods for solving this optimization problem may be used.

A natural candidate for the iterative solution of the maximization problem given by equation (4.18) is SGA, as shown in figure 4.7.

It should be noted that, since the new vector $\mathcal{T}_{\theta+1}$ is a random variable, depending on a sample, there is no use in running the SGA process till full convergence. Instead, in order to obtain some robustness against sampling noise, a fixed number of SGA updates may be used. One particular choice, which is of special interest, is the use of a single gradient ascent update, leading to the update rule

$$\mathcal{T}_{\theta+1} = \mathcal{T}_\theta + \alpha \sum_{s \in S_\theta} q_f(s) \nabla \ln P_{\mathcal{T}_\theta}(s), \qquad (4.19)$$

which is identical to the SGA update [equation (4.13)]. However, as already mentioned, the CE method imposes fewer restrictions on the quality function (e.g.,

procedure SGAforCrossEntropy

$\quad T' \leftarrow T_\theta$

$\quad$ **while** (termination condition not met) **do**

$\quad\quad T' \leftarrow T' + \alpha \sum_{s \in S_\theta} q_f(s) \nabla \ln P_{T'}(s)$

$\quad$ **end-while**

$\quad$ **return** $T_{\theta+1} = T'$

end-procedure

Figure 4.7
High-level pseudo-code for the SGAforCrossEntropy procedure. The procedure starts by setting $T' = T_\theta$; other starting points are possible, but this is the most natural one, since we may expect $T_{\theta+1}$ to be close to T_θ. In case the **while** loop is iterated only once, the procedure coincides with equation (4.19).

allowing it to change over time), hence the resulting algorithm may be seen as a generalization of SGA.

As with SGA, in order to have an efficient algorithm, a model is needed, for which the calculation of the derivatives can be carried out in reasonable time. In the next section, we show that this is indeed possible for the models typically used in ACO.

4.4.3 ACO, SGA, and CE

So far we have limited our discussion to the generic approaches for updating the model. However, this is only one of the components needed in any model-based search algorithm. In the following, we show how ACO implements the other component, that is, the probabilistic model.

As we said, the particular type of probabilistic model used by ACO algorithms is the coupling of the structure called construction graph with a set of stochastic procedures called artificial ants. The artificial ants build solutions in an iterative manner using local information stored in the construction graph. In this section we present the pheromone updates derived from the SGA algorithm and the CE method.

The SGA Update in ACO
In section 4.4.2 an update rule for the stochastic gradient was derived:

$$T_{\theta+1} = T_\theta + \alpha \sum_{s \in S_\theta} q_f(s) \nabla \ln P_{T_\theta}(s), \tag{4.20}$$

where S_θ is the sample at iteration θ.

As was shown by Meuleau & Dorigo (2002), in case the distribution is implicitly defined by an ACO-type construction process, parameterized by the vector of the pheromone values, T, the gradient $\nabla \ln P_T(s)$ can be efficiently calculated. The following calculation (Zlochin et al., 2001) is a generalization of the one in Meuleau & Dorigo (2002).

From the definition of AntSolutionConstruction, it follows that, for a solution $s = \langle c_1, c_2, \ldots, c_{|s|} \rangle$ built by ant k, we have

$$P_T(s) = \prod_{h=1}^{|s|-1} P_T(c_{h+1} \mid \text{pref}_h(s)), \tag{4.21}$$

where $\text{pref}_h(s)$ is the h-prefix of s (i.e., the sequence formed by the first h components of s), and consequently

$$\nabla \ln P_T(s) = \sum_{h=1}^{|s|-1} \nabla \ln P_T(c_{h+1} \mid \text{pref}_h(s)). \tag{4.22}$$

Finally, given a pair of components $(i, j) \in C^2$, using equation (4.1), it can be verified that:

- if $i = c_h$ and $j = c_{h+1}$, then

$$\frac{\partial}{\partial \tau_{ij}} \{\ln P_T(c_{h+1} \mid \text{pref}_h(s))\} = \frac{\partial}{\partial \tau_{ij}} \left\{ \ln \left(F(\tau_{ij}) \Big/ \sum_{(i,y) \in \mathcal{N}_i^k} F(\tau_{i,y}) \right) \right\}$$

$$= \frac{\partial}{\partial \tau_{ij}} \left\{ \ln F(\tau_{ij}) - \ln \sum_{(i,y) \in \mathcal{N}_i^k} F(\tau_{i,y}) \right\}$$

$$= F'(\tau_{ij})/F(\tau_{ij}) - F'(\tau_{ij}) \Big/ \sum_{(i,y) \in \mathcal{N}_i^k} F(\tau_{i,y})$$

$$= \left\{ 1 - F(\tau_{ij}) \Big/ \sum_{(i,y) \in \mathcal{N}_i^k} F(\tau_{i,y}) \right\} \frac{F'(\tau_{ij})}{F(\tau_{ij})}$$

$$= \{1 - P_T(j \mid \text{pref}_h(s))\} D(\tau_{ij}),$$

where $D(\cdot) = F'(\cdot)/F(\cdot)$ and the subscripts of F were omitted for the clarity of presentation;

- if $i = c_h$ and $j \neq c_{h+1}$, then (by a similar argument)

$$\frac{\partial \ln(P_T(c_{h+1} \mid \text{pref}_h(s)))}{\partial \tau_{ij}} = -P_T(j \mid \text{pref}_h(s)) D(\tau_{ij});$$

procedure SGAPheromoneUpdate

    foreach $s \in S_\theta$ **do**

        foreach $(i, j) \in s$ **do**

            $\tau_{ij} \leftarrow \tau_{ij} + \alpha q_f(s) D(\tau_{ij})$

        end-foreach

    end-foreach

    foreach $s = \langle c_1, \ldots, c_h, \ldots \rangle \in S_\theta$ **do**

        foreach $(i, j) : i = c_h, 1 \le h < |s|$ **do**

            $\tau_{ij} \leftarrow \tau_{ij} - \alpha q_f(s) P_T(j \,|\, \mathrm{pref}_h(s)) D(\tau_{ij})$

        end-foreach

    end-foreach

end-procedure

Figure 4.8
High-level pseudo-code for the SGAPheromoneUpdate procedure. The first two nested **for** loops add pheromones to all the connections *used* to build the solution (i.e., those connections that belong to the built solution). The second two nested **for** loops decrease the pheromone on all the connections *considered* during solution construction [i.e., those connections that contributed to the computation of the probabilities given by equation (4.1) during solution construction]. Note that the internal loop is over all (i, j) because those j which were not considered during solution construction are automatically excluded, since for them it holds $P_T(j \,|\, \mathrm{pref}_h(s)) = 0$.

- if $i \ne c_h$, then $P_T(c_{h+1} \,|\, \mathrm{pref}_h(s))$ is independent of τ_{ij} and therefore

$$\frac{\partial \ln(P_T(c_{h+1} \,|\, \mathrm{pref}_h(s)))}{\partial \tau_{ij}} = 0.$$

By combining these results, the SGAPheromoneUpdate procedure, shown in figure 4.8, is derived. In practice, any connection (i, j) *used* in the construction of a solution is reinforced by an amount $\alpha q_f(s) D(\tau_{ij})$, and any connection *considered* during the construction, has its pheromone values evaporated by an amount given by $\alpha q_f(s) P_T(j \,|\, \mathrm{pref}_h(s)) D(\tau_{ij})$. Here, with *used* connections we indicate those connections that belong to the built solution, whereas with *considered* connections we indicate those that contributed to the computation of the probabilities given by equation (4.1), during solution construction.

Note that, if the solutions are allowed to contain loops, a connection may be updated more than once for the same solution.

In order to guarantee stability of the resulting algorithm, it is desirable to have a bounded gradient $\nabla \ln P_T(s)$. This means that a function F, for which $D(\cdot) = F'(\cdot)/F(\cdot)$ is bounded, should be used. Meuleau & Dorigo (2002) suggest using $F(\cdot) = \exp(\cdot)$, which leads to $D(\cdot) \equiv 1$. It should be further noted that if, in addition,

$q_f = 1/f$ and $\alpha = 1$, the reinforcement part becomes $1/f$ as in the original AS (see chapter 3, section 3.3.1).

The CE Update in ACO

As we have shown in section 4.4.2, the CE method requires solving the following intermediate problem:

$$P_{\theta+1} = \underset{P \in \mathcal{M}}{\mathrm{argmax}} \sum_{s \in S_\theta} q_f(s) \ln P(s). \tag{4.23}$$

Let us now consider this problem in more detail in case of an ACO-type probabilistic model. Since at the maximum the gradient must be 0, we have

$$\sum_{s \in S_\theta} q_f(s) \nabla \ln P_T(s) = 0. \tag{4.24}$$

In some relatively simple cases, for example, when the solution s is represented by an unconstrained string of bits of length n, that is, $s = (s_1, \ldots, s_i, \ldots, s_n)$, and there is a single parameter τ_i for the i-th position in the string, such that $P_T(s) = \prod_i p_{\tau_i}(s_i)$, the equation system [equation (4.24)] reduces to a set of independent equations:

$$\frac{d \ln p_{\tau_i}}{d\tau_i} \sum_{\substack{s \in S_\theta \\ s_i=1}} q_f(s) = -\frac{d \ln(1 - p_{\tau_i})}{d\tau_i} \sum_{\substack{s \in S_\theta \\ s_i=0}} q_f(s), \quad i = 1, \ldots, n, \tag{4.25}$$

which often may be solved analytically. For example, for $p_{\tau_i} = \tau_i$ it can be verified that the solution of equation (4.25) is simply

$$p_{\tau_i} = \tau_i = \frac{\sum_{s \in S_\theta} q_f(s)s_i}{\sum_{s \in S_\theta} q_f(s)}, \tag{4.26}$$

and, in fact, a similar solution also applies to a more general class of Markov chain models (Rubinstein, 2001).

Now, since the pheromone trails τ_i in equation (4.26) are random variables, whose values depend on the particular sample, we may wish to make our algorithm more robust by introducing some conservatism into the update. For example, rather than discarding the old pheromone values, the new values may be taken to be a convex combination of the old values and the solution to equation (4.26):

$$\tau_i \leftarrow (1 - \rho)\tau_i + \rho \frac{\sum_{s \in S_\theta} q_f(s)s_i}{\sum_{s \in S_\theta} q_f(s)}. \tag{4.27}$$

The resulting update is identical to the one used in the hyper-cube framework for ACO (see chapter 3, section 3.4.3).

However, for many cases of interest, equation (4.24) is coupled and an analytic solution is unavailable. Nevertheless, in the actual implementations of the CE method the update was of the form given by equation (4.26), with some brief remarks about using equation (4.27) (Rubinstein, 2001), which may be considered as an approximation to the exact solution of the CE minimization problem [equation (4.18)].

Since, in general, the exact solution is not available, an iterative scheme such as gradient ascent could be employed. As we have shown in the previous section, the gradient of the log-probability may be calculated as follows:

- if $i = c_h$ and $j = c_{h+1}$, then

$$\frac{\partial \ln(P_\mathcal{T}(c_{h+1} \mid \mathrm{pref}_h(s)))}{\partial \tau_{ij}} = (1 - P_\mathcal{T}(j \mid \mathrm{pref}_h(s)))D(\tau_{ij});$$

- if $i = c_h$ and $j \neq c_{h+1}$, then

$$\frac{\partial \ln(P_\mathcal{T}(c_{h+1} \mid \mathrm{pref}_h(s)))}{\partial \tau_{ij}} = -P_\mathcal{T}(j \mid \mathrm{pref}_h(s))D(\tau_{ij});$$

- if $i \neq c_h$, then

$$\frac{\partial \ln(P_\mathcal{T}(c_{h+1} \mid \mathrm{pref}_h(s)))}{\partial \tau_{ij}} = 0;$$

and these values may be plugged into any general iterative solution scheme of the CE minimization problem, for example, the one described by equation (4.19).

To conclude, we have shown that if we use equation (4.26) as a (possibly approximate) solution of equation (4.18), the same equations as used in the hyper-cube framework for ACO algorithms are derived. If otherwise we use a single-step gradient ascent for solving equation (4.18), we obtain a generalization of the SGA update, in which the quality function is permitted to change over time.

4.5 Bibliographical Remarks

The first convergence proof for an ACO algorithm, called Graph-based Ant System (GBAS), was provided by Gutjahr (2000). GBAS is very similar to $\mathrm{ACO}_{bs, \tau_{min}}$ except that (1) $\tau_{min} = 0$ and (2) the pheromone update rule changes the pheromones only when, in the current iteration, a solution at least as good as the best found so far is generated. For GBAS, Gutjahr proves convergence in solution by showing that a fixed ant constructs the optimal solution to the given problem instance with a probability larger or equal to $1 - \varepsilon$. In particular, he proved that for each $\varepsilon > 0$ it holds that (1) for a fixed ρ and for a sufficiently large number of artificial ants, the probability P that a fixed ant constructs the optimal solution at iteration θ is $P \geq 1 - \varepsilon$ for all $\theta \geq \theta_0$, with $\theta_0 = \theta_0(\varepsilon)$; (2) for a fixed number of ants and for an evaporation rate ρ sufficiently close to zero, the probability P that a fixed ant constructs the optimal solution at iteration θ is $P \geq 1 - \varepsilon$ for all $\theta \geq \theta_0$, with $\theta_0 = \theta_0(\varepsilon)$. Although the theorem has the great merit to be the first theoretical work on ACO, its main limitation is that the proof only applies to GBAS, an ACO algorithm which has never been implemented and for which therefore no experimental results are available.

As a next step, Stützle & Dorigo (2002) proved the convergence in value result for $ACO_{bs, \tau_{min}}$, presented in theorem 4.1. They also proved some additional results related to convergence in solution: they provided bounds on the probability of constructing the optimal solution if fixed lower pheromone trail limits are used. As stated in section 4.3.5, the main importance of this result is that it applies to two of the experimentally most successful ACO algorithms, $\mathcal{MMAS}$ and ACS (see chapter 3, sections 3.3 and 3.4, respectively).

The first proof of convergence in solution, similar to the one given by theorems 4.2 and 4.3, was given by Gutjahr for variants of GBAS, which were called GBAS/tdlb (for time-dependent lower pheromone bound) and GBAS/tdev (for time-dependent evaporation rate). The first variant, GBAS/tdlb, uses a bound on the lower pheromone trail limits very similar to the one used in theorem 4.2. Differently, in GBAS/tdev it is the pheromone evaporation rate that is varied during the run of the algorithm: for proving that GBAS/tdev converges in solution, pheromone evaporation is decreased slowly, and in the limit it tends to zero.

The first work showing the relationship between AS and SGA was by Meuleau & Dorigo (2002). In section 4.4.3 we presented an extension of that work by Zlochin et al. (2001) and Dorigo et al. (2002). The CE method is an extension of the MIMIC algorithm proposed by De Bonet et al. (1997) and developed by Rubinstein (1999, 2001). The relationship between ACO and CE was first formally described by Zlochin et al. (2001).

4.6   Things to Remember

- It is possible to prove asymptotic convergence for particular subsets of ACO algorithms. In particular, asymptotic convergence in value was proved for ACS and

$\mathcal{MMAS}$, two of the experimentally best-performing ACO algorithms, while convergence in solution was proved for GBAS and for $\text{ACO}_{bs,\,\tau_{min}(\theta)}$. Proving convergence in value intuitively means proving that the algorithm generates at least once the optimal solution. Proving convergence in solution can be interpreted as proving that the algorithm reaches a situation in which it generates over and over the same optimal solution.

- Convergence proofs tell us that the bias introduced in the stochastic algorithm does not rule out the possibility of generating an optimal solution. They do not say anything about the speed of convergence, that is, the computational time required to find an optimal solution.

- ACO algorithms belong to the class of MBS algorithms. In MBS algorithms candidate solutions are generated using a parameterized probabilistic model. This probabilistic model is updated using the previously seen solutions in such a way that the search will concentrate in the regions containing high-quality solutions. The construction graph, together with the artificial ant procedures, defines the probabilistic model, which is parameterized by the pheromone trails.

- When interpreting ACO as an instance of MBS, it is possible to use methods such as SGA and CE minimization to define rules to update pheromone trails. In this view, AS can be seen as an algorithm that performs an approximate SGA in the space of pheromone trails.

4.7   Thought and Computer Exercises

Exercise 4.1   Prove that theorem 4.2 can be extended to the case in which the function $F(\tau_{ij})$, defined in section 4.3, is a grade p polynomial of the form $F(\tau_{ij}) = a_0\tau_{ij}^p + a_1\tau_{ij}^{p-1} + \cdots + a_h\tau_{ij}^{p-h} + \cdots + a_p$, with $a_0 > 0$, $a_h \geq 0$, $0 < h < p$ and $a_p = 0$.

Exercise 4.2   In section 4.4.2 we wrote that even if the distribution P_θ belongs to the family $\mathcal{M}$, the distribution $\hat{P}$ as defined by equation (4.14), does not necessarily remain in $\mathcal{M}$. Give an example showing that this is indeed the case.

Exercise 4.3   In section 4.4.3, an SGA update for ACO algorithms was derived. Try to rederive the same equations in the more specific case of the TSP. A solution can be found in Meuleau & Dorigo (2002).

Exercise 4.4   Convergence in solution can also be proved for $\text{ACO}_{gb,\,\rho(\theta)}$, an ACO algorithm that differs from $\text{ACO}_{bs,\,\tau_{min}(\theta)}$ in that its pheromone evaporation rate is modified at run time and its lower pheromone trail limits are set to 0. In particular, assume that until some iteration $\theta_0 \geq 1$, a fixed pheromone evaporation ρ is applied, and that from $\theta > \theta_0$ on we have

$$\rho_\theta \leq 1 - \frac{\log \theta}{\log(\theta + 1)}, \quad \forall \theta > \theta_0,$$

and that

$$\sum_{\theta=1}^{\infty} \rho_\theta = +\infty.$$

Prove convergence in solution for this algorithm along the lines of the convergence proof given in section 4.3.2 for $\text{ACO}_{bs,\tau_{min}(\theta)}$.

Hint: You may have a look at the paper by Gutjahr (2002).

5 Ant Colony Optimization for $\mathcal{NP}$-Hard Problems

We shall refer to a problem as intractable if it is so hard that no polynomial time algorithm can possibly solve it.
—*Computers and Intractability*, Michael R. Garey & David S. Johnson, 1979

This chapter gives an overview of selected applications of ACO to different $\mathcal{NP}$-hard optimization problems. The chapter is intended to serve as a guide to how ACO algorithms can be adapted to solve a variety of well-known combinatorial optimization problems rather than being an exhaustive enumeration of all possible ACO applications available in the literature. Our main focus is on presenting and discussing interesting applications that either present a different perspective on how to apply ACO algorithms or for which very good results have been obtained.

Most of the problems considered fall into one of the following categories: routing, assignment, scheduling, and subset problems. For each of these categories a full section is devoted to explain how ACO has been applied to the corresponding category. We then review applications of ACO to a few additional $\mathcal{NP}$-hard problems such as shortest common supersequence, bin packing, protein folding, and constraint satisfaction—problems that do not easily fit in the above-mentioned categories—and to problems typically found in machine learning, such as the learning of classification rules and of the structure of Bayesian networks.

For each problem we describe the construction graph, how constraints are handled, the way pheromone trails and heuristic information are defined, how solutions are constructed, the pheromone trail update procedure, and the computational results achieved. Additionally, when available in published papers, we give details about the local search used with ACO.

We conclude the chapter with a discussion of some "ACO application principles" that can be used by practitioners as a guide when trying to apply ACO to an $\mathcal{NP}$-hard problem not yet considered in the literature.

5.1 Routing Problems

In this section we consider routing problems, that is, problems in which one or more agents have to visit a predefined set of locations and whose objective function depends on the ordering in which the locations are visited. The problems we discuss are the sequential ordering problem and the vehicle routing problem.

5.1.1 Sequential Ordering

The sequential ordering problem consists in finding a minimum weight Hamiltonian path on a directed graph with weights on arcs and nodes subject to precedence constraints. The SOP can be formulated as a generalization of the asymmetric TSP as follows:

- A solution connects a starting node to a final node by a Hamiltonian path (in the asymmetric TSP a solution is a Hamiltonian circuit).

- Weights are assigned only to arcs. If given, node weights can be removed from the original definition by redefining the weight c_{ij} of arc (i, j) by adding the node weight p_j of node j to each arc incident to j, resulting in new arc weights $c'_{ij} = c_{ij} + p_j$.

- Precedence constraints are defined among the nodes. If a node j has to precede node i in the path, this is represented by assigning to arc (i, j) the weight $c'_{ij} = -1$; hence, if $c'_{ij} \geq 0$, then the weight gives the cost of arc (i, j), while if $c'_{ij} = -1$, then node j must precede, not necessarily immediately, node i.

   The application of ACO to the SOP is particularly interesting because a straightforward extension of one of the best-performing ACO algorithms for the TSP, ACS, turns out to have world-class performance on a problem that, although closely connected to the TSP, cannot be solved efficiently by the best available exact algorithms. In fact, the main adaptations necessary to apply ACS to the SOP are minor modifications in the solution construction procedure and the implementation of a new and efficient local search for the SOP. The resulting algorithm, called HAS–SOP (Hybrid AS–SOP) is currently the best available algorithm for the SOP (Gambardella & Dorigo, 2000).

Construction graph   The set of components contains the set of all the n nodes plus the start node, node 0, and the final node, node $n + 1$. As usual, the set of components is fully connected. Solutions are Hamiltonian paths that start at node 0, end at node $n + 1$, and that comply with all the precedence constraints.

Constraints   In the SOP the constraints require that all nodes of the graph be visited once and only once and that all precedence constraints be satisfied. The constraints are taken into account in the definition of the ants' feasible neighborhood at construction time, as explained below.

Pheromone trails   A pheromone trail τ_{ij} indicates the desirability of choosing node j when an ant is at node i. This is the same definition as in the TSP application (see chapter 2, section 2.3.2).

Heuristic information   The heuristic information is the same as for the TSP, that is, $\eta_{ij} = 1/c'_{ij}$ when $c'_{ij} \neq -1$, and $\eta_{ij} = 0$ otherwise.

Solution construction   All ants are initially put on node 0 and build a Hamiltonian path that connects node 0 to node $n + 1$. Ants build solutions by choosing probabilistically the next node from their feasible neighborhood. In practice, ant k located on node i chooses the node j, $j \in \mathcal{N}_i^k$, to move to with a probability given by ACS's pseudorandom proportional rule [equation (3.10)]. The feasible neighborhood $\mathcal{N}_i^k$ contains all nodes j that ant k has not visited yet and that, if added to ant k's partial solution, do not violate any precedence constraint.

A particularity of HAS–SOP is the value given to the parameter q_0 in equation (3.10), which is set to $q_0 = 1 - s/n$, where the parameter s, $1 \leq s \leq n$, gives the expected number of nodes to be chosen according to the probabilistic part of equation (3.10); the parameter s allows defining q_0 independently of the problem size.

Pheromone update HAS–SOP uses the same local and global pheromone update procedures as ACS for the TSP (see chapter 3, section 3.4.1, for details).

Local search The local search is the most innovative part of HAS–SOP. It is a specific 3-opt procedure, called SOP-3-exchange, which can handle efficiently multiple precedence constraints during the local search without increasing computational complexity. For a detailed description of the SOP-3-exchange procedure, see the original paper by Gambardella & Dorigo (2000).

Results Computational results obtained with HAS–SOP are excellent. HAS–SOP was compared to state-of-the-art algorithms for the SOP, the best-performing one being a genetic algorithm called MPO/AI that was explicitly designed to solve sequencing problems. Since MPO/AI was found to be significantly inferior to HAS–SOP, MPO/AI was extended with the same SOP-3-exchange local search used by HAS–SOP. MPO/AI plus SOP-3-exchange and HAS–SOP were compared using as benchmark all the SOP instances available in the TSPLIB (Reinelt, 1991), accessible at www.iwr.uni-heidelberg.de/iwr/comopt/soft/TSPLIB95/. Also, in this case HAS–SOP outperformed MPO/AI plus SOP-3-exchange, and it was able to find new upper bounds for twelve TSPLIB instances. The most recent information on HAS–SOP can be found at www.idsia.ch/~luca/has-sop_start.htm, maintained by Luca M. Gambardella.

5.1.2 Vehicle Routing

The vehicle routing problem (VRP) is a central problem in distribution management (Toth & Vigo, 2001). In the capacitated VRP (CVRP) n customers have to be served from one central depot, which is typically identified by the index 0. Each customer i has a non-negative demand b_i of the same merchandise and for each pair of customers (i, j) a travel time d_{ij} between the two customers is given. The customers are served by a fleet of vehicles of equal capacity B. The goal in the CVRP is to find a set of routes that minimizes the total travel time such that (1) each customer is served once by exactly one vehicle, (2) the route of each vehicle starts and ends at the depot, and (3) the total demand covered by each vehicle does not exceed its capacity B.

The CVRP is an $\mathcal{NP}$-hard problem because it contains the TSP as a subproblem. In practice, it is much more difficult to solve than the TSP, the main reason being that it consists of two nested problems. The first is a bin-packing problem where the goal is to pack the customers into an a priori unknown number of routes (bins). Then, for each of the routes a shortest tour visiting all the customers assigned to the route has to be found, which involves solving a TSP.

The CVRP is the most basic form of a VRP. It is therefore not surprising that it was also the first VRP to be tackled by an ACO approach. Bullnheimer et al. (1999b) presented an adaptation of AS_{rank} (AS_{rank}-CVRP), which was later improved by Reimann, Stummer, & Doerner (2002b) (AS_{rank}-CVRPsav).

Several extensions of the basic CVRP exist, the most studied being the VRP with time window constraints (VRPTW). In this version of the VRP, each customer i has a time window $[e_i, l_i]$ during which she must be served (a time window $[e_0, l_0]$ is associated with the depot); here, e_i is the earliest possible service time for customer i and l_i is the latest possible time.

The objective function in the VRPTW is different from the one in the CVRP. The VRPTW has two objectives: (1) to minimize the number of vehicles used and (2) to minimize the total travel time. The two objectives are ordered hierarchically, that is, a solution with fewer vehicles is always preferred over a solution with more vehicles but a smaller total travel time. Two solutions with a same number of vehicles are rated according to their total tour duration. Currently, the most successful ACO algorithm for the VRPTW is MACS-VRPTW (Gambardella, Taillard, & Agazzi, 1999a), where MACS stays for multiple ACS. MACS-VRPTW uses two parallel, interacting ant colonies, one for each objective. The central idea is that one colony, called ACS-VEI, is trying to minimize the number of vehicles (routes) to be used, while the other colony, called ACS-TIME, is trying to minimize, for a given number of vehicles, the traveling time. MACS-VRPTW uses ACS for these single-objective problems (for a description of ACS, see chapter 3, section 3.4.1). The two algorithms run in parallel, with ACS-TIME using v_{min} vehicles and ACS-VEI searching for a feasible solution with $v_{min} - 1$ vehicles. Each time ACS-VEI finds a feasible solution with $v_{min} - 1$ vehicles, ACS-TIME and ACS-VEI are restarted with the new, reduced number of vehicles.

In the following, we briefly describe AS_{rank}-CVRP, AS_{rank}-CVRPsav, and MACS-VRPTW.

Construction graph   In AS_{rank}-CVRP and AS_{rank}-CVRPsav, the construction graph comprises one component for each of the customers and one component for the depot. MACS-VRPTW uses multiple copies of the depot; the number of copies (including the original depot) is equal to the number of vehicles that are currently in use. The distances between the copies of the depot are zero. As usual, the components are fully connected.

Constraints   The constraints in the CVRP require that each customer be visited exactly once and that the vehicle capacities not be exceeded. In the VRPTW, additionally the time window constraints need to be satisfied.

Pheromone trails   In each algorithm, pheromone trails τ_{ij} are associated only with connections. The pheromone trails refer to the desirability of visiting customer j directly after i. ACS-VEI and ACS-TIME use two different sets of pheromones.

Heuristic information AS_{rank}-CVRP and AS_{rank}-CVRPsav base the heuristic information on the savings heuristic (Clarke & Wright, 1964). To explain how this heuristic information is defined, consider first the savings algorithm for the CVRP. It starts from a solution with one separate tour per customer, that is, with n tours. For each pair (i, j) of customers a saving $s_{ij} = d_{i0} + d_{0j} - d_{ij}$ is computed, where the index 0 denotes the depot (see figure 5.1). The savings algorithm combines customers into tours following the usual greedy strategy.

AS_{rank}-CVRP uses as heuristic information a parameterized saving (Paessens, 1988) given by

$$\eta_{ij} = d_{i0} + d_{0j} - g \cdot d_{ij} + f \cdot |d_{i0} - d_{0j}|. \tag{5.1}$$

Good settings were reported to be $g = f = 2$. AS_{rank}-CVRPsav uses the original saving definition, that is, it sets $\eta_{ij} = s_{ij}$.

The heuristic information η_{ij} in MACS-VRPTW is defined as a function of the travel time d_{ij}, of the time window $[e_i, l_i]$, and of the number ns_j of times a node j was

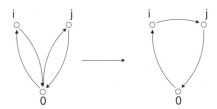

Figure 5.1
The savings algorithm for a situation with two customers i and j and a depot 0. On the left, customers i and j are connected to the depot with two separated tours of total length $d_{0i} + d_{i0} + d_{0j} + d_{j0}$. On the right the two customers are connected to the depot with a single tour of length $d_{0i} + d_{ij} + d_{j0}$. The saving $s_{ij} = d_{i0} + d_{0j} - d_{ij}$ is given by the difference between the lengths of the two tours on the left and the tour on the right.

not included in an ant's solution in previous iterations of the algorithm (in fact, ACS-VEI may build candidate solutions that do not include all customers). For details, see Gambardella et al. (1999a).

Solution construction   In AS_{rank}-CVRP, ants build solutions using the same probabilistic rule as in AS [equation (3.2)]. During solution construction, AS_{rank}-CVRP ants choose the next customer among the feasible ones from a candidate list of length $n/4$, where n is the number of customers. (Candidate lists were defined in chapter 3, section 3.4.1. They are discussed in more detail in section 5.7.7.) If no customer can be added without making the tour infeasible, the tour is closed by returning to the depot. A new tour is then started if there are unvisited customers left.

AS_{rank}-CVRPsav follows the main steps of the savings algorithm in the solution construction. It starts with n individual tours and then merges tours as long as feasible. Then, instead of the deterministic choices based on the s_{ij} values as done in the savings algorithm, the ants in AS_{rank}-CVRPsav choose the next two tours to be merged based on the random proportional action choice rule of AS [equation (3.2)]. At each construction step, AS_{rank}-CVRPsav chooses customers from a candidate list that consists of the pairs (i, j) corresponding to the $n/4$ largest savings s_{ij}.

In MACS-VRPTW, both colonies, ACS-TIME and ACS-VEI, use the same solution construction procedure which is similar to the one used by ACS. An ant starts from a randomly chosen copy of the depot and at each step either adds a customer that does not violate the time window constraints and the capacity constraints, or returns to the depot (this means that ants are allowed to move to a still unvisited copy of the depot even if they could still add unvisited customers to their partial solution without violating any constraint). If no customer can be added, the ant returns to one of the copies of the depot. If, after the construction of a solution is completed (i.e., all the depot copies have been used), there remain some unscheduled customers, MACS-VRPTW tries to include them in the incomplete solution. To do so, it uses an insertion procedure which considers customers in order of nonincreasing demand and inserts them, if possible, at a position such that the travel time is minimized.

Pheromone update   The pheromone update in AS_{rank}-CVRP and AS_{rank}-CVRPsav follows the pheromone update rule used in AS_{rank} [equation (3.8)].

ACS-TIME and ACS-VEI of MACS-VRPTW use the global and local pheromone update rules of ACS [equations (3.11) and (3.12)], with a caveat. In fact, ACS-VEI typically generates infeasible solutions that visit fewer than n customers. Accordingly, the objective function optimized by ACS-VEI is the maximization of the number of customers visited. In fact, if a solution that visits all n customers is found, this corresponds to a solution with one vehicle less than the previous best-so-far solution. Therefore, the global pheromone update rule updates pheromones belonging to the solution that visited the largest number of customers. However, according to Gambardella et al. (1999a), it is possible to greatly enhance the algorithm's performance by letting the global pheromone update rule also update pheromones belonging to the best-so-far (feasible) solution provided by ACS-TIME.

Local search AS_{rank}-CVRP applies a 2-opt local search for the TSP to improve the routes generated by the ants. AS_{rank}-CVRPsav first applies a local search based on an exchange move between tours, where each exchange move exchanges two customers from two (different) routes; it then improves the resulting tours by applying 2-opt. MACS-VRPTW uses a more sophisticated local search that is based on *cross*-exchange moves taken from Taillard, Badeau, Gendreau, Guertin, & Potvin (1997).

Results AS_{rank}-CVRP and AS_{rank}-CVRPsav were applied only to the CVRP. In general, AS_{rank}-CVRPsav was performing much better than AS_{rank}-CVRP. AS_{rank}-CVRPsav was compared in Reimann et al. (2002b) and in a recent follow-up paper (Reimann, Doerner, & Hartl, 2004) to several tabu search algorithms for this problem, showing that it performs better than earlier tabu search algorithms.

   MACS-VRPTW was, at the time of its publication, one of the best performing metaheuristics for the VRPTW and it was able to improve the best-known solutions for a number of well-known benchmark instances, both with and without time windows. Only recently has the approach presented in Bräysy (2003) achieved competitive results.

Remarks   ACO algorithms have also been applied to a number of other VRP variants, including a pickup and delivery problem under time window constraints in a hub network (Dawid, Doerner, Hartl, & Reimann, 2002) or the VRP with backhauls and time window constraints (Reimann, Doerner, & Hartl, 2002a). The main differences in these applications to extensions of the basic VRPs we described consist in (1) the way the heuristic information and the pheromone trails are computed or used, and (2) the details of the solution construction procedure. Recently, Reimann, Doerner, & Hartl (2003) proposed an algorithm called "Unified Ant System" that was applied to four VRP variants obtained by considering both, one, or none of the two characteristics, "time windows" and "backhauls." (The only differences among variants concern the use of different heuristic information.)

5.2   Assignment Problems

The task in assignment problems is to assign a set of items (objects, activities, etc.) to a given number of resources (locations, agents, etc.) subject to some constraints. Assignments can, in general, be represented as a mapping from a set $\mathcal{I}$ to a set $\mathcal{J}$, and the objective function to minimize is a function of the assignments done.

To apply the ACO metaheuristic to assignment problems, a first step is to map the problem on a construction graph $G_C = (C, L)$, where C is the set of components (usually the components consist of all the items and all the resources plus possibly some additional dummy nodes) and L is the set of connections that fully connects the graph. The construction procedure allows the ants to perform walks on the construction graph that correspond to assigning items to resources.

For the practical application of the ACO metaheuristic to assignment problems it is convenient to distinguish between two types of decision. The first refers to the assignment order of the items, that is, the order in which the different items are assigned to resources. The second decision refers to the actual assignment, that is, the choice of the resource to which an item is assigned. Pheromone trails and heuristic information may be associated with both decisions. In the first case, pheromone trails and heuristic information can be used to decide on an appropriate assignment order. In the second case, the pheromone trail τ_{ij} and the heuristic information η_{ij} associated with the pair (i, j), where i is an item and j a resource, determine the desirability of assigning item i to resource j.

All ACO algorithms for assignment problems have to take these two decisions into account. In all the applications of ACO to assignment problems that we are aware of, pheromone trails are used only for one of these two decisions. Typically, the pheromone trails refer to the second one, the assignment step. For the first step, deciding about the assignment order, most of the algorithms either use some heuristically derived order or a random order.

5.2.1   Quadratic Assignment

The quadratic assignment problem is an important problem in theory and practice. Many practical problems such as backboard wiring (Steinberg, 1961), campus and hospital layout (Dickey & Hopkins, 1972; Elshafei, 1977), typewriter keyboard design (Burkard & Offermann, 1977), and many others can be formulated as QAPs. The QAP can best be described as the problem of assigning a set of facilities to a set of locations with given distances between the locations and given flows between the facilities. The goal is to place the facilities on locations in such a way that the sum of the products between flows and distances is minimized.

More formally, given n facilities and n locations, two $n \times n$ matrices $A = [a_{ij}]$ and $B = [b_{rs}]$, where a_{ij} is the distance between locations i and j and b_{rs} is the flow between facilities r and s, and the objective function

$$f(\pi) = \sum_{i=1}^{n} \sum_{j=1}^{n} b_{ij} a_{\pi_i \pi_j}, \tag{5.2}$$

where π_i gives the location of facility i in the current solution $\pi \in S(n)$, then the goal in the QAP is to find an assignment of facilities to locations that minimizes the objective function. The term $b_{ij}a_{\pi_i\pi_j}$ describes the cost contribution of simultaneously assigning facility i to location π_i and facility j to location π_j.

The QAP is an $\mathcal{NP}$-hard optimization problem (Sahni & Gonzalez, 1976). It is considered one of the hardest combinatorial optimization problems in practice: the largest nontrivial QAP instance in QAPLIB, a benchmark library for the QAP accessible at www.seas.upenn.edu/qaplib/, solved to optimality at the time of writing, has dimension $n = 36$ (Brixius & Anstreicher, 2001; Nyström, 1999). The relatively poor performance of exact algorithms explains the interest in metaheuristic approaches when the practical solution of a QAP is required. Therefore, it is not surprising that the QAP has attracted a large number of research efforts in ACO. ACO algorithms for the QAP comprise AS (Maniezzo et al., 1994; Maniezzo & Colorni, 1999), $\mathcal{MMAS}$ (Stützle, 1997b; Stützle & Hoos, 2000), and ANTS (Maniezzo, 1999). Overall, these research efforts have led to high-performing ACO algorithms; in fact, the most recent ACO algorithms are among the best-performing metaheuristics for the QAP. In addition, other ant-based algorithms like Hybrid AS (HAS) (Gambardella et al., 1999b) and FANT (Fast Ant System) (Taillard, 1998) were applied to the QAP. Note, however, that HAS and FANT, although inspired by early research efforts on AS, are not ACO algorithms because they depart in essential aspects from the structure of the ACO metaheuristic.

In the following, we describe the application of AS (AS-QAP), $\mathcal{MMAS}$ ($\mathcal{MMAS}$-QAP), and ANTS (ANTS-QAP) to the QAP. Some of the main design choices, such as the definition of the construction graph and of the pheromone trails, are the same for the three algorithms. Therefore, only significant differences among the three algorithms are indicated, where necessary. A more detailed description of applications of ACO and, more in general, of ant algorithms to the QAP can be found in Stützle & Dorigo (1999a).

Construction graph The set of components C comprises all facilities and locations. The connections L fully connect the components. A feasible solution is an assignment consisting of n pairs (i, j) between facilities and locations, with each facility and each location being used exactly once. There are no explicit costs assigned to the couplings.

Constraints The only constraint is that a feasible solution for the QAP assigns each facility to exactly one location and vice versa. This constraint can be easily enforced in the ants' walk by building only couplings between still unassigned facilities and locations.

Pheromone trails   The pheromone trails τ_{ij} refer to the desirability of assigning facility i to location j (or the other way round, the two choices being equivalent).

Heuristic information $\mathcal{MMAS}$-QAP does not use any heuristic information, whereas AS-QAP and ANTS-QAP do.

In AS-QAP two vectors d and f are computed in which the i-th components d_i and f_i represent respectively the sum of the distances from location i to all other locations and the sum of the flows from facility i to all other facilities. The lower d_i, the distance potential of location i, the more central is the location; the higher f_i, the flow potential of facility i, the more important the facility. Flow potentials are used to order facilities (see "Solution construction" below), while the inverse of distance potentials $\eta_j = 1/d_j$ is used as a heuristic value to bias location choice. The motivation for using this type of heuristic information is that, intuitively, good solutions will place facilities with high flow potential on locations with low distance potential.

ANTS-QAP uses lower bounds on the completion of a partial solution to derive the heuristic desirability η_{ij} of adding a specific pair (i, j). The lower bound is computed by tentatively adding the pair to the current partial solution and by estimating the cost of a complete solution containing that coupling by means of the LBD lower bound (Maniezzo, 1999). This lower bound has the advantage of having a computational complexity of $\mathcal{O}(n)$. The lower bound estimate gives the heuristic information: the lower the estimate, the more attractive is the addition of a specific coupling. For details on the lower bound computation see Maniezzo (1999). Note that in an earlier variant of the ANTS-QAP algorithm (Maniezzo & Colorni, 1999), the well-known Gilmore-Lawler lower bound (GLB) (Gilmore, 1962; Lawler, 1963) was applied at each step; however, the GLB requires a computation time of order $\mathcal{O}(n^3)$.

Solution construction   AS-QAP sorts the facilities in nonincreasing order of flow potentials and at each construction step an ant k assigns the next, still unassigned, facility i to a free location j using the action choice rule of AS [equation (3.2)].

The only difference between $\mathcal{MMAS}$-QAP and AS-QAP concerning solution construction is that $\mathcal{MMAS}$-QAP sorts the facilities randomly and, as said before, does not use any heuristic information.

In ANTS-QAP, the lower bound is computed once at the start of the algorithm. Along with the lower bound computation one gets the values of the dual variables corresponding to the constraints when formulating the QAP as an integer programming problem (see Maniezzo [1999] for details). These values are used to define the order in which locations are assigned. The action choice rule is the same as that used by the ANTS algorithm [equation (3.13)].

Pheromone update   The pheromones for all three algorithms are updated following the corresponding rules defined for each of these algorithms (see chapter 3 for details).

Box 5.1
About Local Search in ACO

Very often the best ACO algorithms are those that combine two components: solution construction by artificial ants and a local search procedure (remember that, within the general definition of the ACO metaheuristic given in chapter 2, section 2.1, local search is a particular type of the so-called *daemon actions*). In general, the choice of which local search procedure to use is not only problem-specific, but may also depend on the problem instances considered. As an example, we consider two well-known QAP instances, `tai50a` and `tai50b`, and we compare two $\mathcal{MMAS}$-QAP algorithms using as local search two different procedures: a best-improvement 2-opt and short runs of a tabu search. The resulting two algorithms, which are compared in the figure below, are called $\mathcal{MMAS}$-QAP$_{2\text{-opt}}$ and $\mathcal{MMAS}$-QAP$_{TS}$, respectively. The results are averaged over ten runs for each algorithm.

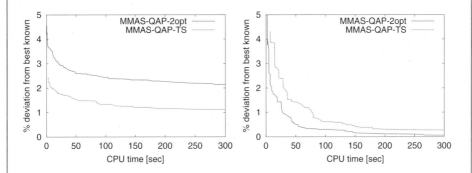

As can be observed in the figure, which local search performs better once coupled to $\mathcal{MMAS}$-QAP depends on the instance type: for `tai50a`, on the left, $\mathcal{MMAS}$-QAP$_{TS}$ performs significantly better than $\mathcal{MMAS}$-QAP$_{2\text{-opt}}$, whereas, for `tai50b`, on the right, $\mathcal{MMAS}$-QAP$_{2\text{-opt}}$ outperforms $\mathcal{MMAS}$-QAP$_{TS}$.

Local search All three ACO algorithms were combined with a 2-opt local search procedure for the QAP. $\mathcal{MMAS}$-QAP was also tested with a local search procedure based on short runs of a tabu search algorithm. It was found that whether the 2-opt or the tabu search should be preferred is a function of the particular instance class; for more details, see box 5.1.

Results ACO algorithms were experimentally shown to be among the best available algorithms for structured real-life, and for large, randomly generated real-life-like QAP instances. Their excellent performance is confirmed in a variety of studies (Gambardella, 1999b; Maniezzo, 1999; Stützle & Hoos, 2000; Stützle & Dorigo, 1999a). Interestingly, ANTS-QAP was also shown to outperform tabu search algorithms (Taillard, 1991) for a class of hard, randomly generated instances. Although the dimension of the considered random instances was limited to $n = 40$, this is a very noteworthy result, because tabu search algorithms typically perform much better than other metaheuristics on these types of instances (Taillard, 1995).

Among ACO algorithms, ANTS-QAP and $\mathcal{MMAS}$-QAP appear to perform significantly better than AS-QAP (Stützle & Dorigo, 1999a), which confirms the observation made for the TSP application that the more recent and more sophisticated ACO algorithms strongly improve over AS performance.

5.2.2   Generalized Assignment

In the generalized assignment problem, a set of tasks has to be assigned to a set of agents in such a way that a cost function is minimized. Each agent j has only a limited capacity a_j and each task i consumes, when assigned to agent j, a quantity b_{ij} of the agent's capacity. Also, the cost d_{ij} of assigning task i to agent j is given. The objective then is to find a feasible task assignment of minimum cost. The GAP was described in chapter 2, section 2.3.3, where an outline was given of how, in principle, ACO algorithms can be applied to it; we refer the reader to that description for more details. The first ACO application to the GAP was presented by Lourenço & Serra (1998) and is based on $\mathcal{MMAS}$ ($\mathcal{MMAS}$-GAP).

Construction graph   The set of components is given by $C = I \cup J$ and it is fully connected; the construction graph is identical to that described in chapter 2, section 2.3.3.

Constraints   The problem constraints may lead to a situation in which a partial assignment cannot be extended to a full assignment that satisfies all the agents' capacity constraints. $\mathcal{MMAS}$-GAP deals with this problem by allowing construction of infeasible solutions.

Pheromone trails   The pheromone trail τ_{ij} represents the desirability of assigning task i to agent j.

Heuristic information $\mathcal{MMAS}$-GAP uses heuristic information only for the pheromone initialization, but not while constructing solutions. In $\mathcal{MMAS}$-GAP, pheromone trails are initialized using the heuristic information; their initial value is set to $\tau_0 = 1/d_{ij}$.

Solution construction   Solutions are constructed by iteratively assigning tasks to agents. At each construction step, first the next task to be assigned is chosen randomly; then the chosen task is assigned to an agent applying the pseudorandom proportional action choice rule of ACS [see equation (3.10)]. In the solution construction, care is taken in assigning tasks only to agents that still have enough spare capacity. Only if no agent has enough spare capacity to accept the task is the task assigned randomly to any of the agents, generating in this way an infeasible assignment.

Pheromone update   After each iteration, the iteration-best solution deposits pheromone. The way pheromones are updated in $\mathcal{MMAS}$-GAP shows a particularity: the amount of pheromone deposited depends only on the feasibility status of a solution, and not on the solution quality. If a solution is feasible, a constant quantity of 0.05 units of pheromone is deposited, otherwise 0.01 units are deposited.

Local search   Several local search algorithms, including a simple iterative improvement algorithm, a tabu search, and an ejection chain approach (Glover, 1996; Glover & Laguna, 1997), were tested.

Particularities $\mathcal{MMAS}$-GAP does not use a colony of ants: in each iteration only one ant constructs a solution and deposits pheromone. In fact, this corresponds to the parameter setting $m = 1$ in ACO. Such a parameter setting can result in a faster convergence of the ACO algorithm to good solutions, but it may result in worse solution quality for long computation times. For a discussion of how the number of ants influences the performance of ACO algorithms, see section 5.7.6.

A further particularity of $\mathcal{MMAS}$-GAP is that the amount of pheromone deposited by an ant depends only on whether its solution is feasible or not (see "Pheromone update" above), that is, the amount of pheromone deposited does not depend on how good a feasible solution is.

Results $\mathcal{MMAS}$-GAP was shown to perform better than a GRASP algorithm that used the same local search (Lourenço & Serra, 1998). A comparison of the computational results obtained with $\mathcal{MMAS}$-GAP with those obtained by other metaheuristics showed that, at the time the research was done, $\mathcal{MMAS}$-GAP could reach state-of-the-art performance. However, since that time, better algorithms for the GAP have been proposed; the best algorithm currently available is that of Yagiura, Ibaraki, & Glover (2004).

5.2.3   Frequency Assignment

In the frequency assignment problem (FAP) are given a set of links, a set of frequencies, and channel separation constraints that for each pair of links give a minimum distance to be maintained between the frequencies assigned to the links. There exist a number of different variants of the FAP (for an overview, see Aardal, van Hoesel, Koster, Mannino, & Sassano, 2001). Maniezzo & Carbonaro (2000) applied the ANTS algorithm to a version of the FAP in which, given a maximum number of frequencies, the objective is to minimize the sum of the costs of violating the channel separation constraints plus the costs of modifying the frequencies of links that have a preassigned frequency.

Construction graph   The set of components C comprises the set of links and the set of available frequencies; as usual, the construction graph is fully connected.

Constraints   The only constraint for the solution construction is that a frequency must be assigned to each link. Violations of the channel separation constraint are penalized by the objective function.

Pheromone trails   Pheromone trails are associated with components representing the links. A pheromone trail τ_{ij} indicates the desirability of assigning a frequency j to link i.

Heuristic information   At each construction step of an ant, a lower bound based on an adaptation of the orientation model (Borndörfer, Eisenblätter, Grötschel, & Martin, 1998a) is computed and used as heuristic information.

Solution construction   Solutions are constructed iteratively by assigning frequencies to links, using the probabilistic decision policy of ANTS [equation (3.13)].

Pheromone update   The pheromone update rule of ANTS is applied [equation (3.15)].

Local search   Each constructed solution is locally optimized using an iterative descent algorithm that tries to improve the objective function by modifying at each step the assignment of frequencies to links.

Results   ANTS was compared to reimplementations of two simulated annealing algorithms (Hurkens & Tiourine, 1995; Smith, Hurley, & Thiel, 1998), to a tabu search, and to a constructive algorithm based on the DSATUR heuristic, originally designed for the graph coloring problem (GCP) (Brelaz, 1979). Experimental results were presented for a set of benchmark instances (CELAR, GRAPH, and PHILA-DELPHIA problems), which were adapted to the FAP formulation used in Maniezzo & Carbonaro (2000). ANTS performed particularly well on the CELAR and GRAPH instances.

5.2.4   Other ACO Applications to Assignment Problems

Graph Coloring Problem

A number of other ACO applications to assignment-type problems have been proposed. One of the first is an approach based on AS to the GCP by Costa & Hertz (1997). Given an undirected graph $G = (N, A)$, the goal in the GCP is to find the minimum number of colors to assign to nodes such that no pair of adjacent nodes is assigned the same color. In their ACO algorithm, Costa and Hertz use pheromones to indicate the desirability of assigning the same color to two nodes. For ants' solution construction, they adapted the heuristics used in two well-known constructive algorithms, the DSATUR heuristic (Brelaz, 1979) and the Recursive Largest First (RLF) heuristic (Leighton, 1979). They experimentally compared eight variants of their ACO algorithm on a set of randomly generated GCP instances with up to 300 nodes. With good parameter settings, all the considered variants significantly improved over the underlying DSATUR and RLF heuristics, with those based on the DSATUR heuristic yielding the overall best results. The performance of this last ACO algorithm, in general, appears to be way behind good graph coloring algorithms such as tabu search algorithms (Dorne & Hao, 1999) or several hybrid approaches (Galinier & Hao, 1999; Paquete & Stützle, 2002). However, a more successful ACO approach to the GCP could certainly be obtained by employing a local search and by using ACO algorithms that are more advanced than AS.

University Course Timetabling Problem
In the university course timetabling problem (UCTP) one is given a set of time slots, a set of events, a set of rooms, a set of features, a set of students, and two types of constraints: hard and soft constraints. Hard constraints have to be satisfied by any feasible solution, while soft constraints do not concern the feasibility of a solution but determine its quality. The goal is to assign the events to the time slots and to the rooms so that all hard constraints are satisfied and an objective function, whose value depends on the number of violated soft constraints, is optimized. The only UCTP attacked by ACO algorithms that we are aware of was proposed within the research activities of the European project "Metaheuristics Network" (for details, see www.metaheuristics.org). Two ACO algorithms were implemented, the most successful of these being an adaptation of $\mathcal{MMAS}$ to the UCTP ($\mathcal{MMAS}$-UCTP) (Socha et al., 2002, 2003). In $\mathcal{MMAS}$-UCTP the pheromone trail τ_{ij} refers to the desirability of assigning an event i to a time slot j; no heuristic information is used. Solutions are constructed by first preordering the events and then assigning the events to time slots using the probabilistic action choice rule of AS. Once the solution construction is completed, the iteration-best solution is improved by a local search procedure. $\mathcal{MMAS}$-UCTP, when compared to the other metaheuristics tested in the research done in the "Metaheuristics Network," obtained good results and showed particularly good performance on the largest instances (Socha et al., 2003).

5.3   Scheduling Problems

Scheduling, in the widest sense, is concerned with the allocation of scarce resources to tasks over time. Scheduling problems are central to production and manufacturing

Table 5.1
Available ACO algorithms for scheduling problems discussed in the text

Problem	Main references
JSP	Colorni, Dorigo, Maniezzo, & Trubian (1994)
OSP	Pfahringer (1996); Blum (2003b)
PFSP	Stützle (1997a, 1998a)
SMTTP	Bauer, Bullnheimer, Hartl, & Strauss (2000)
SMTWTP	den Besten, Stützle, & Dorigo (2000); Merkle & Middendorf (2000)
RCPSP	Merkle, Middendorf, & Schmeck (2000a, 2002)
GSP	Blum (2002a, 2003a)
SMTTPSDST	Gagné, Price, & Gravel (2002)

JSP is the job shop problem, OSP is the open-shop problem, PFSP is the permutation flow shop problem, SMTTP is the single-machine total tardiness problem, SMTWTP is the single-machine total weighted tardiness problem, RCPSP is the resource-constrained project scheduling problem, GSP is the group shop scheduling problem, and SMTTPSDST is the single-machine total tardiness problem with sequence dependent setup times. Details on the ACO algorithms for these problems are given in the text.

industries, but also arise in a variety of other settings. In the following, we mainly focus on shop scheduling problems, where jobs have to be processed on one or several machines such that some objective function is optimized. In case jobs have to be processed on more than one machine, the task to be performed on a machine for completing a job is called an operation. For all the machine-scheduling models considered in the following it holds that (1) the processing times of all jobs and operations are fixed and known beforehand and (2) the processing of jobs and operations cannot be interrupted (scheduling without preemption). For a general introduction to scheduling, see Brucker (1998) or Pinedo (1995).

Scheduling problems play a central role in ACO research, and many different types of scheduling problems have been attacked with ACO algorithms (see table 5.1). The performance, however, varies across problems. For some problems, such as the single-machine total weighted tardiness problem (SMTWTP), the open shop problem, and the resource constrained project scheduling problem, ACO is among the best-performing approaches. For other, classic scheduling problems, however, like the permutation flow shop problem and the job shop problem, the computational results obtained so far are far behind the state of the art.

The construction graph for scheduling problems is typically represented by the set of jobs (for single-machine problems) or operations. However, often it is convenient to add to the construction graph nodes that represent positions in a sequence that jobs (operations) can take, and to view sequences as assignments of jobs (operations) to these positions. This is important, because in many scheduling problems the absolute position of a job in a sequence is important. However, there exists some computational evidence that for some problems the relative ordering of jobs in the sequence may be more important (Blum & Sampels, 2002a).

5.3.1 Single-Machine Total Weighted Tardiness Scheduling

In the single-machine total weighted tardiness problem n jobs have to be processed sequentially on a single machine, without interruption. Each job has an associated processing time p_j, a weight w_j, and a due date d_j, and all jobs are available for processing at time zero. The tardiness of job j is defined as $T_j = \max\{0, CT_j - d_j\}$, where CT_j is its completion time in the current job sequence. The goal in the SMTWTP is to find a job sequence, that is, a permutation of the job indices, that minimizes the sum of the weighted tardiness, given by $\sum_{j=1}^{n} w_i T_i$. The unweighted case, in which all the jobs have the same weight, is called the single-machine total tardiness problem (SMTTP). It is well known that the SMTWTP is harder to solve than the SMTTP. This is true from a theoretical perspective, because the SMTTP can be solved in pseudopolynomial time (Lawler, 1977), while the SMTWTP with no restrictions on the weights is $\mathcal{NP}$-hard in the strong sense (Lenstra, Rinnooy Kan, & Brucker, 1977). But it is also true from the experimental perspective: while SMTWTP instances with more than 50 jobs often cannot be solved to optimality with state-of-the-art branch & bound algorithms (Abdul-Razaq, Potts, & Wassenhove, 1990; Crauwels, Potts, & Wassenhove, 1998), the best available branch & bound algorithms solve SMTTP instances with up to 500 jobs (Szwarc, Grosso, & Della Croce, 2001).

ACO algorithms have been developed for both the SMTTP and the SMTWTP. First, Bauer et al. (2000) applied ACS to the SMTTP (ACS-SMTTP), then den Besten et al. (2000) and Merkle & Middendorf (2003a) in parallel developed ACS applications to the SMTWTP, referred to respectively as ACS-SMTWTP-BSD and ACS-SMTWTP-MM in the rest of this section. These ACO algorithms are very similar to each other and share many characteristics.

Construction graph The set of components C consists of the n jobs and the n positions to which the jobs are assigned. The set L of arcs fully connects the graph.

Constraints The only constraint that has to be enforced is that all jobs have to be scheduled.

Pheromone trails The pheromone trails τ_{ij} refer to the desirability of scheduling a job j as the i-th job, that is, the desirability of assigning job j to position i.

Heuristic information In ACS-SMTTP two priority rules were tested to define two different types of heuristic information. The rules are (1) the earliest due date rule, which puts the jobs in nondecreasing order of the due dates d_j, and (2) the modified due date rule (Bauer et al., 2000), which puts the jobs in nondecreasing order of the modified due dates given by $mdd_j = \max\{\hat{p} + p_j, d_j\}$, where $\hat{p}$ is the sum of the processing times of the already sequenced jobs.

ACS-SMTWTP-BSD also considered, in addition to the earliest due date and to the modified due date rules, the apparent urgency priority rule, which puts the jobs in nondecreasing order of apparent urgency (Morton, Rachamadugu, & Vepsalainen, 1984), defined as

$$au_j = (w_j/p_j) \cdot \exp(-(\max\{d_j - CT_j, 0\})/k\bar{p}),$$

where $\bar{p}$ is the average processing time of the remaining jobs, and k is a parameter set as proposed in Potts & Wassenhove (1991). In each case, the heuristic information was defined as $\eta_{ij} = 1/h_j$, where h_j is either d_j, mdd_j, or au_j, depending on the priority rule used.

ACS-SMTWTP-MM used a variation of Bauer and colleagues' modified due date rule. This new rule was defined as $vmdd_j = \max\{\hat{p} + p_j, d_j\} - \hat{p}$. The heuristic information was set to be $\eta_{ij} = w_j/vmdd_j$. This variation is based on the observation that, as $\hat{p}$ increases, the values mdd_j become large, and the differences between the heuristic values for the remaining jobs become small; hence, ants can no longer differentiate effectively between the alternatives based on the heuristic values. This problem is reduced by using $vmdd_j$. Finally, in particular situations, it can be shown that a good policy is to schedule the remaining jobs in a deterministic order. If such a situation occurs, ants in ACS-SMTWTP-MM follow this deterministic rule.

Solution construction and pheromone update   The ants construct a sequence by first choosing a job for the first position, then a job for the second position, and so on until all jobs are scheduled. The action choice and the pheromone update rules are those of ACS, except for two details in ACS-SMTWTP-BSD and ACS-SMTWTP-MM: the first uses appropriately defined candidate lists, while the second uses the *pheromone summation* rule (which is explained in depth in box 5.2).

Local search   In ACS-SMTTP a best-improvement local search that considers all possible exchanges between pairs of jobs was applied to the best ant after each iteration. ACS-SMTWTP-MM applied a truncated first-improvement strategy that checked exactly once for each pair of jobs whether an exchange of their positions led to an improved solution. ACS-SMTWTP-BSD combined ACS with a powerful local search based on variable neighborhood descent (Hansen & Mladenović, 1999).

Results   Of the three approaches, ACS-SMTWTP-BSD obtained the best performance results. This algorithm was able to obtain on all benchmark instances from

Box 5.2
The Pheromone Summation Rule

In permutation scheduling applications, pheromone trails τ_{ij} typically refer to the desirability of assigning job j to position i. Now, assume that, because of the stochastic nature of the algorithm, although job j has a high pheromone value τ_{ij}, it happens that it is job h, with a low τ_{ih}, which is assigned to position i of the schedule. Then, in many scheduling problems, as is the case, for example, in the SMTWTP, it may be advantageous to assign job j to a position close to position i. However, if for positions $l > i$ the pheromone trails τ_{lj} happen to be low, it is probable that the job gets sequenced toward the end of the schedule, far away from position i, leading to highly suboptimal schedules. Unfortunately, the situation in which the τ_{lj}'s are low for positions $l > i$ may easily occur. For example, this is the case if no solution which assigns job j to a position $l > i$ has yet been found, or if such solutions were found many iterations before, so that the corresponding pheromone values have decreased because of evaporation. An elegant solution to this problem has been proposed by Merkle and Middendorf: the use of the so-called *summation rule*.

The summation rule consists of choosing the job j to assign to position i using the sum of all the τ_{hj}'s, with $h \leq i$. In this way, if it happens that, notwithstanding a high value of pheromone τ_{ij}, job j is not allocated to position i, the high value τ_{ij} continues to influence the probability of allocating job j to the position $i + 1$. In this way, job j has a high probability of being assigned to a position close to i.

In ACS-SMTWTP-MM (see section 5.3.1), when using the summation rule, an ant chooses with probability q_0 to assign to position i a job that maximizes

$$\left[\sum_{k=1}^{i} \tau_{kj}\right]^{\alpha} [\eta_{ij}]^{\beta}, \tag{5.3}$$

while with probability $1 - q_0$ the job j is chosen according to a probability given by

$$p_{ij} = \frac{\left[\sum_{k=1}^{i} \tau_{kj}\right]^{\alpha} [\eta_{ij}]^{\beta}}{\sum_{h \in \mathcal{N}_i}\left(\left[\sum_{k=1}^{i} \tau_{kh}\right]^{\alpha} [\eta_{ih}]^{\beta}\right)}, \tag{5.4}$$

where $\mathcal{N}_i$ is the set of still unscheduled jobs. As said, the summation rule was experimentally shown to lead to improved computational results for the SMTWTP (as well as for the SMTTP). In further experiments, Merkle and Middendorf used a weighted summation rule as well as combinations of the standard way of using pheromones with the weighted summation rule.

In the weighted summation rule, equation (5.4) becomes

$$p_{ij} = \frac{\left[\sum_{k=1}^{i} \gamma^{i-k} \tau_{kj}\right]^{\alpha} [\eta_{ij}]^{\beta}}{\sum_{h \in \mathcal{N}_i}\left(\left[\sum_{k=1}^{i} \gamma^{i-k} \tau_{kh}\right]^{\alpha} [\eta_{ih}]^{\beta}\right)}, \tag{5.5}$$

where the parameter γ, $\gamma > 0$, determines the influence of pheromone trails corresponding to earlier positions. Setting $\gamma = 1$ gives the (unweighted) summation rule, a value $\gamma < 1$ gives less influence to pheromone trails corresponding to earlier decisions, while a value $\gamma > 1$ increases their influence.

Merkle and Middendorf have combined the standard way of using pheromones with the weighted summation rule by computing pheromone trails τ'_{ij} as follows:

$$\tau'_{ij} = c \cdot x_i \cdot \tau_{ij} + (1 - c) \cdot y_i \cdot \sum_{k=1}^{i} \gamma^{i-k} \tau_{kj}, \tag{5.6}$$

Box 5.2
(continued)

---

where $x_i = \sum_{h \in \mathcal{N}_i} \sum_{k=1}^{i} \gamma^{i-k} \tau_{kh}$ and $y_i = \sum_{h \in \mathcal{N}_i} \tau_{ih}$ are factors to adjust for the different range of values in the standard and in the summation rule and the parameter c adjusts the relative influence of the local and the summation rule: for $c = 1$ is obtained the standard rule, while for $c = 0$ is obtained the pure, weighted summation rule. Then, the probability of choosing a job j for position i is computed using the usual equation:

$$p_{ij} = \frac{[\tau'_{ij}]^{\alpha} [\eta_{ij}]^{\beta}}{\sum_{h \in \mathcal{N}_i} ([\tau'_{ih}]^{\alpha} [\eta_{ih}]^{\beta})}. \tag{5.7}$$

An experimental study of ACO algorithms for the SMTWTP and the resource-constrained project scheduling problem (RCPSP), a scheduling problem that is presented in section 5.3.3, showed that by setting appropriate values for the parameters γ and c it is possible to obtain much better results in comparison to those obtained with either the standard rule or with a pure (unweighted) summation rule (Merkle, Middendorf, & Schmeck, 2000b).

Finally, we would like to emphasize that the usefulness of the summation rule depends on the property that in good schedules the positions of jobs are similar. If this is not the case, pheromone evaluation based on the summation rule may fail. In fact, Merkle & Middendorf (2002a) defined the single-machine total earliness problem with multiple due dates, for which such property does not hold and where the summation rule fails. For that problem, they showed that good performance can be achieved by a *relative pheromone evaluation rule* that normalizes the pheromone value τ_{ij} with the relative amount of the pheromones on the remaining positions for job j.

ORLIB (mscmga.ms.ic.ac.uk/jeb/orlib/wtinfo.html) the best-known solutions, which are conjectured to be the optimal ones. ACS-SMTWTP-BSD is currently among the best available algorithms for the SMTWTP; however, the iterated Dynasearch approaches of Congram et al. (2002) and Grosso, Della Croce, & Tadei (2004) appear to be faster, reaching the same level of performance in a shorter computation time.

Good results were also reported for ACS-SMTWTP-MM, although it could not reach the same solution quality of ACS-SMTWTP-BSD; the main reason is certainly the less powerful local search algorithms used by Merkle and Middendorf. Experiments with ACS-SMTWTP-MM applied to the SMTTP showed a significantly better performance when compared to ACS-SMTTP, the main reason being the use of more sophisticated heuristic information and the use of the summation rule (Merkle & Middendorf, 2003a).

5.3.2   Job Shop, Open Shop, and Group Shop Scheduling

In job shop, open shop, and group shop scheduling problems we are given a finite set $\mathcal{O}$ of operations that is partitioned into a set of subsets $\mathcal{M} = \{M_1, \ldots, M_m\}$, where each M_i corresponds to the operations to be processed by machine i, and into a set of subsets $\mathcal{J} = \{J_1, \ldots, J_n\}$, where each set J_j corresponds to the operations belonging

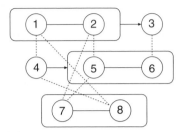

Figure 5.2
Disjunctive graph representation (Roy & Sussmann, 1964) of a simple group shop scheduling problem instance with eight operations. The nodes in the graph correspond to the operations, groups having more than one node are indicated by boxes. We have $\mathcal{O} = \{1,\ldots,8\}$, $\mathcal{J} = \{J_1 = \{1,2,3\}, J_2 = \{4,5,6\}, J_3 = \{7,8\}\}$, $\mathcal{M} = \{M_1 = \{1,4,8\}, M_2 = \{2,5,7\}, M_3 = \{3,6\}\}$, and $\mathcal{G} = \{G_1 = \{1,2\}, G_2 = \{3\}, G_3 = \{4\}, G_4 = \{5,6\}, G_5 = \{7,8\}\}$. There are directed arcs between the groups belonging to the same job, and operations within one group are connected by an undirected arc. Additionally, there are undirected (dashed) arcs between all pairs of operations to be processed on a same machine. A feasible solution can be obtained by directing the undirected arcs so that there is no cycle in the resulting graph.

to job j. Each operation is assigned a non-negative processing time and preemption is not allowed.

The job shop, open shop, and group shop scheduling problems differ only in the order that is imposed on the operations belonging to a job. In the job shop problem (JSP), precedence constraints among all operations of a job exist and they induce a total ordering of the operations of each job. On the contrary, in the open shop problem (OSP), there are no precedence constraints, that is, any ordering of the operations is allowed as long as only one operation of a job is processed at a time. In the group shop scheduling problem (GSP), the operations of each job are additionally partitioned into groups. The operations within one group can be processed in any order, but the groups of a job are totally ordered; therefore, the order of the groups within a job induces a partial order among the operations. The GSP is the most general of these problems, as it contains the JSP and the OSP as special cases. In the JSP each group contains only one operation and every OSP instance can be seen as a GSP instance with only one group per job. An example of a simple GSP instance is given in figure 5.2.

So far, ACO algorithms have been applied to the above-mentioned scheduling problems with the minimization of the completion time of the last operation, also called makespan, as objective function. Colorni et al. (1994) were the first to attack one of these problems: they applied AS to the JSP (AS-JSP-CDMT). This first approach was followed by an application of Ant-Q (Gambardella & Dorigo, 1995; Dorigo & Gambardella, 1996) to the OSP (AntQ-OSP) by Pfahringer (1996), and by

$\mathcal{MMAS}$-HC-GSP, an $\mathcal{MMAS}$ algorithm for the GSP proposed by Blum (2002a, 2003a). In all these approaches the solution construction is based on schedule generation methods. These are algorithms that generate at each construction step a set of operations that can be added to the partial schedule maintaining feasibility. Then, according to some heuristic, one of the operations is chosen and appended to the schedule. More details about schedule generation methods are given in box 5.3. The main difference between the algorithms concerns the way pheromone trails are defined, the heuristic information chosen, the type of ACO algorithm used, and the use of local search.

Construction graph   The construction graph contains one node for every operation plus two additional nodes that represent a source and a destination node. As usual, it is fully connected.

Constraints   The constraints in the problem require that the precedence constraints between the operations are met and that all operations must be scheduled exactly once.

Pheromone trails   The available ACO algorithms use different ways of defining the pheromone trails. AS-JSP-CDMT and AntQ-OSP use a pheromone representation intended to learn a predecessor relationship. In this case, a pheromone trail τ_{ij} exists between every pair of operations and between the source and all operations; τ_{ij} gives the desirability of choosing operation j directly after operation i. $\mathcal{MMAS}$-HC-GSP applies a pheromone model where a pheromone trail τ_{ij} is assigned to related operations; operations are related either if they are in the same group or if they are processed on the same machine. In this case, τ_{ij} refers to the desirability of scheduling operation j after, but not necessarily *immediately* after, operation i.

Heuristic information   Several types of heuristic information are applied in the various algorithms. AS-JSP-CDMT uses the *most work remaining* heuristic (Haupt, 1989) that computes for each job j the total processing time of the operations still to be scheduled (mwr_j) and the heuristic information is then set to $\eta_j = mwr_j$. AntQ-OSP showed best performance using the earliest start heuristic, which favors operations with minimal valid starting time with respect to the partial schedule; in this case we have $\eta_i = 1/est_i$, where est_i is the earliest possible starting time of operation i. The earliest start heuristic is also used in $\mathcal{MMAS}$-HC-GSP, where the heuristic information is computed based on the value of $1/est_i$, which is then normalized such that for all eligible operations the heuristic information is in the interval $[0, 1]$.

Solution construction   At the start all ants are put in the source node and the construction phase terminates when all ants have reached the destination. Solutions are constructed using a schedule generation method that restricts the set of eligible

Box 5.3
Schedule Generation Methods

Schedule generation methods are probably the most frequently applied constructive heuristics for solving shop scheduling problems in practice. As a side effect, schedule generation methods build a sequence s containing all the operations of $\mathcal{O}$ exactly once. (Note that a sequence s unambiguously defines a solution to an instance of a shop scheduling problem.) For the problem instance given in figure 5.2, for example, the sequence (1 2 4 8 5 7 6 3) defines group order $1 \preceq 2$ in group G_1, $5 \preceq 6$ in group G_4 and $8 \preceq 7$ in group G_5. It also defines the order in which the operations are processed on the machines.

From a high-level perspective, a schedule generation method works as depicted below. In the algorithm we denote by $o' \preceq_{pred} o$ the fact that the problem constraints enforce that an operation o' has to be processed before operation o; $s[i]$ denotes the operations assigned at position i in sequence s.

procedure ScheduleGenerationMethod
    $\mathcal{O}_{rem} \leftarrow \mathcal{O}$
    $s \leftarrow [\ ]$
    for $i = 1$ **to** $|\mathcal{O}|$ **do**
        $S \leftarrow \{o \in \mathcal{O}_{rem} \,|\, \nexists o' \in \mathcal{O}_{rem} \text{ with } o' \preceq_{pred} o\}$
        $S' \leftarrow$ GenerateCandidateOperations(S)
        $o^* \leftarrow$ ChooseOperation(S')
        $s[i] \leftarrow o^*$
        $\mathcal{O}_{rem} \leftarrow \mathcal{O}_{rem} \backslash \{o^*\}$
    end-for

The main procedures defining a schedule generation method are GenerateCandidateOperations and ChooseOperation.

There are two main ways of implementing the GenerateCandidateOperations procedure. The proposal of Giffler & Thompson (1960) is to compute first the earliest completion times of all the operations in the set S. Then, one of the machines with minimal completion time ect_{min} is chosen and the set S' is the set of all operations in S which need to be processed on the chosen machine and whose earliest possible starting time is smaller than ect_{min}. A second typical approach is to compute first the earliest possible starting time est_{min} of all operations in S. Then, S' consists of all operations in S which can start at est_{min}.

Over the years quite a lot of research has been devoted to finding rules to be used in the ChooseOperation procedure. These rules, which are commonly called priority rules or dispatching rules, are most often applied in a deterministic way, although examples of probabilistic use can be found. None of the rules proposed in the literature can be considered to be the "best-performing" priority rule, as their relative performance depends strongly on the structure of the problem instance to be solved. A survey of priority rules can be found in Haupt (1989). In all the current ACO applications to GSP problems the choice of the next operation from the set S' is biased by both heuristic information and pheromone trails.

operations to a set $\mathcal{N}(s^k)$, where s^k is the k-th ant's partial solution. Each eligible operation $i \in \mathcal{N}(s^k)$ is rated according to the pheromone trail and the heuristic information and added according to the rules applicable in the corresponding ACO algorithm. Only $\mathcal{MMAS}$-HC-GSP departs in some details from the standard action choice rule, because of the different meaning of the pheromone trails: If an operation does not have any related and unscheduled operation left, it is chosen deterministically. Otherwise an ant chooses the next operation i with a probability

$$
p_i^k = \begin{cases} \dfrac{\min_{j \in R_u(i)} \tau_{ij}[\eta_i]^\beta}{\sum_{l \in \mathcal{N}(s_k)}(\min_{j \in R_u(l)} \tau_{lj}[\eta_i]^\beta)}, & \text{if } i \in \mathcal{N}(s^k); \\ 0, & \text{otherwise;} \end{cases}
\tag{5.8}
$$

where $R_u(i)$ is the set of operations that are related to i but not yet in the partial schedule constructed so far.

Pheromone update   The pheromone update follows the basic rules of the various ACO algorithms; note that $\mathcal{MMAS}$-HC-GSP includes the ideas of $\mathcal{MMAS}$ in the hyper-cube framework for ACO and therefore only one ant is depositing pheromone after each iteration. We refer the reader to the original sources for details on the pheromone update.

Local search   AS-JSP-CDMT and AntQ-OSP do not use local search. Concerning $\mathcal{MMAS}$-HC-GSP, several variants were tested, some of them including local search. The best performance was reported for a variant that first improves all solutions constructed by the ants, applying an iterative improvement algorithm, and then applies to the best local optimum a tabu search algorithm which is terminated after $|\mathcal{O}|/2$ iterations. The local search algorithm adapts the JSP neighborhood introduced by Nowicki & Smutnicki (1996a) to the GSP.

Particularities   AS-JSP-CDMT and AntQ-OSP are rather straightforward applications of ACO algorithms. Only $\mathcal{MMAS}$-HC-GSP includes some particular features, apart from the local search, that try to enhance performance. In particular, $\mathcal{MMAS}$-HC-GSP maintains an *elitist list* of solutions comprising recent, very good solutions and iterates through intensification and diversification phases of the search. Diversification is achieved through the application of pheromone reinitialization. Intensification is achieved by letting the restart-best solution or some ant of the elitist list deposit pheromone. Which ants are allowed to deposit pheromone is a function of the value assumed by a variable called *convergence factor* and of the state of the search with respect to convergence. Solutions which are member of the elitist list are occasionally removed from the list to free space for different high-quality solutions. A further feature is that at each iteration an ant in $\mathcal{MMAS}$-HC-GSP randomly chooses, before starting solution construction, one of two different mechanisms for generating the set $\mathcal{N}(s^k)$ of eligible operations.

Results   The computational results obtained with AS-JSP-CDMT and AntQ-OSP are not satisfactory. This is particularly true for AS-JSP-CDMT, which is far behind the performance of the currently best algorithms for the JSP (Nowicki & Smutnicki, 1996a; Balas & Vazacopoulos, 1998; Grabowski & Wodecki, 2001). AntQ-OSP was compared to an earlier evolutionary algorithm (Fang, Ross, & Corne, 1994) on small-size OSP instances, reaching a similar level of performance. However, it appears to be quite far from current state-of-the-art approaches for the OSP (Liaw, 2000). Blum extensively tested $\mathcal{MMAS}$-HC-GSP on a large number of GSP instances, including JSP and OSP instances (Blum, 2002b). $\mathcal{MMAS}$-HC-GSP resulted in being the current best-performing ACO algorithm for these types of scheduling problems, in particular for those instances that are neither pure JSP nor pure OSP problems. However, even when applied to the "pure" versions, $\mathcal{MMAS}$-HC-GSP has a very good performance. In fact, for the OSP it has a performance similar to that of the state-of-the-art algorithm, a hybrid genetic algorithm proposed by Liaw (2000); and for the JSP it is only beaten by neighborhood search based methods (Balas & Vazacopoulos, 1998; Nowicki & Smutnicki, 1996a; Grabowski & Wodecki, 2001). Blum also compared $\mathcal{MMAS}$-HC-GSP with a tabu search approach for the GSP. The result was that the tabu search approach appears to be slightly better than $\mathcal{MMAS}$-HC-GSP for GSP instances close to JSP instances, while $\mathcal{MMAS}$-HC-GSP outperforms the tabu search approach on instances that are closer to OSP instances.

Remarks   Recently, Blum (2003b) has proposed a hybridization between an ACO algorithm and Beam search. In this approach, called Beam-ACO, at each algorithm iteration each ant builds more than one solution, adopting a procedure which is a probabilistic version of Beam search. The results obtained are very promising: Beam-ACO seems to outperform the best algorithms for the OSP.

5.3.3   Resource-Constrained Project Scheduling

The resource-constrained project scheduling problem (RCPSP) is a general scheduling problem in which the set of activities $\mathcal{A}$ of a project must be scheduled minimizing the makespan, subject to resource constraints and to precedence constraints among the activities. More formally, in the RCPSP one is given a set of activities $\mathcal{A} = \{a_0, a_1, \ldots, a_{n+1}\}$, with given precedence constraints among the activities, and a set of resource types $\mathcal{R} = \{r_1, \ldots, r_l\}$, each resource r_i having an associated capacity rc_i. Every activity a_i has associated a processing time p_i and resource requirements $ar_{i1}, \ldots, ar_{il}$, where ar_{ij} is the resource requirement for resource j. The goal in the

RCPSP is to assign to each activity a start time such that the makespan is minimized, subject to the constraints that all precedence constraints are satisfied (i.e., an activity a_i must not start before all its predecessors have finished being processed), and at every instant the resource constraints are met. The RCPSP is the most general of the scheduling problems treated so far with ACO, since, as can easily be verified, it contains the GSP (as well as the permutation flow shop problem, which is briefly discussed in section 5.3.4) as a special case.

The RCPSP is $\mathcal{NP}$-hard and it has attracted a large number of researchers in recent years. A general overview of the RCPSP is given in Brucker et al. (1999) and a comparison of RCPSP heuristics available up to the year 1998 is reported in Hartmann & Kolisch (2000). The ACO algorithm for the RCPSP (EAS-RCPSP), by Merkle, Middendorf, & Schmeck (2002), is currently among the best-performing metaheuristic approaches for the RCPSP.

Construction graph   The construction graph comprises a node for each of the activities to be scheduled plus two dummy nodes corresponding to the (dummy) start activity and the (dummy) end activity.

Constraints   The constraints require that each activity is scheduled, all the precedence constraints among the activities are satisfied, and all resource constraints are met.

Pheromone trails   The pheromone trails τ_{ij} refer to the desirability of scheduling activity j as the i-th activity.

Heuristic information   The heuristic information η_{ij} indicates the heuristic desirability of scheduling activity j as the i-th activity. Best performance in EAS-RCPSP was obtained by using $\eta_{ij} = \max_{l \in \mathcal{N}} LS_l - LS_j + 1$, where LS_j is the latest possible start time of activity j, which can be computed before solving the problem, using an upper bound on the makespan (Elmaghraby, 1977), and $\mathcal{N}$ is the set of activities that are eligible given a partial schedule. The heuristic information is based on a normalized version of the latest start time (nLST) heuristic. The reason for using a normalized version is that the relative differences between the start times become small if only a few activities are eligible in the final phases of the generation of the activity list; this problem is reduced when normalizing the absolute start time values, which is done by using their difference to the maximum latest start time.

Several other possible ways of defining the heuristic information were considered in EAS-RCPSP, but the one based on the nLST heuristic gave the best performance.

Solution construction   The ants in EAS-RCPSP construct a schedule using schedule generation methods that take the resource requirements in the RCPSP into account (see box 5.3 for a general outline of schedule generation methods). The ants start their solution construction by first assigning activity 0, the (dummy) start activity, to position zero. At construction step i, an eligible activity j to be scheduled as the i-th activity is chosen according to the action choice rule of AS as implemented by equation (3.2), where the pheromone trails τ_{ij} are defined according to equation (5.6); that is, EAS-RCPSP uses a combination of the local pheromone evaluation method and the weighted pheromone summation rule [equation (5.5)], as explained in box 5.2.

Two different ways of defining the set of eligible activities $\mathcal{N}$ at each construction step were examined in EAS-RCPSP. The best performance was reported for a *serial schedule generation scheme* (Kolisch & Hartmann, 1999), where $\mathcal{N}$ comprises all the activities that have not been scheduled so far and for which all predecessors have already been scheduled.

Pheromone update   In EAS-RCPSP only the best-found schedule and the iteration-best schedule are allowed to deposit an amount of $(2 \cdot CT)^{-1}$ units of pheromone, where CT is the makespan of the best-so-far or the iteration-best schedule. Before the pheromone deposit, all pheromone trails are lowered by using pheromone evaporation.

Local search   A first-improvement 2-opt local search algorithm that exchanges activities at positions i and j, $i \neq j$, is used.

Particularities   There are several particular features in EAS-RCPSP. First, as already mentioned when describing solution construction, a combination of the standard way of using pheromones and the weighted summation rule is used. Second, the parameter β, which regulates the influence of the heuristic information in the solution construction, and the parameter ρ, which determines the amount of pheromone evaporation, are modified at run time. Third, occasional forgetting of the best-so-far solution is implemented. If the best-so-far ant was not improved for i_{max} iterations, it is replaced by the iteration-best ant. This avoids early convergence to a specific solution. Fourth, forward and backward ants are used. While forward ants construct solutions starting from the start activity, backward ants construct the activity list starting from the end activity and moving on the reversed RCPSP instance (i.e., the same instance with all the precedence constraints reversed). Forward and backward ants use separate pheromone matrices. After a fixed number of iterations (a parameter of the algorithm), the two colonies compare their results and the one with the better results continues the search.

Results   EAS-RCPSP achieved excellent performance. Two different experiments were run to evaluate EAS-RCPSP. The first experimental setup corresponds to the case in which each algorithm is allowed to evaluate a maximum of 5000 schedules. Under these experimental conditions, EAS-RCPSP was compared with a number of other (heuristic) algorithms, including all algorithms tested in Hartmann & Kolisch (2000) and a genetic algorithm presented in Hartmann & Kolisch (1999), and it was found to perform significantly better than the competitors. In these experiments, local search was applied only to the last best-so-far solution, obtained when the ACO algorithm stops. In a second type of experiment, local search was applied to the best solution after every iteration and the ACO algorithm was stopped after 20000 iterations. In this case, for 278 out of 600 test instances, a solution was found that was at least as good as the known best solution and for 130 of the largest instances, a new best solution could be found.

5.3.4 Other ACO Applications to Scheduling Problems

Single-Machine Total Tardiness Problem with Sequence-Dependent Setup Times
Gagné et al. (2002) applied ACS to the single-machine total tardiness problem with sequence-dependent setup times (SMTTPSDST). As in the SMTTP, the objective in the SMTTPSDST is to minimize the total tardiness of the jobs (see section 5.3.1 for ACO approaches to the SMTTP problem and its weighted version). SMTTPSDST differs from SMTTP in that in the former there are sequence-dependent setup times t_{ij}, where t_{ij} is the time that would be required to start the processing of job j if it followed immediately job i in the sequence. The pheromone trails τ_{ij} refer in the SMTTPSDST case to the desirability that job j follows directly job i. A particularity of the ACO approach to the SMTTPSDST is that three types of heuristic information are combined in the solution construction. The three types of heuristic information are based on the setup times, the slack times (the slack time is the maximum between the time span in which a job would be completed before its due date and zero), and a lower bound on the total tardiness for completion of the sequence. Each of the three heuristic information types is then normalized into the interval $[0, 1]$. Additionally, a candidate list is generated; it comprises the *cand* still unscheduled jobs with the smallest slack times, where *cand* is a parameter. The job to be appended is chosen based on the pseudorandom proportional action choice rule of ACS [equation (3.10)]. Once a solution is completed, it is improved by a local search algorithm. Computational results have shown that the proposed ACS algorithm performs better than competitors (Tan & Narashiman, 1997; Rubin & Ragatz, 1995) on the largest available benchmark instances.

Permutation Flow Shop Problem
In the permutation flow shop problem (PFSP) we are given n jobs that have to be processed on m machines following the same order $m_1, \ldots, m_m$. The processing times

are fixed, non-negative, and may be 0 if a job is not processed on some machine. The objective is to find a sequence, that is, a permutation of the numbers $1, \ldots, n$, that minimizes the completion time $CT_{\max}$ of the last job. An application of $\mathcal{MMAS}$ to the PFSP ($\mathcal{MMAS}$-PFSP) was presented by Stützle (1998a). Solutions are constructed as in the SMTWTP case, using the very same interpretation for the pheromone trails, that is, τ_{ij} refers to the desirability of scheduling a job j in position i of the permutation. However, no heuristic information is used. There are two features of this approach that are worth emphasizing. First, the ACO algorithm applied is a hybrid between $\mathcal{MMAS}$ and ACS: from $\mathcal{MMAS}$ it takes the idea of using explicit pheromone trail limits and the pheromone trail update rule (i.e., pheromone trail evaporation and pheromone deposit are applied as usual in $\mathcal{MMAS}$); from ACS it takes the aggressive pseudorandom proportional action choice rule. Second, $\mathcal{MMAS}$-PFSP uses only one ant in each iteration. An experimental evaluation of $\mathcal{MMAS}$-PFSP was done only for short run times. It was found to outperform earlier proposed simulated annealing and tabu search algorithms. However, $\mathcal{MMAS}$-PFSP performance is way behind the efficient tabu search algorithm of Nowicki & Smutnicki (1996b) for the PFSP. Recently, an extension of $\mathcal{MMAS}$-PFSP using the pheromone summation rule (see box 5.2) was proposed by Rajendran & Ziegler (2003), achieving a somewhat better performance, but still not reaching the level of current state-of-the-art algorithms.

5.4   Subset Problems

In subset problems, a solution to the problem under consideration is represented as a subset of the set of available items (components) subject to problem-specific constraints. Obviously, many problems not considered in this section could be interpreted as subset problems. For example, in the TSP a solution may be seen as consisting of a subset of the set of available arcs. However, these problems are often represented, more conveniently, using other representations, such as a permutation of the graph nodes, cities, in the TSP case. In the following, we consider some paradigmatic subset problems that arise in a variety of applications. Despite the interest in these problems, so far there have been only a few attempts to solve subset problems by ACO.

When compared to the applications presented so far, there are two main particularities involved with ACO applications to subset problems. First, in subset problems one is not particularly interested in an ordering of the components. Therefore, the most recently incorporated item need not be necessarily considered when selecting the next item to be added to the solution under construction. As a result, in subset problems pheromone trails are typically associated with components and not with connections. Second, the number of components in the solutions built by different ants may be different, so that the solution construction phase ends only when all the ants have completed their solutions.

5.4.1   Set Covering

In the set covering problem (SCP) we are given an $m \times n$ matrix $A = [a_{ij}]$ in which all the matrix elements are either 0 or 1. Additionally, each column is given a non-negative cost b_j. We say that a column j covers a row i if $a_{ij} = 1$. The goal in the SCP is to choose a subset of the columns of minimal weight that covers every row. Let $\mathcal{J}$ denote a subset of the columns and y_j be a binary variable which is 1, if $j \in \mathcal{J}$, and 0 otherwise. The SCP can be defined formally as follows:

$$\min f(y) = \sum_{j=1}^{n} b_j \cdot y_j, \tag{5.9}$$

subject to

$$\sum_{j=1}^{n} a_{ij} \cdot y_j \geq 1, \quad i = 1\ldots, m, \tag{5.10}$$

$$y_j \in \{0, 1\}, \quad j = 1, \ldots, n, \tag{5.11}$$

where the constraints given by equation (5.10) enforce that each row is covered by at least one column.

At least two ACO approaches to the SCP have been proposed: one by Leguiza-món & Michalewicz (2000), which we refer to as AS-SCP-LM, and another by Hadji et al. (2000), which we refer to as AS-SCP-HRTB; the two approaches are similar in many respects.

Construction graph   The set C of components of the construction graph comprises the set of columns plus a dummy node on which all ants are put in the first construction step. The construction graph is, as usual, fully connected. Solutions are subsets of the set of components.

Constraints   The constraints say that each component can be visited by an ant at most once and that a solution has to cover all the rows.

Pheromone trails   Pheromone trails are associated with components; the pheromone trail τ_j associated with component j measures the desirability of including component j in a solution.

Heuristic information   In both ACO algorithms (Leguizamón & Michalewicz, 2000; Hadji et al., 2000), the heuristic information is computed as a function of an ant's partial solution. Let e_j be the *cover value* of column j, that is, the number of additional rows covered when adding column j to the current partial solution. Then, b_j/e_j gives the cost per additional covered row when adding column j to the solution under construction. The heuristic information η_j is given by $\eta_j = e_j/b_j$.

Solution construction   In AS-SCP-LM each ant starts with an empty solution and constructs a solution by iteratively adding components until all rows are covered. The action choice rule for choosing the next component is an adaptation of the probabilistic action choice rule of AS [equation (3.2)] to the case in which pheromones are assigned to components only. A component is then chosen with probability

$$p_i^k(t) = \frac{\tau_i [\eta_i]^\beta}{\sum_{l \in \mathcal{N}^k} \tau_l [\eta_l]^\beta}, \quad \text{if } i \in \mathcal{N}^k, \tag{5.12}$$

where $\mathcal{N}^k$ is the feasible neighborhood of ant k before adding component i; $\mathcal{N}^k$ consists of all columns that cover at least one still uncovered row. An ant has completed a solution when all rows are covered. AS-SCP-HRTB uses essentially the same way of constructing solutions with the only difference that in a first stage an ant adds l randomly chosen components, where l is a parameter; this is done to increase diversification during solution construction. A further difference between AS-SCP-LM and AS-SCP-HRTB is that in the latter approach a postoptimization step is applied, in which each ant removes redundant columns—columns that only cover rows which are also covered by a subset of other columns in the final solution.

Pheromone update   In both AS-SCP-LM and AS-SCP-HRTB, the pheromone update rule used is an adaptation of the AS rule to the case in which pheromones are assigned to components. Hence, equation (3.2), becomes

$$\tau_i \leftarrow (1 - \rho)\tau_i, \quad \forall i \in C, \tag{5.13}$$

and the pheromone update rule of equation (3.3), becomes

$$\tau_i \leftarrow \tau_i + \sum_{k=1}^{m} \Delta \tau_i^k(t). \tag{5.14}$$

In AS-SCP-HRTB the value $\Delta \tau_i^k(t)$ is set to $1/f(s_k)$, where $f(s_k)$ is the objective function value of ant's k solution s_k, if the component i is an element of s_k, and 0 otherwise. AS-SCP-LM uses the same update rule with the only difference that the amount of pheromone deposited is multiplied by the sum of the costs of all columns in the problem definition.

Local search   AS-SCP-HRTB applies a local search to the best solution constructed in the last iteration of the algorithm. This local search procedure follows the local search scheme of Jacobs & Brusco (1995).

Results   The computational results obtained with AS-SCP-LM and AS-SCP-HRTB are good, but they do not compete with current state-of-the-art heuristics for the SCP such as those published in Caprara, Fischetti, & Toth (1999) and in Marchiori & Steenbeek (2000).

Remarks   Recently, also, the related set partitioning problem (SPP) was attacked by an ACO algorithm (Maniezzo & Milandri, 2002). The SPP is the same problem as the SCP, except for the constraints in equation (5.10) that are replaced by equality constraints. However, the SPP appears to be a significantly harder problem than the SCP, because finding a feasible solution is already $\mathcal{N}\mathcal{P}$-hard.

5.4.2   Weight Constrained Graph Tree Partition Problem

In the weight constrained graph tree partition problem (WCGTPP) we are given an undirected graph $G(N, A)$ of n nodes and l arcs. Each arc (i, j) is assigned a cost c_{ij} and each node i is assigned a weight w_i. The goal is to find a minimum cost spanning forest F of p trees such that the weight of each tree falls in a given range $[W^-, W^+]$. The WCGTPP, a problem that arises in the design of telecommunications networks, is $\mathcal{N}\mathcal{P}$-hard and, in general, not approximable (Cordone & Maffioli, 2003). Cordone & Maffioli (2001) attacked the WCGTPP using an ACO algorithm based on ACS (ACS-WCGTPP). Their algorithm has the particularity that it employs pheromone trails of p different "colors," one color for each of the p trees.

Construction graph   There is one component for each node and, as usual, the construction graph is fully connected.

Pheromone trails   For each node j we have p pheromone trails τ_{ij}. The value τ_{ij} indicates the desirability of including node j in tree T_i.

Heuristic information   The heuristic information is computed by considering the minimum cost $\bar{c}_{ij}$ of adding a node j to a tree T_i, that is, $\bar{c}_{ij} = \min_{v \in T_i} c_{vj}$. In addition, a penalty is computed as

$$\pi_{ij} = 1 + \pi \cdot \max\left\{\frac{w_{T_i} + w_j - W^+}{w_{max} - W^+}, \frac{W^- - w_{T_i} - w_j}{W^- - w_{min}}, 0\right\},$$

where w_{max} is the weight of the heaviest possible tree, that is, the tree that spans the whole graph except for the $p - 1$ nodes that have the $p - 1$ lowest weights associated. The value w_{T_i} is the weight of tree T_i and w_{min} is the weight of the lightest tree, that is, the lightest node in G. The value π is a penalty factor that is adjusted at run time (see "Particularities" below). The penalty ranges from 1 (corresponding to a feasible move) to $1 + \pi$ (for the most unbalanced assignment). The heuristic information is set to $\eta_{ij} = 1/(\bar{c}_{ij}\pi_{ij})$.

Constraints The constraints impose that every node has to be assigned to exactly one tree, that is, the trees form a partition of the set of nodes. In addition, each tree has to obey some lower and upper weight limits.

Solution construction Each ant produces a full solution which corresponds to a partition of the graph into p trees. Ant k is initialized with one root node for each tree. Then, at each construction step, one node is added to one of the trees using the pseudorandom proportional rule of ACS [equation (3.10)]. With probability q_0 ant k adds to the tree T_i the node j that maximizes the value $\tau_{ij} \cdot \eta_{ij}$, while with probability $1 - q_0$ the random proportional action choice rule of AS [equation (3.2)] is applied. Hence, at each construction step a simultaneous choice is made about the node to add and the tree to which this node is added.

Pheromone update ACS-WCGTPP uses the local and global pheromone update rules of ACS [equations (3.12) and (3.11), respectively]. The value τ_0 is set to $1/(n \cdot c_{MSF})$, where c_{MSF} is the cost of a minimum spanning forest. The amount of pheromone to be deposited in the global pheromone update rule is determined by $1/(c_{F^{gb}}\pi_{F^{gb}})$, where $c_{F^{gb}}$ and $\pi_{F^{gb}}$ are, respectively, the cost and the penalty associated with F^{gb}, the best spanning forest found so far.

Local search ACS-WCGTPP applies a local search procedure that is called *swinging forest*. This procedure starts by considering moves of one node of a tree to a different tree (neighborhood $\mathcal{N}_1$) and exchanges of pairs of nodes contained in different trees (neighborhood $\mathcal{N}_2$). If a local optimum is reached with respect to $\mathcal{N}_1$ and $\mathcal{N}_2$, additional moves based on splitting and removing trees are considered.

Particularities Apart from using more than one pheromone trail, ACS-WCGTPP modifies penalty factors at computation time as a function of the number of feasible solutions constructed. Let m_f be the number of feasible solutions among the m constructed ones. Then, π is adjusted by

$$\pi \leftarrow \pi \cdot 2^{(m-2m_f)/m}.$$

The effect of this adjustment is either to decrease the penalty factor if many solutions are feasible or to increase the penalty factor if many solutions are infeasible. In the first case, it becomes easier to build infeasible solutions and, therefore, to move to different regions of the search space containing feasible solutions. In the second case, the search is more strongly directed toward feasible solutions.

Of significant importance in the solution process is the choice of the root nodes. In the first iteration, they are chosen so that they cover the graph as uniformly as possible. Then the greedy heuristic is applied to generate a spanning forest and the roots move to the centroids of the trees. This is repeated until the roots stabilize or repeat cyclically. This process is performed $n = |N|$ times using each node as a seed. The best root assignment overall is used to initialize ACS-WCGTPP and only changes when a new best-so-far solution is found.

Results   The WCGTPP was not discussed before in the literature. Therefore, experimental comparisons were limited to different variants of the proposed algorithm. ACS-WCGTPP was tested with and without local search. Three variants using local search were considered: (1) a variant in which the swinging forest procedure was applied only to the best solution returned by the ACS-WCGTPP algorithm; (2) a variant in which the iteration-best solutions were improved by applying a local search in $\mathcal{N}_1$ plus the swinging forest procedure; and (3) a last variant, which is the same as the second one, except that the iteration-best solutions are improved by a local search using neighborhood $\mathcal{N}_1 \cup \mathcal{N}_2$. Overall, the best results with respect to solution quality were obtained by the third variant, which makes the strongest use of the local search.

5.4.3   Arc-Weighted *l*-Cardinality Tree Problem

The *l*-cardinality tree problem is a generalization of the minimum spanning tree problem, which consists in finding a subtree with exactly *l* arcs in a graph $G = (N, A)$ with arc or node weights, such that the sum of the weights is minimal. This problem has a variety of applications such as oil-field leasing (Hamacher & Jörnsten, 1993), facility layout problems (Foulds, Hamacher, & Wilson, 1998), or matrix decomposition (Borndörfer, Ferreira, & Martin, 1998b).

In the following, we focus on the arc-weighted *l*-cardinality tree problem (AW*l*CTP), in which weights are associated only with arcs. The AW*l*CTP is known to be $\mathcal{NP}$-hard (Fischetti et al., 1994; Marathe, Ravi, Ravi, Rosenkrantz, & Sundaram, 1996). Blum & Blesa (2003) proposed an $\mathcal{MMAS}$ algorithm for the AW*l*CTP ($\mathcal{MMAS}$-AW*l*CTP), implemented within the hyper-cube framework for ACO. Their algorithm was compared with a tabu search algorithm and an evolutionary algorithm.

Construction graph   The problem components are the arcs, and the construction graph is fully connected as usual.

Pheromone trails   Each arc (component) (i, j) is assigned a pheromone trail τ_{ij}. Hence, there are $l = |A|$ pheromone trails.

Heuristic information   The heuristic information is based on the inverse of the arc weights, that is, $\eta_{ij} = 1/w_{ij}$, where w_{ij} is the weight of arc (i, j).

Constraints   The only constraints are that a tree with exactly l arcs needs to be obtained, that is, a connected subgraph of G with l arcs and $l + 1$ nodes.

Solution construction   Initially, an arc is chosen randomly with a generic arc (i, j) having a probability of being chosen given by $\tau_{ij}/\sum_{(i,\sigma) \in A} \tau_{i\sigma}$. Let x_h be the h-cardinality tree constructed after h construction steps; we have $x_1 = \langle (i, j) \rangle$. Let $\mathcal{N}(x_h)$ be the set of arcs that (1) are not yet in x_h and (2) that have exactly one endpoint in common with some arc in x_h. At construction step $h + 1$, $h + 1 \leq l$, an ant k chooses an arc of $\mathcal{N}(x_h)$ using the pseudorandom proportional action choice rule of ACS [equation (3.10)], that is, with probability $q_0 = 0.8$ it chooses the arc $(i, j) \in \mathcal{N}(x_h)$ that maximizes the ratio τ_{ij}/w_{ij}; otherwise a random proportional probabilistic choice is made.

Pheromone update   The pheromone update uses the update rules of $\mathcal{MMAS}$, adapted to the hyper-cube framework. The choice of which ant is allowed to deposit pheromone is made between the iteration-best, the best-so-far, and the restart-best ant, as a function of a *convergence factor*. The update rule is rather flexible in the sense that it allows for interpolations between the iteration-best and the restart-best solution, by giving weights to these solutions in the pheromone deposit.

Local search   Each solution is improved using a best improvement local search based on a neighborhood structure that comprises all neighboring solutions that can be obtained by deleting one arc of an l-cardinality tree, resulting in an $l - 1$-cardinality tree x_{l-1}, and adding a different arc taken of the set $\mathcal{N}(x_{l-1})$. The iteration-best solution was further locally optimized by applying $2l$ iterations of a tabu search algorithm.

Results $\mathcal{MMAS}$-AWlCTP was compared to a tabu search and an evolutionary algorithm on a large set of benchmark instances for the AWlCTP. In general, none of the algorithms dominated any of the others on the whole range of benchmark instances. However, $\mathcal{MMAS}$-AWlCTP was shown to perform better than the two competitors, when l is smaller than 60% of the number of nodes (the largest possible cardinality of a tree is $n - 1$ for a graph with n nodes), while for larger cardinalities it was inferior to the tabu search and the evolutionary algorithms. For a detailed summary of the computational results, see Blum & Blesa (2003).

5.4.4   Other ACO Applications to Subset Problems

Multiple Knapsack Problem

Recall from section 2.3.4 of chapter 2, that in the multiple knapsack problem we are given a set I of items and a set R of resources. To each item $i \in I$ is associated a profit b_i and a requirement r_{ij} for resource $j \in R$. The goal is to find a subset of the items that maximizes the profit and that meets all the constraints. Constraints are given by limits a_j on the availability of resources j. AS was applied to the MKP by Leguiza-món & Michalewicz (1999) (AS-MKP). Their approach mainly follows the steps already outlined in section 2.3.4, the main difference being the way they compute the heuristic information. AS-MKP uses a dynamic heuristic information, which relates the average consumption of the remaining amount of resources by component i to the profit of adding component i. More concretely, the heuristic information for AS-MKP is defined as follows. Let $s_k(t)$ be the (partial) solution of ant k at construction step t. Then, $u_j(k, t) = \sum_{l \in s_k(t)} r_{lj}$ is the amount of resource j that is consumed at step t by ant k and $v_j(k, t) = a_j - u_j(k, t)$ is the remaining amount of resource j. The value $v_j(k, t)$ is used to define the tightness $w_{ij}(k, t)$ of a component i with respect to resource j: $w_{ij}(k, t) = r_{ij}/v_j(k, t)$. Finally, the average tightness of all constraints with respect to component i is computed as $\bar{w}_i(k, t) = \sum_j w_{ij}(k, t)/l$, where l is the number of resource constraints. The lower this value, the less critical it is to add component i with respect to the resource consumption. Finally, taking into account the profit p_i of adding a particular component i, AS-MKP defines the heuristic information as

$$\eta_i(s_k(t)) = \frac{p_i}{\bar{w}_i(k, t)}.$$

AS-MKP was tested on MKP instances from ORLIB (mscmga.ms.ic.ac.uk/jeb/orlib/mknapinfo.html) and compared favorably to an earlier evolutionary algorithm. However, as was the case for the set covering problem, AS-MKP did not reach state-of-the-art results (such as those obtained by the algorithm of Vasquez & Hao, 2001). The inclusion of a local search might significantly improve the quality of the results obtained.

Maximum Independent Set Problem

In a similar vein, Leguizamón & Michalewicz (1999) attacked the maximum independent set problem (MIS). Given a graph $G = (N, A)$, the objective in MIS is to find a largest subset of nodes such that none of the nodes are connected by an arc. More formally, a subset $N' \subseteq N$ must be found such that for all $i, j \in N'$ it holds that $(i, j) \notin A$ and $|N'|$ is maximal. The main adaptation necessary to extend AS-MKP to

MIS is the heuristic information to be used. The heuristic information chosen by Leguizamón and Michalewicz is $\eta_i = |\mathcal{N}_i^k|$, where $\mathcal{N}_i^k$ is the set of components that can be added to a partial solution s_k after component i is added to s_k. Comparisons to an evolutionary algorithm and a GRASP algorithm showed promising results (Leguizamón, Michalewicz, & Schütz, 2001).

Maximum Clique Problem

Given an undirected graph $G = (N, A)$ with N being the set of nodes and A being the set of arcs, a clique is a subset $N' \subseteq N$ of the set of nodes such that there exists an arc for every pair of distinct nodes in N', that is, for all $i, j \in N'$, $i \neq j$ it holds that $(i, j) \in A$. The goal of the maximum clique problem (MCP) is to find a set N' of maximum size. Fenet & Solnon (2003) have applied $\mathcal{MMAS}$ to the MCP ($\mathcal{MMAS}$-MCP). The pheromone trails τ_{ij} in $\mathcal{MMAS}$-MCP refer to the desirability of assigning nodes i and j to a same clique. Cliques are constructed by first choosing a random node and then, at each step, adding a node that is connected to all the nodes of the clique under construction. The probability of choosing a node i is proportional to $\sum_{j \in N'} \tau_{ij}$, that is, to the sum of the pheromone trails between node i and the nodes already in the partial solution N'. $\mathcal{MMAS}$-MCP does not use any heuristic information, as its use appeared to yield worse performance for long runs. Solution construction ends when there are no more nodes to be added that are connected to all the nodes in N'. Finally, pheromone is deposited on all the arcs of the clique found by the iteration-best ant. $\mathcal{MMAS}$-MCP's performance was compared in detail with the best results obtained by Marchiori (2002) for genetic local search and iterated local search algorithms. The results were that $\mathcal{MMAS}$-MCP found on many benchmark problems better-quality solutions, but, typically, at a higher computational cost. The performance of $\mathcal{MMAS}$-MCP is still way behind the current best algorithm for the MCP, a reactive local search by Battiti & Protasi (2001), but further improvements, like the use of an effective local search, should greatly enhance the results obtainable by $\mathcal{MMAS}$-MCP.

Redundancy Allocation Problem

Liang & Smith (1999) attacked the problem of maximizing the reliability of a system. In particular, they considered a system that is composed of a series of subsystems, where each of the subsystems functions if k out of n components are working; the overall system fails if one of the subsystems fails. The goal then is to assign to each subsystem a set of components from the available ones for each subsystem (each component has an associated reliability, a cost, and a weight), such that the overall system reliability is maximized subject to cost and weight constraints. The ACO algorithm used for this problem, EAS-RA, is based on elitist Ant System (see chapter

3, section 3.3.2). In EAS-RA, solutions to each subsystem are constructed independently of the others. A pheromone trail τ_{ij} refers to the desirability of adding a component j to subsystem i. For the solution construction no heuristic information is taken into account. One important particularity of EAS-RA is that it uses an adaptive penalty function for penalizing violations of the cost and weight constraints. The penalty function is adaptive because with increasing run time of the algorithm, the penalization of a fixed amount of constraint violation increases. The computational results indicate that EAS-RA can find a good system design.

5.5   Application of ACO to Other $\mathcal{NP}$-Hard Problems

In this section we present some additional applications of ACO that do not fit in the previous section but which present some particularly interesting features not present in other ACO implementations.

5.5.1   Shortest Common Supersequence Problem

Given a set L of strings over an alphabet Σ, the shortest common supersequence problem (SCSP) consists in finding a string of minimal length that is a supersequence of each string in L. The string B is a supersequence of a string A if A can be obtained from B by deleting in B zero or more characters. Consider, for example, the set $L = \{bbbaaa, bbaaab, cbaab, cbaaa\}$ over the alphabet $\Sigma = \{a, b, c\}$. A shortest supersequence for L is $cbbbaaab$ (see figure 5.3). The SCSP is an $\mathcal{NP}$-hard problem with applications in DNA analysis or in the design of conveyor belt workstations in machine production processes. Michel & Middendorf (1998, 1999) developed an ACO algorithm for the SCSP (AS-SCSP), which showed very promising results when compared to a number of alternative approaches.

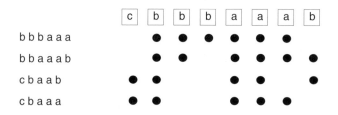

Figure 5.3
Graphical representation of a supersequence (given as the string *cbbbaaab* in the boxes, first line) and of the four strings *bbbaaa, bbaaab, cbaab, cbaaa* from which it is generated. Each bullet indicates the characters in the strings that are covered by the supersequence.

Construction graph   The character at position j in string L_i is denoted by s_{ij}. The components of the construction graph are the s_{ij}'s. The graph is fully connected.

Constraints   The constraints enforce that a true supersequence of the strings in L has to be built. These constraints are implicitly enforced through the construction policy used by the ants.

Pheromone trails   A pheromone trail τ_{ij} is associated with each component s_{ij}. It gives the desirability of choosing character s_{ij} when building the supersequence. All pheromone trails are initialized to one.

Heuristic information   AS-SCSP does not use any heuristic information. However, it uses lookahead in its place, as explained in "Particularities" below.

Solution construction   The construction policy of the ants can be informally described as follows. Ants build supersequences of the strings in L, independently of each other. Each ant k receives a copy of the original set of strings L and initializes its supersequence to the empty string. At the first construction step, ant k adds to the supersequence it is building a character that occurs at the front of at least one string $L_i \in L$ (i.e., it chooses at least one component s_{i1}). The choice of the character to add is based on pheromone trails, as well as on some additional information, as explained in "Particularities" below. Once a character is added, the same character is removed from the front of the strings on which it occurred. Then the procedure is reapplied to the modified set of strings L', until all characters have been removed from all strings and the set L' consists of empty strings.

To describe the solution construction procedure more formally, we need to define for each ant k an indicator vector $v^k = (v_1^k, \ldots, v_l^k)$, with $l = |L|$. Element v_i^k of v^k points to the "front" position $s_{iv_i^k}$ of string L_i—the position that contains the character that is a candidate for inclusion in the supersequence.

Consider, for example, the set $L = \{bbbaaa, bbaaab, cbaab, cbaaa\}$. In this case, the vector $v^k = (2, 2, 3, 3)$ represents a situation in which the first character of the first and second strings, as well as the first two characters of the third and fourth strings, is already included in a supersequence. The characters that are candidates for inclusion in the supersequence are therefore a and b. In fact, $s_{1v_1^k} = s_{12} = b$, $s_{2v_2^k} = s_{22} = b$, $s_{3v_3^k} = s_{33} = a$, and $s_{4v_4^k} = s_{43} = a$.

At the beginning of solution construction, the vector v^k is initialized to $v^k = (1, \ldots, 1)$. The solution construction procedure is completed once the indicator vector has reached the value $v^k = (|L_1| + 1, \ldots, |L_l| + 1)$. As we said above, at each construction step the feasible neighborhood $\mathcal{N}_{v^k}^k$, that is, the set of characters that can be appended to the supersequence under construction, is composed of the characters occurring at the positions pointed by the indicator vector: $\mathcal{N}_{v^k}^k = \{x \in \Sigma \mid \exists i$

such that $x = s_{iv_i^k}$}. The choice of which character in $x \in \mathcal{N}_{v^k}^k$ to append to the super-sequence is done according to the pseudorandom proportional action choice rule of ACS [equation (3.10)], using as the pheromone trail value the sum of the pheromone trails of all the occurrences of x in the l strings:

$$\sum_{i:s_{iv_i^k}=x} \tau_{iv_i^k}. \tag{5.15}$$

Finally, the indicator vector is updated. That is, for $i = 1$ to l:

$$v_i^k = \begin{cases} v_i^k + 1, & \text{if } s_{iv_i^k} = x; \\ v_i^k, & \text{otherwise;} \end{cases} \tag{5.16}$$

where x is the character appended to the supersequence.

Note that equation (5.15) gives a pheromone amount to character x that is the larger (1) the higher is the amount of pheromone on components s_{ij} for which it holds $s_{ij} = s_{iv_i^k} = x$, and (2) the higher is the number of times character x occurs at the current front of the strings in L (i.e., the larger the cardinality of the set $\{s_{ij} \,|\, s_{ij} = s_{iv_i^k} = x\}$). This latter rule (2) reflects the majority merge heuristic that at each step chooses the character that occurs most often at the front of strings (Foulser, Li, & Yang, 1992). Michel and Middendorf considered an additional vari-ant that weighs each pheromone trail with $|L_i| - v_i^k + 1$, giving higher weight to characters occurring in strings in which many characters still need to be matched by the supersequence. This latter choice is inspired by the L-majority merge (LM) heu-ristic that weighs each character with the length of the string L_i and then selects the character with the largest sum of weights of all its occurrences (Branke, Middendorf, & Schneider, 1998). In fact, this latter variant was found to improve the algorithm's performance and was therefore retained in the final version of Michel and Mid-dendorf's ACO algorithm for the SCSP.

Pheromone update   The amount of pheromone ant k deposits is given by

$$\Delta \tau^k = \frac{g(r^k)}{|s^k|},$$

where r^k is the rank of ant k after ordering the ants according to the quality of their solutions in the current iteration, g is some function of the rank, and s^k is the solu-tion built by ant k.

Pheromones are updated as follows. First, a vector $z^k = (z_1^k, \ldots, z_l^k)$, with $l = |L|$, analogous to the one used for solution construction, is defined and initialized to

$z^k = (1, \ldots, 1)$. Element z_i^k of z^k points to the character in string L_i that is a candidate for receiving pheromone. Then, s^k, the supersequence built by ant k, is scanned from the first to the last position. Let x_h denote the h-th character in s^k, $h = 1, \ldots, |s^k|$. At each step of the scanning procedure, first is memorized the set $\mathcal{M}_h^k = \{s_{iz_i^k} \mid s_{iz_i^k} = x_h\}$ of elements in the strings belonging to L that are pointed by the indicator vector and whose value is equal to x_h; note that, by construction of the supersequence, at least one such character x_h exists. Next, the indicator vector is updated. That is, for $i = 1$ to l:

$$z_i^k = \begin{cases} z_i^k + 1, & \text{if } s_{iz_i^k} = x_h; \\ z_i^k, & \text{otherwise.} \end{cases} \tag{5.17}$$

Once the supersequence has been entirely scanned, the amount of pheromone to be deposited by ant k on each visited component can be determined as follows. For each character x_h the amount of pheromone deposited by ant k on component $s_{ij} \in \mathcal{M}_h^k$ is given by

$$\Delta \tau_{ij}^k = \frac{\Delta \tau^k}{|\mathcal{M}_h^k|} \cdot \frac{2(|s_k| - h + 1)}{|s_k|^2 + |s_k|}. \tag{5.18}$$

The left term of the right-hand side of equation (5.18) says that the pheromone for the h-th character in s_k is distributed equally among the components of the strings in L, if x_h occurred in more than one string. The right term of the right-hand side of equation (5.18) is a scaling factor that ensures that the overall sum of pheromones deposited by ant k is equal to $\Delta \tau^k$; additionally, this scaling factor ensures that the earlier a character occurs in s_k, the larger the amount of pheromone it receives. Hence, each character of a string receives an amount of pheromone that depends on how early the character was chosen in the construction process of ant k, how good ant k's solution is, and the number of strings from which the character was chosen in the same construction step.

Once all pheromone trails are evaporated and the above computations are done, the contributions of all ants to the characters' pheromone trails are summed and added to the pheromone trail matrix.

Local search   AS-SCSP does not make use of local search.

Particularities   AS-SCSP differs from standard ACO implementations in at least three main aspects, as explained below.

- First, it uses an optional lookahead function that takes into account the "quality" of the partial solutions (i.e., partial supersequences) that can be reached in the following construction steps. To do so, AS-SCSP tentatively adds the character x to the supersequence and generates the vector $\bar{v}^k$ that is obtained by the tentative addition of x. Then, the maximum amount of pheromone on any of the characters pointed by $\bar{v}^k$ is determined. In AS-SCSP, the lookahead function plays the role of the heuristic information and is be therefore indicated by η, as usual. It is defined by

$$\eta(x, \bar{v}^k) = \max\{\tau_{ij}^k \mid s_{ij} = s_{i\bar{v}_i^k} = x\}.$$

As usual with heuristic information, the value $\eta(x, \bar{v}^k)$ is weighted by an exponent β when using the pseudorandom proportional action choice rule. It is also possible to extend the one-step lookahead to deeper levels.

- Second, at each step of the solution construction procedure, an ant, although adding a single character to the supersequence under construction, can visit in parallel more than one component s_{ij} of the construction graph.

- Last, the AS-SCSP algorithm has been implemented as a parallel algorithm, based on an island model approach, where the whole ant colony is divided into several subpopulations that occasionally exchange solutions. In this approach, two different types of colonies are considered: *forward* and *backward* colonies. The backward colony works on the set $\hat{L}$, obtained from the set L by reversing the order of the strings.

Results   Three variants of AS-SCSP, which used (1) no lookahead, (2) lookahead of depth one, and (3) lookahead of depth two, were compared with the LM heuristic using the same three levels of lookahead and with a genetic algorithm designed for the SCSP (GA-SCSP) (Branke et al., 1998). The comparison was done on three classes of instances: (1) randomly generated strings, (2) strings that are similar to the type of strings arising in a variety of applications, and (3) several special cases, which are known to be hard for the LM heuristic. The computational results showed that the AS-SCSP variants, when compared to LM variants or GA-SCSP, performed particularly well on the instances of classes (2) and (3). The addition of the lookahead proved to increase strongly the solution quality for all instance classes, at the cost of additional computation time. The addition of backward colonies gave substantial improvements only for the special strings. Overall, AS-SCSP proved to be one of the best-performing heuristics for the SCSP; this is true, in particular, for structured instances, which occur in real-world applications.

5.5.2   Bin Packing

In the bin-packing problem (BPP) one is given a set of n items, each item i having a fixed weight w_i, $0 < w_i < W$, and a number of bins of a fixed capacity W. The goal in the BPP is to pack the items in as few bins as possible. The BPP is $\mathcal{NP}$-hard

(Garey & Johnson, 1979) and is an intensively studied problem in the area of approximation algorithms (Coffman, Garey, & Johnson, 1997). Levine & Ducatelle (2003) adapted $\mathcal{MMAS}$ to its solution ($\mathcal{MMAS}$-BPP).

Construction graph   The fully connected construction graph includes one node for each item and each bin used by the algorithm; since the number of bins is not determined before constructing a solution, the number of bins can be set to be the same as the number of items or to some upper bound on the optimal solution. Each node representing an item is assigned the weight w_i of the item and each node representing a bin is assigned its capacity W.

Constraints   The constraints in the BPP say that all items have to be assigned to some bin and that the capacity of any bin cannot be exceeded.

Pheromone trails   In general, in the BPP the pheromone trails τ_{ij} encode the desirability of having item i in a same bin with item j. However, because in the BPP only relatively few different item weights may occur when compared to the number of items, items of a same weight w_i may be grouped in a set W_i. The pheromone representation used by $\mathcal{MMAS}$-BPP takes advantage of this possibility. Let the weight of items in W_i be denoted by $\hat{w}_i$ and $\{\hat{w}_1, \ldots, \hat{w}_c\}$ be the set of the different weights of the items. The pheromone trails τ_{ij} encode the desirability of having an item of weight $\hat{w}_i$ in a same bin with an item of weight $\hat{w}_j$. This representation takes into account the fact that the essential information in the BPP concerns associations between item weights, that is, whether an item of weight $\hat{w}_i$ was packed together with an item of weight $\hat{w}_j$ in one bin, while it is not important which particular item of weight $\hat{w}_i$ was in a bin with some particular item of weight $\hat{w}_j$.

Heuristic information   In $\mathcal{MMAS}$-BPP the bins are filled one after the other, and the heuristic information for an item i is set to $\eta_i = w_i$. The heuristic information is inspired by the first-fit decreasing rule that first orders items according to non-increasing weight and then, starting with the heaviest, places the items into the first bin in which they fit.

Solution construction   Each ant is initialized with a list of all the items to be placed and one empty bin. It starts by filling the available bin and, once none of the remaining items can be added to the bin, it starts again with a new, empty bin and the remaining items. The procedure is then iterated until all items are allocated to bins. The probability that an ant k places an item of weight $\hat{w}_j$ into the current bin b is

$$p_{jb}^k(t) = \frac{\tau_j[\eta_j]^\beta}{\sum_{h \in \mathcal{N}_b^k}(\tau_h[\eta_h]^\beta)}, \quad \text{if } j \in \mathcal{N}_b^k, \tag{5.19}$$

and 0 otherwise. The feasible neighborhood $\mathcal{N}_b^k$ of ant k for bin b consists of all those items that still fit into bin b, while the value τ_j is given by

$$
\tau_j = \begin{cases} \dfrac{\sum_{i \in b} \tau_{ij}}{|b|}, & \text{if } b \neq \varnothing; \\[2mm] 1, & \text{otherwise.} \end{cases} \tag{5.20}
$$

In other words, τ_j is the average of the pheromone values between the item to be added, of weight $\hat{w}_j$, and the items that are already in the bin b; if the bin is still empty, then τ_j is set to 1.

Pheromone update   The pheromone update follows the update rule of the $\mathcal{MMAS}$ algorithm, slightly modified because of the nature of BPP. In particular, τ_{ij} is increased every time items i and j are put in a same bin, resulting in the following update rule:

$$
\tau_{ij} \leftarrow (1 - \rho)\tau_{ij} + m_{ij}f(s^{bs}), \tag{5.21}
$$

where m_{ij} indicates how often items i and j are in a same bin, and s^{bs} is the best-so-far solution. The amount of pheromone deposited is given by

$$
f(s) = \frac{\sum_{i=1}^{N}(F_i/W)^\phi}{N}, \tag{5.22}
$$

where N is the number of bins, F_i is the total weight of the items in bin i, and W is the bin capacity (hence, the factor F_i/W is the *relative fullness* of bin i). The value ϕ is a parameter that determines the relative weight of the relative fullness of the bins and the number of bins. Following Falkenauer (1996), Levine and Ducatelle set $\phi = 2$. The use of this evaluation function allows better differentiation between the various solutions than simply counting the number of bins. In fact, typically, a very large number of solutions use the same number of bins.

Local search   Levine and Ducatelle implemented a local search algorithm that first deletes the l least-filled bins, where l is an empirically determined parameter. Next, for each bin the local search tests whether one or two items in the bin can be exchanged for items currently not allocated, making the bin fuller. Once such an exchange is no longer possible, the remaining items are distributed using the first-fit decreasing heuristic and the local search is repeated. For more details on the local search, see Levine & Ducatelle (2003).

Results $\mathcal{MMAS}$-BPP was applied to randomly generated instances ranging in size from 120 to 8000 items, where each bin has a maximum capacity of 150 and the item weights are generated according to a uniform random distribution over the set $\{20, 21, \ldots, 99, 100\}$. $\mathcal{MMAS}$-BPP was compared with the hybrid grouping genetic algorithm (HGGA) by Falkenauer (1996), which was earlier shown to be among the best-performing algorithms for the BPP, and with Martello and Toth's reduction method (MTP) (Martello & Toth, 1990). The computational results showed that $\mathcal{MMAS}$-BPP performs significantly better than both. In particular, $\mathcal{MMAS}$-BPP obtained much better solution quality than MTP in shorter time, and could obtain slightly better solution quality than HGGA, but with a much lower computation time. $\mathcal{MMAS}$-BPP was also applied to solve a collection of difficult BPP instances provided by Wäscher & Gau (1996): for five of these instances it could find new best-known solutions.

5.5.3 2D-HP Protein Folding

A central problem in bioinformatics is the prediction of a protein's structure based on its amino acid sequence. Protein structures can be determined by rather time-consuming, expensive techniques like magnetic resonance imaging or X-ray crystallography, which require additional preprocessing like the isolation, purification, and crystallization of the protein. Therefore, the prediction of protein structure by algorithmic means is very attractive. However, an accurate algorithmic prediction of a protein structure is difficult because of the requirement of good measures for the quality of candidate structures and of effective optimization techniques for finding optimal structures. Therefore, often simplified models for protein folding are studied.

One such simplified model is the two-dimensional (2D) hydrophobic–polar protein folding problem (2D-HP-PFP) introduced by Lau & Dill (1989). The 2D-HP-PFP problem focuses on the hydrophobic interaction process in protein folding by representing an amino acid sequence by the pattern of hydrophobicity in the sequence. In particular, a sequence can be written as $\{H, P\}^{+}$, where H stands for a hydrophobic amino acid and P represents polar amino acids (polar amino acids are classified as hydrophilic). For convenience, the conformations into which this sequence can fold are restricted to self-avoiding paths on a lattice (a self-avoiding path is a path without intersections); in the 2D-HP-PFP model, a 2D square lattice is considered.

The objective function value $f(c)$ of a conformation c is defined to be the number of hydrophobic amino acids that are adjacent on the lattice and not consecutive in the sequence. The goal then becomes, given an amino acid sequence $s = s_1 s_2 \ldots s_n$ with each $s_i \in \{H, P\}$, to find a conformation that maximizes the objective function. It is known that this problem is $\mathcal{NP}$-hard for square (Crescenzi, Goldman, Papadimitriou, Piccolboni, & Yannakakis, 1998) and cubic lattices (Berger & Leight, 1998).

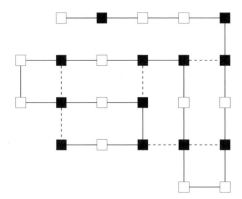

Figure 5.4
Given is a sample conformation of a protein for the 2D-HP-PFP problem on a two-dimensional lattice. White squares represent polar amino acids (P) and black squares represent a hydrophobic amino acid (H). The protein sequence $s = \mathtt{PHPPHHPHPPHPHHPHPPHPHHPH}$ is obtained starting at the leftmost white square in the upper row. The solid lines connect consecutive amino acids in the sequence and the dotted lines represent the adjacent positions of the hydrophobic amino acids.

Figure 5.4 gives an example of conformation of a protein sequence which has an objective function value of 6.

A first ACO algorithm for the 2D-HP-PFP was presented in Shmygelska et al. (2002). This approach represented candidate solutions for the 2D-HP-PFP by sequences of local structure motifs that correspond to relative folding directions. There are three possible such motifs, which are given in figure 5.5. Each of the motifs gives the position of an amino acid relative to its two predecessors; hence, a candidate solution for a protein of length n corresponds to a sequence of local structure motifs of length $n - 2$.

Solutions are constructed by first randomly choosing, according to a uniform distribution, a start position l in the sequence. From position l, an ant extends the partial conformation till position 1 is reached (obtaining a partial conformation for positions $s_1 \ldots s_l$) and then from position l to position n, resulting in a conformation for the full sequence $s_1 \ldots s_n$. When extending the conformation to the right of the sequence, the direction into which s_{i+1} is placed with respect to $s_{i-1} s_i$ is determined using the random proportional action choice rule of AS [equation (3.2)].

Construction graph   The construction graph consists of $3 \cdot (n - 2)$ local structure motifs, one corresponding to each of the positions 2 to $n - 1$ in the sequence. The "3" comes from the fact that there are three different structure motifs. As usual, the construction graph is fully connected.

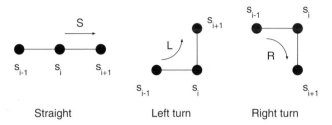

Figure 5.5
Structural motifs that define the solution components for the solution construction and the local search.
Here we consider the case that a sequence is extended from position i to position $i + 1$.

Constraints   The constraints impose that every lattice position can be occupied by at
most one amino acid.

Pheromone trails   Pheromones τ_{ij} refer to the desirability of extending a sequence at
position i to the right by one of the possible folding directions $j \in \{S, L, R\}$, where
these letters stand for *straight* (S), *left turn* (L), and *right turn* (R). Pheromone trails
τ'_{ij} are defined for extending a sequence at position i to the left. Because of symme-
tries in the problem, $\tau'_{iL} = \tau_{iR}$, $\tau'_{iR} = \tau_{iL}$, and $\tau'_{iS} = \tau_{iS}$. The symmetries are due to the
fact that when extending a sequence by a left turn, this corresponds to extending
a sequence to the left by a right turn.

Heuristic information   The heuristic information is computed based on the number
h_{ij} of $H - H$ adjacent positions that would result by placing the next amino acid
using motif $j \in \{S, L, R\}$. The value of h_{ij} is computed by testing all the seven possible neighbor positions of s_{i+1} in the 2D lattice. Figure 5.6 illustrates the positions
that are neighbors of the three possible extensions. The heuristic information is then
taken to be $\eta_{ij} = h_{ij} + 1$ to avoid the heuristic information becoming 0 and specific
placements being excluded from eligibility. Note that if $s_{i+1} = P$, that is, at position
$i + 1$ in the sequence we have a polar amino acid, then we have $h_{ij} = 0$.

Solution construction   As said above, a solution is constructed by extending partial
conformations. When extending the conformation to the right of the sequence, this is
done by using the pheromone trails τ_{ij} plus the heuristic information as usual in the
random proportional action choice rule of AS [equation (3.2)]. When extending the
sequence to the left, the same is done but using instead pheromone trails τ'_{ij}.

One problem that might occur during solution construction is that a partial conformation cannot be extended, because all neighboring positions are already occupied. The measure taken to avoid this is the following. A lookahead procedure is
used to avoid placing an amino acid at a position where all neighboring positions on

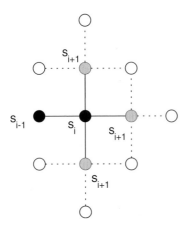

Figure 5.6
Illustration of the neighbor positions to be considered for the three possible locations of sequence element s_{i+1}. The three tentative positions of s_{i+1} are indicated by the shaded circles and the seven neighbor positions to these three positions are indicated by white circles.

the lattice are already occupied. Additionally, once the lookahead procedure has ruled out an otherwise apparently feasible extension, backtracking is invoked that undoes half of the construction steps and then restarts the construction process from the resulting sequence position.

Pheromone update   In the pheromone update, as usual, the pheromone trails evaporate and a number of the best solutions obtained after each iteration are allowed to deposit pheromone. Occasionally, also the best-so-far solution deposits pheromone if no improved solution was found in the current iteration. The amount of pheromone to be deposited is $f(c)/f^*$, where f^* is an approximation of the optimal value. Additional techniques, similar in spirit to the pheromone trail limits of $\mathcal{MMAS}$, are used to avoid search stagnation (see chapter 3, section 3.3.1).

Local search   Two local search algorithms are applied. The first one is based on a *macro-mutation neighborhood* described in Krasnogor, Hart, Smith, & Pelta, 1999), while the second is a one-exchange neighborhood, in which two neighboring solutions differ in exactly one motif. For details on the local search see Shmygelska et al. (2002).

Results   Computational results with this ACO approach have shown that for sequences with length up to 50 the optimal solutions or the best-known solutions can be obtained.

An improved version of this ACO algorithm was proposed recently by Shmygelska & Hoos (2003). It differs from the algorithm described here mainly in the use of a construction process that probabilistically extends partial solutions in both directions, the use of an improved local search algorithm, and of an additional local search phase with a simulated annealing-type acceptance criterion applied to the iteration-best or the best-so-far solution. This latter algorithm obtained computational results that in many cases are very close to, and in some instances improve on, those of current state-of-the-art algorithms for this problem (Hsu, Mehra, Nadler, & Grassberger, 2003).

5.5.4   Constraint Satisfaction

Constraint satisfaction problems (CSPs) are an important class of combinatorial problems with many important practical applications in spatial and temporal reasoning (Dechter, Meiri, & Pearl, 1991; Guesgen & Hertzberg, 1992), graph coloring (Brelaz, 1979), frequency assignment (Gamst, 1986), and scheduling problems (Sadeh, Sycara, & Xiong, 1995; Sadeh & Fox, 1996). See Dechter (2003) for an overview.

A CSP is defined by a triple $(\mathcal{Y}, \mathcal{D}, \mathcal{C})$, where $\mathcal{Y} = \{y_1, \ldots, y_n\}$ is a finite set of variables, $\mathcal{D} = \{D_1, \ldots, D_n\}$ is a finite set of domains, where each variable y_i has an associated domain D_i, and $\mathcal{C} = \{c_1, \ldots, c_m\}$ is a finite set of constraints, that is, relations defined among a subset of the variables that restrict the set of values that these variables can take simultaneously. Depending on the domains and the type of constraints, several classes of CSPs are obtained. In the following, we focus our attention on finite discrete CSPs, that is, CSPs in which each domain is discrete and finite. Additionally, in the following presentation we restrict ourselves to the case in which all the constraints are binary, that is, they involve two variables. However, the presented algorithm can be applied also to nonbinary constraints, because any nonbinary CSP instance can be transformed into a binary CSP instance in a rather straightforward way (see Dechter & Pearl, 1989; Rossi, Petrie, & Dhar, 1990; and Bacchus, Chen, van Beek, & Walsh, 2002).

A variable instantiation is the assignment of a value $d \in D_i$ to a variable y_i, which is denoted by the pair $\langle y_i, d \rangle$, and a candidate solution s consists of a variable instantiation for all variables. A candidate solution s violates a constraint $c_i \in \mathcal{C}$ if the values assigned simultaneously to the variables involved in c_i do not satisfy the constraint c_i. The goal in a CSP is to find a candidate solution that satisfies all constraints or, if no such candidate solution exists, to prove it. ACO algorithms can be used to find satisfying variable instantiations, but they cannot prove that there is no satisfying instantiation if none exists.

In the following, we present a general ACO algorithm applicable to CSPs, which is based on $\mathcal{MMAS}$ ($\mathcal{MMAS}$-CSP) (Solnon, 2002). Similar to many other approximate algorithms for CSPs, $\mathcal{MMAS}$-CSP tries to minimize the number of violated constraints. This means that $\mathcal{MMAS}$-CSP is actually solving a MAX-CSP problem (Freuder & Wallace, 1992), where the goal is to maximize the number of satisfied constraints or, equivalently, to minimize the number of unsatisfied constraints. If a solution to the MAX-CSP is found that satisfies all constraints, then also a satisfying solution for the CSP instance is found.

Construction graph   For each possible variable-value instantiation $\langle y_i, d \rangle$ we have one node in the construction graph and for any pair of nodes corresponding to different variables there is an arc.

Constraints   The constraints (for generating a candidate solution) in the CSP enforce that each variable is assigned exactly one value of its domain.

Pheromone trails   Pheromone trails are associated with every arc in the construction graph. A pheromone trail $\tau_{\langle i,d \rangle \langle j,e \rangle}$ associated with the arc $(\langle y_i, d \rangle, \langle y_j, e \rangle)$ between nodes $\langle y_i, d \rangle$ and $\langle y_j, e \rangle$ intuitively represents the desirability of assigning the value d to y_i and e to y_j, simultaneously.

Heuristic information   The heuristic information is defined to be $\eta_{\langle y_j, e \rangle}(s_k) = 1/(1 + f(s_k \cup \langle y_j, e \rangle) - f(s_k))$, where $s_k \cup \langle y_j, e \rangle$ is the partial solution after adding component $\langle y_j, e \rangle$ to ant k's partial solution s_k and $f(s_k)$ is the number of constraints violated by s_k. In other words, the heuristic information is inversely proportional to the number of newly violated constraints after assigning the value e to variable y_j.

Solution construction   Solutions are constructed in a way analogous to the one followed in assignment problems (see section 5.2). Each ant iteratively constructs solutions by assigning values to variables. At each construction step, an ant has to make two decisions: (1) which variable to assign a value next and (2) which value to assign to the chosen variable. In $\mathcal{MMAS}$-CSP, the first decision is made based on the *smallest domain ordering* heuristic that selects the unassigned variable with the smallest number of consistent values with respect to the partial assignment already built (in other words, for each variable the number of values of its domain that are consistent with the partial assignment is computed and the variable that has the smallest number is chosen). The second decision, the choice of value to assign, is done using AS's action choice rule [equation (3.2)], but using instead of the τ_{ij}'s the sum of the pheromone values on arcs connecting $\langle y_j, e \rangle$ to the already instantiated variables:

$$\sum_{\langle y_i, d \rangle \in s_k} \tau_{\langle y_i, d \rangle \langle y_j, e \rangle}.$$

This sum gives the desirability of assigning the value e to variable y_j.

Pheromone update The pheromone update follows the general rules of $\mathcal{MMAS}$ with the exception that more than one ant may deposit pheromone. In particular, in $\mathcal{MMAS}$-CSP all iteration-best ants deposit an amount $1/f(s_k)$ of pheromone, favoring in this way assignments that violate few constraints. Pheromone is deposited on all arcs between any pair of visited nodes on the construction graph.

Local search $\mathcal{MMAS}$-CSP improves candidate solutions using the min-conflicts heuristic (MCH) (Minton, Johnston, Philips, & Laird, 1992). At each local search step MCH chooses randomly, according to a uniform distribution, a variable that is involved in some violated constraint and then assigns to it a new value that minimizes the number of constraint violations. The local search is stopped after $n = |\mathcal{Y}|$ steps without improvement.

Particularities $\mathcal{MMAS}$-CSP uses a preprocessing step that is motivated by the following observations. If ρ, the parameter regulating pheromone evaporation, is very low, then all pheromone values will be very similar in the first cycles of $\mathcal{MMAS}$ because the pheromone trails are all initialized to a large value τ_{max}. In turn, if the differences between the absolute values of the pheromones are very small, then the probabilities of the various available extensions of a partial solution are close to a uniform distribution (if we do not take into account the heuristic information). Hence, in the first iterations of $\mathcal{MMAS}$, pheromones do not give strong guidance toward good solutions. However, good solutions can easily be found by local search algorithms. $\mathcal{MMAS}$-CSP exploits this observation by using a preprocessing phase, where a local search is applied to candidate solutions that are generated using only heuristic information in the solution construction [this can be achieved by setting $\alpha = 0$ in equation (3.2)]. In $\mathcal{MMAS}$-CSP this process is repeated until the average cost of the n_b best solutions seen so far does not improve anymore by more than a given ε or until a candidate solution satisfying all constraints is found. The n_b best solutions found in this preprocessing phase are then used to initialize the pheromone trails. In the $\mathcal{MMAS}$-CSP case, this preprocessing step was reported to considerably improve performance.

Results $\mathcal{MMAS}$-CSP was tested on hard, randomly generated binary CSPs taken from the phase transition region (Smith & Dyer, 1996) and compared with an extension of MCH using random walk (WMCH) (Wallace & Freuder, 1996), and with a local search procedure that is currently one of the best-performing MCH variants for CSPs (Wallace, 1996). The experimental results showed that, while WMCH performed better than $\mathcal{MMAS}$-CSP for instances with a low number of constraints, for a higher number of constraints $\mathcal{MMAS}$-CSP was the best. However, it is not known how $\mathcal{MMAS}$-CSP compares to other local search algorithms that are better-performing than WMCH; in particular, it would be desirable to compare it to the tabu search algorithm of Galinier & Hao (1997), which appears to be one of the best available local search algorithms for binary CSPs. In any case, $\mathcal{MMAS}$-CSP is one of the first approaches showing that ACO algorithms can be applied to solve constraint satisfaction problems successfully.

Remarks   There have been a few other approaches attacking CSPs. Solnon (2000) applied an ACO algorithm to constraint satisfaction problems with a global permutation constraint that enforces that all the variables are assigned different values, the number of values being the same as the number of variables. This ACO algorithm was shown to yield promising results for the car sequencing problem (Gottlieb, Puchta, & Solnon, 2003). Schoofs & Naudts (2000) introduced an ACO algorithm for binary CSPs that includes preprocessing techniques that can also sometimes show that an instance is infeasible. A few approaches were also presented for the satisfiability problem in propositional logic (SAT) that can be seen as a particular case of CSP, where all variables' domains comprise exactly two values (Pimont & Solnon, 2000; Lokketangen, 2000). However, these approaches so far have not achieved good performance when compared to current state-of-the-art algorithms for SAT.

Recently, two applications of ACO to the MAX-SAT problem were presented. MAX-SAT is an extension of the SAT problem in which instead of searching for a feasible assignment, one searches for the assignment with the lowest number of unsatisfied clauses. Roli, Blum, & Dorigo (2001) applied an ACO algorithm to MAX-SAT and tested different ways of defining pheromone trails. They found that the best way is to let pheromones indicate the desirability of having a particular truth-value assignment. Zlochin & Dorigo (2002) compare ACO with a number of other model-based approaches and show that, on this particular problem, all the model-based algorithms reach a similar level of performance.

5.6   Machine Learning Problems

Many problems in the field of machine learning can be cast as optimization problems and be solved by either classic exact algorithms or by heuristic approaches. Examples are the generation of classification rules, data clustering, regression, and several others. ACO algorithms have been applied to a few such problems so far, opening the application of this algorithmic technique to a new range of important problems.

Here we present two such applications, the learning of classification rules and the learning of the structure of a Bayesian network. Note that there exist a few more applications of ant algorithms, a class of algorithms that is loosely inspired by various behavioral patterns of ants, to other machine learning–type problems, most noteworthy to clustering problems (Lumer & Faieta, 1994; Kuntz, Snyers, & Layzell, 1999). An overview of these approaches is given in section 7.3 of chapter 7.

5.6.1   Learning of Classification Rules

Parpinelli, Lopes, & Freitas (2002a,b) designed an ACO algorithm called Ant-miner for a particular data mining task: the generation of classification rules. The problem they consider is the following. Given a set of attributes $A = \{a_1, \ldots, a_n\}$ (the domain of each attribute being a set $V_i = \{v_{i1}, \ldots v_{if_i}\}$ of f_i values), a set of l classes $B = \{b_1, \ldots, b_l\}$, and a training set $TS = \{ts_1, \ldots ts_h\}$, where each ts_i is a case, the task is to learn a set of IF-THEN rules, each rule taking the form

$$\text{IF } \langle term_1 \wedge term_2 \wedge \cdots \rangle \quad \text{THEN } \langle b_i \rangle. \tag{5.23}$$

The IF part of the rule is called the antecedent, the THEN part is called the consequent and gives the class predicted by the rule. Each term in the antecedent is a triple (a, o, v), where v is a value in the domain of attribute a, and o is an operator. Ant-miner allows only for discrete attributes and the only possible operator is "=", the equality operator. If continuous attributes are to be used, their domains need to be discretized in a preprocessing step.

Ant-miner builds an ordered set of IF-THEN rules by iteratively appending rules of the form given by equation (5.23) to the rule set. Each rule to be appended is determined by running an ACO algorithm that is similar to AS but that uses only one ant.

To construct a rule, the ant starts with an empty rule and iteratively adds terms to the rule using the probabilistic action choice rule of AS given by equation (3.2) in chapter 3. Once rule construction has finished, first the rule is simplified by a pruning process, the class predicted by this rule is chosen to be the one that covers the largest number of cases in the training set, and then the pheromone trails are updated. The ACO algorithm stops to generate new rules either when it enters stagnation or when a maximum number of rule constructions has been reached. In a final step, the best rule *rule*$_{best}$ found in this process, where best is measured according to some evaluation function $f(\cdot)$, is added to the set of rules and the training cases that are correctly classified by the rule *rule*$_{best}$ are removed from *TS*.

procedure Ant-miner
    $TS \leftarrow$ InitializeTrainingSet
    $rule\_list \leftarrow [\ ]$
    while $(|TS| > uc)$ **do**   % uc is the number of uncovered cases
       $\tau \leftarrow$ InitializePheromones
       $i \leftarrow 1$
       $rule_{best} \leftarrow \varnothing$
       while (termination condition not met) **do**
          $rule_i \leftarrow$ ConstructRule
          $rule_i \leftarrow$ PruneRule$(rule_i)$
          UpdatePheromones
          if $(f(rule_i) > f(rule_{best}))$
             $rule_{best} \leftarrow rule_i$
          end-if
          $i \leftarrow i + 1$
       end-while
       $rule\_list \leftarrow [rule\_list, rule_{best}]$
       $TS \leftarrow TS \backslash CoveredCases(rule_{best})$
    end-while
end-procedure *Ant-miner*

Figure 5.7
High-level pseudo-code for Ant-miner.

This process of adding new rules to the rule set is repeated until *TS* contains fewer than *uc* cases, where *uc* is a parameter of the algorithm. An algorithmic outline of Ant-miner is given in figure 5.7.

Construction graph   The construction graph contains one node for every possible term plus a dummy start node. The construction graph is fully connected.

Constraints   The set of constraints enforces that one attribute can be used at most once by each rule and that the minimum number of attributes in a rule is one.

Pheromone trails   The pheromone trail τ_{ij} indicates the desirability of adding the term $(a_i, =, v_{ij})$ to the rule; the term $(a_i, =, v_{ij})$ says that the attribute a_i takes value v_{ij}.

Heuristic information   The heuristic information is based on a measure of the entropy associated with each term $(a_i, =, v_{ij})$ that can be added to a rule. In particular, the entropy is defined as

$$h(B \mid a_i = v_{ij}) = -\sum_{b=1}^{l} P(b \mid a_i = v_{ij}) \cdot \log_2 P(b \mid a_i = v_{ij}), \qquad (5.24)$$

where B is the set of classes and $P(b \mid a_i = v_{ij})$ is the empirical probability of observing class b when having observed that attribute a_i takes value v_{ij}. The higher the entropy, the more uniformly distributed are the classes and the smaller should be the probability that the ant adds the term $(a_i, =, v_{ij})$ to its rule. Ant-miner uses a normalization of the entropy values to derive the heuristic information that is given by

$$\eta_{ij} = \frac{\log_2 l - h(B \mid a_i = v_{ij})}{\sum_{i=1}^{n} x_i \cdot \sum_{j=1}^{f_i} \log_2 l - h(B \mid a_i = v_{ij})}. \qquad (5.25)$$

Here, x_i is 1, if the attribute a_i is not yet used by the ant and 0 otherwise. In the heuristic information, the term $\log_2 l$ is used as a scaling factor, because one can show that it holds $0 \le h(B \mid a_i = v_{ij}) \le \log_2 l$. Hence, the heuristic information gives preference to terms that discriminate well between different classes. The heuristic information for all attribute-value pairs can be precomputed and stored in a table, because $h(B \mid a_i = v_{ij})$ is independent of the rule in which the term occurs.

Solution construction As said above, a rule is constructed by iteratively adding terms. At each step of the rule construction, the probability of adding term $(a_i, =, v_{ij})$ is given by

$$p_{ij} = \frac{\tau_{ij} \eta_{ij}}{\sum_{i=1}^{n} (x_i \sum_{j=1}^{f_i} (\tau_{ij} \eta_{ij}))}, \quad \text{if } (a_i, =, v_{ij}) \in \mathcal{N}. \qquad (5.26)$$

Here, $\mathcal{N}$ is the feasible neighborhood; it comprises all terms except (1) those that contain an attribute that is already used in a partial rule (to avoid invalid rules such as IF $\langle (\text{sex} = \text{male}) \wedge (\text{sex} = \text{female}) \rangle$), and (2) those that would make the resulting rule cover less than mc cases of the training set. Once the rule antecedent is completed, a rule-pruning procedure is applied, and then the class to be predicted by the rule is set to be the one that covers the largest number of training cases satisfying the rule antecedent.

Pheromone update The solution quality $f(rule)$ of a rule is computed as

$$f(rule) = \frac{TP}{TP + FN} \cdot \frac{TN}{FP + TN}, \qquad (5.27)$$

where TP and FP are, respectively, the number of true and false positives, and TN and FN are, respectively, the number of true and false negatives. The left term in the product gives the fraction of positive examples that are recognized as being positive, while the right term gives the fraction of negative examples that are correctly classified as negative ones. The larger these fractions, the higher is $f(rule)$. The pheromone update is performed in two steps. First, an amount of pheromone given by $\tau_{ij} \cdot f(rule)$ is added to each of the terms in the rule. Then pheromone values are normalized: each τ_{ij} is divided by the sum over all τ_{ij}. This implicit form of pheromone evaporation keeps the total amount of pheromone always equal to 1. The net result of the overall pheromone update is that the value τ_{ij} of terms $(a_i, =, v_{ij})$ that do not receive pheromone decreases, while the other terms have their amount of pheromone increased.

Local search   Ant-miner does not use any local search.

Results   Ant-miner was compared to CN2 (Clark & Niblett, 1989; Clark & Boswell, 1991), a classification rule discovery algorithm that uses a strategy for generating rule sets similar to that of Ant-miner. The comparison was done using six data sets from the UCI Machine Learning Repository that is accessible at www.ics.uci.edu/~mlearn/MLRepository.html. The results were analyzed according to the predictive accuracy of the rule sets and the simplicity of the discovered rule set, which is measured by the number of rules in the rule set and the average number of terms per rule. While Ant-miner had a better predictive accuracy than CN2 on four of the data sets and a worse one on only one of the data sets, the most interesting result is that Ant-miner returned much simpler rules than CN2. For more details on the computational results and a study of the influence of several design choices on Ant-miner performance, see Parpinelli et al. (2002b). Similar conclusions could also be drawn from a comparison of Ant-miner to C4.5 (Quinlan, 1993a), a well-known decision tree algorithm (Parpinelli et al., 2002a).

5.6.2   Learning the Structure of Bayesian Networks

Bayesian networks give a graphical representation of knowledge in probabilistic domains, which is becoming increasingly important in artificial intelligence (Pearl, 1998; Jensen, 2001). A Bayesian network is a directed, acyclic graph $G = (N, A)$, where the set of nodes N is in one-to-one correspondence to a set of random variables $X = \{x_1, \ldots, x_n\}$ and the set of arcs A represents direct dependence relationships between the variables. With each of the variables, a set of conditional probability distributions $P(x_i \mid Par(x_i))$ is associated, where $Par(x_i)$ denotes the parents of node x_i, that is, those nodes x_j for which there exists a directed arc (j, i) from j to i. The joint probability distribution across the variables can then be written as

$$P(x_1, x_2, \ldots, x_n) = \prod_{i=1}^{n} P(x_i \mid Par(x_i)) \qquad (5.28)$$

The problem of learning the structure of a Bayesian network is to find a directed acyclic graph, that is, a Bayesian network, that best matches a training set $TS = \{ts_1, \ldots ts_n\}$, where each element of TS consists of values for the random variables that are defined to describe the situation of interest. A commonly used technique for this task is to evaluate the quality of a Bayesian network using a scoring function $f(\cdot)$. In this case, the goal of the design of a Bayesian network is to find a graph that maximizes the scoring function. De Campos, Fernández-Luna, Gámez, & Puerta (2002a) presented an ACS algorithm for this task (ACS-BN). Their algorithm starts from a graph without any arcs and iteratively adds (directed) arcs subject to the constraint that the resulting graph must be acyclic. Hence, ACS-BN searches in the search space defined by directed acyclic graphs. The quality of the graph (Bayesian network) and the heuristic information are based on a transformation of the K2 metric (Cooper & Herskovits, 1992).

Construction graph There is one node for each of the n^2 directed arcs between every pair of variables. Each node ij of the construction graph corresponds to a directed arc (i, j). As usual, the construction graph is fully connected.

Constraints The only constraint is that the final graph (i.e., the Bayesian network) is required to be an acyclic graph.

Pheromone trails A pheromone trail τ_{ij} is associated with every node ij, that is, with every directed arc (i, j).

Heuristic information The heuristic information η_{ij} is set equal to the improvement of the scoring function obtained by adding a directed arc (i, j) to the solution under construction. The scoring function depends on an ant's partial solution. For details on how it is derived, see de Campos et al. (2002a).

Solution construction ACS-BN iteratively adds directed arcs to the Bayesian network as long as there exist arcs for which the heuristic information is positive; a negative heuristic information for an arc indicates that adding this arc would worsen the value of the scoring function. At each construction step, first a candidate list of arcs is formed. An arc (i, j) is included in the candidate list if (1) it is not already included in G, (2) its addition to the graph does not create cycles in the graph, and (3) it holds that $\eta_{ij} > 0$. The selection of the next arc to be added is then done in the candidate list using the pseudorandom proportional rule of ACS [equation (3.10)]. The solution construction is stopped when the candidate list becomes empty.

Pheromone update   The pheromone trails in ACS-BN are initialized by first running the K2SN heuristic (de Campos & Puerta, 2001) and using the resulting score for the definition of the initial pheromone level. The pheromone update follows strictly the rules of ACS.

Local search   ACS-BN makes use of the HCST iterative improvement algorithm (Heckerman, Geiger, & Chickering, 1995), where the neighborhood of a graph G consists of all those graphs that can be obtained from G by deleting, adding, or reversing one arc. Two variants of ACS-BN were presented that differ only in the frequency of applying local search. A first variant applied local search only to the solutions constructed in the last iteration of the algorithm, while a second variant applied local search every ten iterations and in the last iteration to all ants.

Results   ACS-BN was compared to an iterative improvement algorithm, to two model-based search algorithms (Baluja & Caruana, 1995; Mühlenbein, 1998), and to an ILS algorithm on three benchmark instances. The result was that ACS-BN outperformed all the competitors.

Remarks   A different possibility for constructing Bayesian networks is to search for a good sequence of the variables and to build the graph taking into account this sequence. De Campos et al. (2002b) presented an ACO that searches in the space of variable sequences and compared its performance to ACS-BN on two instances. The result was that ACS-BN had a slightly better performance.

Reasoning processes in Bayesian networks are often carried out on a secondary graph, known as the *junction tree*. Gámez & Puerta (2002) attacked the problem of finding a best elimination sequence of the nodes in the graph that arises in the generation of a junction tree. This problem was tackled using ACS; experimental tests showed that, when compared to greedy construction heuristics, as well as to genetic algorithms based on the work of Larrañaga, Kuijpers, Poza, & Murga (1997), ACS typically reached the best average performance and was faster than the genetic algorithms (Gámez & Puerta, 2002).

5.6.3   Other ACO Applications to Machine Learning Problems

Casillas et al. (2000) applied ACO to the problem of learning rules in a fuzzy system. They formulate this problem as a combinatorial optimization problem based on the cooperative rules methodology and show that it can be transformed into an assignment problem that is similar to a quadratic assignment problem. To tackle this problem, they applied AS and ACS and tested two variants of ACS, one without local search and another applying local search to the iteration-best ant. Experimental results were presented on some benchmark problems, as well as on problems stemming from real-world applications. Compared to other greedy techniques and other meta-heuristic algorithms (Casillas et al., 2002; Cordón & Herrera, 2000; Nozaki, Ishi-buchi, & Tanaka, 1997; Wang & Mendel, 1992), the ACO approaches were found to give, with one exception, the best results with respect to the generalization behavior and, especially in the case of ACS, good solutions were found very quickly.

5.7 Application Principles of ACO

Despite being a rather recent metaheuristic, ACO algorithms have been applied to a large number of different combinatorial optimization problems, as shown by the list of applications presented earlier. Based on this experience, we have identified some basic issues that play an important role in several of these applications. These are discussed in the following subsections.

5.7.1 Construction Graph

One of the first issues when applying ACO to a combinatorial optimization problem is the definition of the *construction graph*. Most of the time this is done implicitly, and the construction function graph used depends on the construction heuristic used by the artificial ants. So, in all the cases considered in this book, the choice of the components C (see chapter 2, section 2.2.1), which define the construction graph, is always the most "natural" with respect to the considered problem and to the chosen construction heuristic.

Nevertheless, it is possible to provide a general procedure (Blum & Dorigo, 2004) that applies to any combinatorial optimization problem, and that is based on the following definition:

Definition 5.1. *A combinatorial optimization problem* $(\mathcal{S}, f, \Omega)$ *is defined by*

- *a set of discrete variables* X_i *with values* $x_i \in D_i = \{d_1^i, \ldots, d_{|D_i|}^i\}$, $i = 1, \ldots, n$;
- *a set* Ω *of constraints among variables;*
- *an objective function to be minimized* $f : D_1 \times \cdots \times D_n \to \mathbb{R}$;
- *the set of all the possible feasible assignments:*

$\mathcal{S} = \{s = \{(X_1, x_1), \ldots, (X_n, x_n)\} \mid x_i \in D_i, s \text{ satisfies all the constraints}\}.$

A solution $s^* \in \mathcal{S}$ *is called a globally optimal solution if* $f(s^*) \leq f(s)$, $\forall s \in \mathcal{S}$. *The set of all globally optimal solutions is denoted by* $\mathcal{S}^* \subseteq \mathcal{S}$. *To solve a combinatorial optimization problem one has to find a solution* $s^* \in \mathcal{S}^*$.

Given the above definition, we call *component* the combination of a decision variable X_i and one of its domain values $x_i \in D_i$. The construction graph is the completely connected graph of all the solution components. Unlike the more general description given in section 2.2.1, pheromone trails can only be assigned to components.

It is easy to see that, in many cases, the construction graph obtained using the above definition is equivalent to the one that has been implicitly used in the literature. Consider, for example, a TSP with n cities: If the discrete variables X_i, $i = 1, \ldots, n$ are the n cities and their values x_i are the integer numbers $j = 1, \ldots, n$, $i \neq j$, then for each city i there are $n - 1$ pheromone values corresponding to the $n - 1$ connections to the other cities.

5.7.2   Pheromone Trails Definition

An initial, very important choice when applying ACO is the definition of the intended meaning of the pheromone trails. Let us explain this issue with an example. When applying ACO to the TSP, the standard interpretation of a pheromone trail τ_{ij}, used in all published ACO applications to the TSP, is that it refers to the desirability of visiting city j directly after a city i. That is, it provides some information on the desirability of the relative positioning of city i and j. Yet, another possibility, not working so well in practice, would be to interpret τ_{ij} as the desirability of visiting city i as the j-th city in a tour, that is, the desirability of the city's absolute positioning. Conversely, when applying ACO to the SMTWTP (see section 5.3.1) better results are obtained when using the absolute position interpretation of the pheromone trails, that is, the interpretation in which τ_{ij} represents the desirability of putting job j on the i-th position (den Besten, 2000). This is intuitively due to the different role that permutations have in the two problems. In the TSP, permutations are cyclic, that is, only the relative order of the solution components is important and a permutation $\pi = (1\ 2 \ldots n)$ has the same tour length as the permutation $\pi' = (n\ 1\ 2 \ldots n - 1)$—it represents the same tour. Therefore, a relative position-based pheromone trail is the appropriate choice. On the contrary, in the SMTWTP (as well as in many other scheduling problems), π and π' represent two different solutions with most probably very different costs. Hence, in the SMTWTP the absolute position-based pheromone trails are a better choice. Nevertheless, it should be noted that, in principle, both choices are possible, because any solution of the search space can be generated with both representations.

The influence of the way pheromones are defined has also been studied in a few other researches. One example is the study of Blum and Sampels on different possibilities of how to define pheromone trails for the group shop scheduling problem (Blum & Sampels, 2002a). They examined representations where pheromones refer to absolute positions (same as in the SMTWTP case), to predecessor relations (similar to the TSP case), or to relations among machines (as presented for $\mathcal{MMAS}$-HC-GSP in section 5.3.2). What they found was that the last-named representation gave by far better results than the first two ways of defining pheromones.

The definition of the pheromone trails is crucial and a poor choice at this stage of the algorithm design will result in poor performance. Fortunately, for many problems the intuitive choice is also a very good one, as was the case for the previous example applications. Yet, sometimes the use of the pheromones can be somewhat more involved, which is, for example, the case with the ACO application to the shortest common supersequence problem (Michel & Middendorf, 1999).

5.7.3 Balancing Exploration and Exploitation

Any effective metaheuristic algorithm has to achieve an appropriate balance between the exploitation of the search experience gathered so far and the exploration of unvisited or relatively unexplored search space regions. In ACO, several ways exist of achieving such a balance, typically through the management of the pheromone trails. In fact, the pheromone trails induce a probability distribution over the search space and determine which parts of the search space are effectively sampled, that is, in which part of the search space the constructed solutions are located with higher frequency. Note that, depending on the distribution of the pheromone trails, the sampling distribution can vary from a uniform distribution to a degenerate distribution which assigns probability 1 to a solution and 0 probability to all the others. In fact, this latter situation corresponds to stagnation of the search, as explained in chapter 3, section 3.3.1.

The simplest way to exploit the ants' search experience is to make the pheromone update a function of the solution quality achieved by each particular ant. Yet, this bias alone is often too weak to obtain good performance, as was shown experimentally on the TSP (Stützle, 1999; Stützle & Hoos, 2000). Therefore, in many ACO algorithms an *elitist strategy* was introduced whereby the best solutions found during the search contribute strongly to pheromone trail updating.

A stronger exploitation of the "learned" pheromone trails can also be achieved during solution construction by applying the pseudorandom proportional rule of ACS, as explained in chapter 3, section 3.4.1. Search space exploration is achieved in ACO primarily by the ants' randomized solution construction. Let us consider for a moment an ACO algorithm that does not use heuristic information (this can be easily achieved by setting $\beta = 0$). In this case, the pheromone updating activity of the ants will cause a shift from the initial, rather uniform sampling of the search space to a sampling focused on specific search space regions. Hence, exploration of the search space will be higher in the initial iterations of the algorithm, and will decrease as the computation goes on. Obviously, attention must be paid to avoid too strong a focus on apparently good regions of the search space, which can cause the ACO algorithm to enter a stagnation situation.

There are several ways to avoid such stagnation situations, thus maintaining a reasonable level of exploration of the search space. For example, in ACS the ants use a local pheromone update rule during the solution construction to make the path they have taken less desirable for successive ants and, thus, to diversify search. $\mathcal{MMAS}$ introduces an explicit lower limit on the pheromone trail level so that a minimal level of exploration is always guaranteed. $\mathcal{MMAS}$ also uses a reinitialization of the pheromone trails, which is a way of enforcing search space exploration. Experience has shown that pheromone trail reinitialization, when combined with appropriate choices for the pheromone trail update (Stützle & Hoos, 2000), can be very useful to refocus the search on a different search space region.

In fact, many powerful ACO applications include schedules of how to choose between iteration-best, best-so-far, and the restart-best solution for the pheromone update. (The restart-best solution is the best solution found since the last time the pheromone trails were initialized.) The first studies on such strategies were done by Stützle (1999) and Stützle & Hoos (2000) and considerable improvements over only using iteration-best or best-so-far update were observed. Recently, several high-performing ACO implementations used these or similar features (Blum, 2003a; Blum & Blesa, 2003; Merkle et al., 2002).

Finally, an important, though somewhat neglected, role in the balance of exploration and exploitation is that of the parameters α and β, which determine the relative influence of pheromone trail and heuristic information. Consider first the influence of parameter α. For $\alpha > 0$, the larger the value of α, the stronger the exploitation of the search experience; for $\alpha = 0$ the pheromone trails are not taken into account at all; and for $\alpha < 0$ the most probable choices taken by the ants are those that are less desirable from the point of view of pheromone trails. Hence, varying α could be used to shift from exploration to exploitation and vice versa. The parameter β determines the influence of the heuristic information in a similar way. In fact, systematic variations of α and β could, similarly to what is done in the strategic oscillations approach (Glover, 1990), be part of a simple and useful strategy to balance exploration and exploitation.

A first approach in that direction is followed by Merkle et al. (2002) in their ACO algorithm for the RCPSP (see section 5.3.3). They decrease the value of β from some initial value to zero and they showed that such a schedule for β led to significantly better results than any fixed value for the parameter they tested.

5.7.4   Heuristic Information

The possibility of using heuristic information to direct the ants' probabilistic solution construction is important because it gives the possibility of exploiting problem-specific knowledge. *Static* and *dynamic* heuristic information are the main types of heuristic information used by ACO algorithms. In the static case, the values η are computed once at initialization time and then remain unchanged throughout the whole algorithm's run. An example is the use, in the TSP applications, of the length d_{ij} of the arc connecting cities i and j to define the heuristic information $\eta_{ij} = 1/d_{ij}$. Static heuristic information has the advantage that (1) it is often easy to compute and, in any case, it has to be computed only once, at initialization time, and (2) at each iteration of the ACO algorithm, a table can be precomputed with the values of $\tau_{ij}(t)\eta_{ij}^{\beta}$, which can result in significant savings of computation time. In the dynamic case, the heuristic information depends on the partial solution constructed so far and therefore has to be computed at each step of an ant's walk. This determines a higher computational cost that may be compensated by the higher accuracy of the computed heuristic values. For example, in the ACO application to the SMTWTP it was found that the use of dynamic heuristic information based on the modified due date or on the apparent urgency heuristic (see section 5.3.1) resulted in a better overall performance than the one obtained using only static heuristic information (based on the earliest due date heuristic).

Another way of computing heuristic information was introduced in the ANTS algorithm (see chapter 3, section 3.4.2; see also Maniezzo, 1999), where it is computed using lower bounds on the solution cost of the completion of an ant's partial solution. This method has the advantage that it facilitates the exclusion of certain choices because they lead to solutions that are worse than the best found so far. It allows therefore the combination of knowledge on the calculation of lower bounds from mathematical programming with the ACO paradigm. Nevertheless, a disadvantage is that the computation of the lower bounds can be time-consuming, especially because they have to be calculated at each construction step by each ant.

Finally, it should be noted that, although the use of heuristic information is important for a generic ACO algorithm, its importance is often strongly diminished if local search is used to improve solutions. This is because local search takes into account information about the cost of improving solutions in a more direct way. Fortunately, this means that ACO algorithms can also achieve, in combination with a local search algorithm, very good performance for problems for which it is difficult to define a priori a very informative heuristic information.

5.7.5   ACO and Local Search

In many applications to $\mathcal{NP}$-hard combinatorial optimization problems, ACO algorithms perform best when coupled with local search algorithms (which is, in fact, a particular type of daemon action of the ACO metaheuristic). Local search algorithms locally optimize the ants' solutions and these locally optimized solutions are used in the pheromone update.

The use of local search in ACO algorithms can be very interesting as the two approaches are complementary. In fact, ACO algorithms perform a rather coarse-grained search, and the solutions they produce can then be locally optimized by an adequate local search algorithm. The coupling can therefore greatly improve the quality of the solutions generated by the ants.

On the other side, generating initial solutions for local search algorithms is not an easy task. For example, it has been shown that, for most problems, repeating local searches from randomly generated initial solutions is not efficient (see, e.g., Johnson & McGeoch, 1997). In practice, ants probabilistically combine solution components which are part of the best locally optimal solutions found so far and generate new, promising initial solutions for the local search. Experimentally, it has been found that such a combination of a probabilistic, adaptive construction heuristic with local search can yield excellent results (Boese, Kahng, & Muddu, 1994; Dorigo & Gambardella, 1997b; Stützle & Hoos, 1997).

It is important to note that when using local search a choice must be made concerning pheromone trail update: either pheromone is added to the components or connections of the locally optimal solution, or to the starting solutions for the local search. The quasi-totality of published research has used the first approach. Although it could be interesting to investigate the second approach, some recent experimental results suggest that its performance is worse.

Despite the fact that the use of local search algorithms has been shown to be crucial to achieving best performance in many ACO applications, it should be noted that ACO algorithms also show very good performance where local search algorithms cannot be applied easily. Such examples are the applications to network routing described in chapter 6 and the application to the shortest common supersequence problem discussed in section 5.5.1 of this chapter (Michel & Middendorf, 1999).

5.7.6 Number of Ants

Why use a colony of ants instead of one single ant? In fact, although a single ant is capable of generating a solution, efficiency considerations suggest that the use of a colony of ants is often a desirable choice. This is particularly true for geographically distributed problems (see network routing applications described in chapter 6), because the differential path length effect exploited by ants in the solution of this class of problems can only arise in the presence of a colony of ants. It is also interesting to note that in routing problems ants solve many shortest-path problems in parallel (one between each pair of nodes) and a colony of ants must be used for each of these problems.

On the other hand, in the case of combinatorial optimization problems, the differential length effect is not exploited and the use of m ants, $m > 1$, that build r solutions each (i.e., the ACO algorithm is run for r iterations) could be equivalent to the use of one ant that generates $m \cdot r$ solutions. Nevertheless, experimental evidence suggests that, in the great majority of situations, ACO algorithms perform better when the number m of ants is set to a value $m > 1$.

In general, the best value for m is a function of the particular ACO algorithm chosen as well as of the class of problems being attacked, and most of the times it must be set experimentally. Fortunately, ACO algorithms seem to be rather robust with respect to the actual number of ants used.

5.7.7 Candidate Lists

One possible difficulty encountered by ACO algorithms is when they are applied to problems with a large neighborhood in the solution construction. In fact, a large neighborhood means the ants have a large number of possible moves from which to choose, which determines an increase in the computation time.

In such situations, the computation time can be kept within reasonable limits by the use of candidate lists. Candidate lists constitute a small set of promising neighbors of the current partial solution. They are created using a priori available knowledge on the problem, if available, or dynamically generated information. Their use allows ACO algorithms to focus on the more interesting components, strongly reducing the dimension of the search space.

As an example, consider the ACO application to the TSP. For the TSP it is known that very often optimal solutions can be found within a surprisingly small subgraph consisting of all the cities and of those arcs that connect each city to only a few of its nearest neighbors. For example, for the TSPLIB instance `pr2392.tsp` with 2392 cities an optimal solution can be found within a subgraph constructed using for each city only the eight nearest neighbors (Reinelt, 1994). This knowledge can be used for defining candidate lists, which was first done in the context of ACO algorithms in Gambardella & Dorigo (1996). A candidate list includes for each city its cl nearest neighbors. During solution construction an ant tries to choose the city to move to only from among the cities in the candidate list. Only if all these cities have already been visited can the ant choose from the cities not in the candidate list.

So far, in ACO algorithms the use of candidate lists or similar approaches is still rather unexplored. Inspiration from other techniques like tabu search (Glover & Laguna, 1997) or GRASP (Feo & Resende, 1995), where strong use of candidate lists is made, could be useful for the development of effective candidate list strategies for ACO.

5.7.8   Steps to Solve a Problem by ACO

From the currently known ACO applications we can identify some guidelines for attacking problems by ACO. These guidelines can be summarized in the following six design tasks:

1. Represent the problem in the form of sets of components and transitions or by means of a weighted graph, on which ants build solutions.

2. Define appropriately the meaning of the pheromone trails τ_{rs}, that is, the type of decision they bias. This is a crucial step in the implementation of an ACO algorithm and, often, a good definition of the pheromone trails is not a trivial task and typically requires insight into the problem to be solved.

3. Define appropriately the heuristic preference for each decision that an ant has to take while constructing a solution, that is, define the heuristic information η_{rs} associated with each component or transition. Notice that heuristic information is crucial for good performance if local search algorithms are not available or cannot be applied.

4. If possible, implement an efficient local search algorithm for the problem to be solved, because the results of many ACO applications to $\mathcal{NP}$-hard combinatorial optimization problems show that the best performance is achieved when coupling ACO with local optimizers (Dorigo & Di Caro, 1999b; Dorigo & Stützle, 2002).

5. Choose a specific ACO algorithm (those currently available were described in chapter 3) and apply it to the problem being solved, taking the previous aspects into account.

6. Tune the parameters of the ACO algorithm. A good starting point for parameter tuning is to use parameter settings that were found to be good when applying the

ACO algorithm to similar problems or to a variety of other problems. An alternative to time-consuming personal involvement in the tuning task is to use automatic procedures for parameter tuning (Birattari et al., 2002b).

It should be clear that the above steps can only give a very rough guide to the implementation of ACO algorithms. In addition, often the implementation is an iterative process, where with some further insight into the problem and the behavior of the algorithm, some choices taken initially need to be revised. Finally, we want to insist on the fact that probably the most important of these steps are the first four, because a poor choice at this stage typically can not be overcome with pure parameter fine-tuning.

5.8 Bibliographical Remarks

The first combinatorial problem tackled by an ACO algorithm was the traveling salesman problem. Since the first application of AS in Dorigo's PhD dissertation in 1992, the TSP became a common test bed for several contributions proposing ACO algorithms that perform better than AS (Dorigo & Gambardella, 1997b; Stützle & Hoos, 2000; Bullnheimer et al., 1999c; Cordón et al., 2000); these contributions and their application to the TSP have been described in chapter 3. After these first applications, a large number of different problems were attacked by ACO algorithms. As with other metaheuristics for $\mathcal{NP}$-hard combinatorial optimization problems, the next wave of applications was directed at a number of other academic benchmark problems.

The first of these were the quadratic assignment problem and the job shop problem, in 1994. As for the TSP, also in this case the first ACO algorithm to be applied was AS (Maniezzo et al., 1994; Colorni et al., 1994). This choice was dictated by the fact that, at that time, AS was still the only ACO algorithm available.

After these first studies, the QAP continued to receive significant attention in research efforts striving to improve ACO algorithms (for an overview, see section 5.2.1 or the overview article by Stützle & Dorigo, 1999a). Differently, the JSP received somewhat less attention in the following years, and only recently have researchers started again to attack it with ACO algorithms (Teich, Fischer, Vogel, & Fischer, 2001; Blum, 2002a, 2003a). One reason for this may be that ACO algorithms for the QAP quickly reached world-class performance, while the early applications of ACO to the JSP were much less successful.

However, there is a significant temporal gap between the first publications about AS in the early '90s (Dorigo, Maniezzo, & Colorni, 1991a,b; Dorigo, 1992; Colorni,

Dorigo, & Maniezzo, 1992a,b) and the moment at which the methodology started to be widely known. In fact, research on ACO started flourishing only after the first journal publication about AS (Dorigo et al., 1996). In 1996, Pfahringer presented a widely unknown application to the open shop problem of the Ant-Q algorithm, a predecessor of ACS developed by Gambardella & Dorigo (1995) (see also chapter 3, section 3.4.1, and Dorigo & Gambardella, 1996). From 1997, the variety of ACO applications increased steadily (in part, these researches were published in conference proceedings or journals only in later years; however, technical reports were available much earlier). These include classic vehicle routing problems (Bullnheimer et al., 1999b), sequential ordering (Gambardella & Dorigo, 2000), flow shop scheduling (Stützle, 1998a), and graph coloring (Costa & Hertz, 1997) problems. For some of these applications, excellent computational results were reported, which is especially true for the ACS application to the sequential ordering problem of Gambardella and Dorigo. Slightly later, ACO was applied to the shortest common supersequence problem (Michel & Middendorf, 1998, 1999) and the generalized assignment problem (Lourenço & Serra, 1998). From then on, there was an explosion in the number of different problems attacked, as can be appreciated by browsing the proceedings of the first three workshops, "From Ant Colonies to Artificial Ants" (Dorigo, Middendorf, & Stützle, 2000b; Dorigo et al., 2002a) or journal special issues (Dorigo, Stützle, & Di Caro, 2000c; Dorigo, Gambardella, Middendorf, & Stützle, 2002b). Of interest, as discussed in chapter 7, ACO algorithms are now moving to the real world, with the recent development of interesting applications to solve industrial routing, scheduling, and sequencing problems.

To make the historical overview of the development of ACO applications complete, we should mention the applications to the routing problem in telecommunications networks, starting in 1996 with the work of Schoonderwoerd et al. (1996) and the work on AntNet by Di Caro & Dorigo (1998c). These applications play an important role in the development of ACO and the particularly successful AntNet algorithm is presented in detail in chapter 6.

5.9   Things to Remember

▪ ACO has been applied to many different problems. In this chapter we have presented some of these applications, focusing on the most interesting ones with respect to the goal of illustrating how to adapt ACO to efficiently solve $\mathcal{NP}$-hard combinatorial problems.

▪ Currently, ACO algorithms achieve state-of-the-art performance for several application problems. These include the sequential ordering problem, the resource-constrained project scheduling problem, the quadratic assignment problem, the vehicle routing problem with time window constraints, the bin-packing problem, the shortest common supersequence problem, and the single-machine total weighted tardiness problem. For many other problems they produce results very close to those of the currently best-performing algorithms.

▪ There exist a variety of problems for which other algorithms appear to be superior to ACO algorithms. Examples are the job shop problem and the graph coloring problem. It is an interesting research question to understand for which types of problems this is the case and which are the problems that are particularly suited for ACO algorithms.

▪ A number of application principles and guidelines that suggest how to develop successful ACO applications have been derived exploiting the experience gained so far by ACO researchers.

5.10 Computer Exercises

Exercise 5.1 Try to reach a state-of-the-art ACO algorithm for the permutation flow shop problem (PFSP) by improving over the existing $\mathcal{MMAS}$-PFSP algorithm. Some guidelines on how to improve over $\mathcal{MMAS}$-PFSP can be the following.

1. Implement a basic ACO algorithm for the PFSP; implement basic solution construction heuristics, for example, the NEH heuristic (Nawaz, Enscore, & Ham, 1983).

2. Implement an efficient local search algorithm for the PFSP. For details on how to implement an efficient local search procedure for the PFSP, see Nowicki & Smutnicki (1996b), who describe a tabu search algorithm for the PFSP. Reimplement also the original tabu search algorithm of Nowicki and Smutnicki; the reimplementation can be used to benchmark the performance of the ACO algorithm (take care that you reach the same level of performance as the original TS algorithm).

3. Try to enhance the solution construction by the summation rule taken from ACS-SMTWTP-MM (Merkle & Middendorf, 2003a). Note that improved performance may be obtained by appropriately combining the summation rule with a local evaluation rule.

4. Perform preliminary tests combining the ACO algorithm with the local search. Note that short TS runs may result in better overall performance of the ACO algorithm, similar to what is observed for the QAP (see also box 5.1).

5. If the previous steps did not result in excellent performance, try to improve the ACO algorithm by considering additional diversification/intensification techniques.

Exercise 5.2   As we said in section 5.2.4, concerning the graph coloring problem, a more successful ACO approach to the GCP could be obtained by exploiting a local search and by using ACO algorithms that are more advanced than AS. The exercise is then:

1. Reimplement the ACO algorithm proposed by Costa & Hertz (1997) and add different types of local search such as those proposed in Fleurent & Ferland (1996) and in Johnson, Aragon, McGeoch, & Schevon, 1991).

2. Try to solve the graph coloring problem using either ACS or $\mathcal{MMAS}$. Compare your results with those of Costa & Hertz (1997). Add a local search to your algorithms.

Exercise 5.3   Reimplement AS-MKP (see discussion in section 5.4.4), then implement a local search for the multiple knapsack problem and add it to AS-MKP. Compare the results with those of Leguizamón & Michalewicz (1999) and Vasquez & Hao (2001).

Exercise 5.4   Reimplement $\mathcal{MMAS}$-MCP (see section 5.4.4), implement the reactive local search by Battiti & Protasi (2001), and use it as local search for $\mathcal{MMAS}$-MCP. Compare the results obtained with the new ACO algorithm with those with the original algorithm of Battiti and Protasi.

Exercise 5.5   Reimplement $\mathcal{MMAS}$-CSP (see section 5.5.4) and compare it to the TS algorithm of Galinier & Hao (1997).

6 AntNet: An ACO Algorithm for Data Network Routing

An estimated lower bound on the size of the indexable Web is 320 million pages.
—Steve Lawrence and C. Lee Giles *Science, 280*, 1998

Number of web pages indexed by Google in June 2003: more than 3 billion.
—www.google.com

In this chapter we discuss AntNet, an ACO algorithm designed to help solve the routing problem in telecommunications networks. Network routing refers to the activities necessary to guide information in its travel from source to destination nodes. It is an important and difficult problem. Important because it has a strong influence on the overall network performance. Difficult because networks' characteristics, such as traffic load and network topology, may vary stochastically and in a time-varying way. It is in particular these characteristics of the problem, in addition to the physical distributedness of the overall problem on a real network, that make ACO algorithms a particularly promising method for its solution. In fact, the ACO processing paradigm is a good match for the distributed and nonstationary (in topology and traffic patterns) nature of the problem, presents a high level of redundancy and fault-tolerance, and can handle multiple objectives and constraints in a flexible way.

Although AntNet is not the only ACO algorithm developed for routing problems, and not even the historically first one, we focus on it because it is the sole algorithm to have reached, at least at the experimental/simulation level at which it was tested, state-of-the-art performance. We give a detailed description of AntNet's data structures and control procedures, and a brief overview of the results obtained using a network simulation environment.

6.1 The Routing Problem

Communications networks can be classified as either circuit-switched or packet-switched. The typical example of a circuit-switched network is the telephone network, in which a virtual or physical circuit is set up at the communication start and remains the same for the communication duration. Differently, in packet-switched networks, also called *data networks*, each data packet can, in principle, follow a different route, and no fixed virtual circuits are established. In this case the typical examples are local area computer networks and the Internet.

Arguably, the main function of a data network, on which we focus in this chapter, is to assure the efficient distribution of information among its users. This can be achieved through the exploitation of an adequate network control system. One of the most important components of such a system, in conjunction with the admission, flow, and congestion control components, is routing (Walrand & Varaiya, 1996). Routing refers to the distributed activity of building and using *routing tables*. The routing table is a common component of all routing algorithms: it holds the information used by the algorithm to make the local forwarding decisions. The type of information it contains and the way this information is used and updated strongly depend on the algorithm's characteristics. One routing table is maintained by each node in the network: it tells the node's incoming data packets which among the outgoing links to use to continue their travel toward their destination node. One of the most distinctive aspects of the network routing problem is the nonstationarity of the problem's characteristics. In particular, the characteristics of traffic over a network change all the time, and in some important cases (e.g., the Internet) the traffic can fluctuate in ways difficult to predict. Additionally, the nodes and links of a network can suddenly go out of service, and new nodes and links can be added at any moment. Therefore, network routing is very different from the $\mathcal{NP}$-hard problems we encountered in previous chapters. In fact, although in some simplified situations it is possible to reduce the routing problem to a standard combinatorial optimization problem, in realistic settings the dynamics of the traffic, and therefore of the costs associated with network links, is such that it might even be impossible to give a formal definition of what an optimal solution is.

6.1.1   A Broad Classification of Routing Algorithms

Routing algorithms can be broadly classified as centralized versus distributed and as static versus adaptive.

In *centralized* algorithms, a main controller is responsible for updating all the node's routing tables and for making every routing decision. Centralized algorithms can be used only in particular cases and for small networks. In general, the delays necessary to gather information about the network status and to broadcast the decisions and the updates make them infeasible in practice. Moreover, centralized systems are not fault-tolerant: if the main controller does not work properly, all the network is affected. In contrast, in *distributed* routing, the computation of paths is shared among the network nodes, which exchange the necessary information. The distributed paradigm is currently used in the great majority of networks.

In *static* routing, the path taken by a data packet is determined only on the basis of its source and destination, without regard to the current network traffic. The path chosen is usually the minimum cost path according to some cost criterion, and can be changed only to account for faulty links or nodes. *Adaptive* routing is, in principle, more attractive, because it can adapt the routing policy to time and spatially varying traffic conditions. As a drawback, adaptive algorithms can cause oscillations in the selection of paths. This can cause circular paths, as well as large fluctuations in measured performance (Bertsekas & Gallager, 1992).

Another interesting way of looking at routing algorithms is from an optimization perspective. In this case the main choice is between optimal routing and shortest path routing.

Optimal routing has a network-wide perspective and its goal is to optimize a function of all individual link flows (usually this function is a sum of link costs assigned on the basis of average packet delays) (Bertsekas & Gallager, 1992).

Shortest-path routing has a source-destination pair perspective: there is no global cost function to optimize. Its objective is to determine the shortest path (minimum cost) between two nodes, where the link costs are computed (statically or adaptively) according to some statistical description of the traffic flow crossing the links. Considering the different content stored in each routing table, shortest-path algorithms can be further subdivided into two classes called distance-vector and link-state (Steenstrup, 1995).

Distance-vector algorithms make use of routing tables consisting of a set of triples of the form (*destination, estimated distance, and next hop*), defined for all the destinations in the network and for all the neighbor nodes of the considered switch. In this case, the required topologic information is represented by the list of identifiers of the reachable nodes. The average per node memory occupation is in the order of $O(\varphi \cdot n)$, where φ is the average connectivity degree (i.e., the average number of neighbor nodes considered over all the nodes) and n is the number of nodes in the network. The algorithm works in an iterative, asynchronous, and distributed way. The information that every node sends to its neighbors is the list of its last estimates of the distances (intended as costs) from itself to all the other nodes in the network. After receiving this information from a neighbor node j, the receiving node i updates its table of distance estimates overwriting the entry corresponding to node j with the received values. Routing decisions at node i are made choosing as the next hop node the one satisfying the expression $\arg \min_{j \in \mathcal{N}_i} \{d_{ij} + D_j\}$, where d_{ij} is the assigned cost to the link connecting node i with its neighbor j and D_j is the estimated shortest distance from node j to the destination. It can be shown that this algorithm converges in finite time to the shortest paths with respect to the used metric if no link cost changes after a given time (Bellman, 1958; Ford & Fulkerson, 1962; Bertsekas & Gallager, 1992); this algorithm is also known as distributed *Bellman-Ford*.

Link-state algorithms make use of routing tables containing much more information than that used in distance-vector algorithms. In fact, at the core of link-state algorithms there is a distributed and replicated database. This database is essentially a dynamic map of the whole network, describing the details of all its components and their current interconnections. Using this database as input, each node calculates its best paths using an appropriate algorithm such as Dijkstra's (Dijkstra, 1959), and then uses knowledge about these best paths to build the routing tables. The memory requirement for each node in this case is $\mathcal{O}(n^2)$. In the most common form of link-state algorithm, each node acts autonomously, broadcasting information about its link costs and states and computing shortest paths from itself to all the destinations on the basis of its local link cost estimates and of the estimates received from other nodes. Each routing information packet is broadcast to all the neighbor nodes which in turn send the packet to their neighbors, and so on. A distributed flooding mechanism (Bertsekas & Gallager, 1992) supervises this information transmission, trying to minimize the number of retransmissions.

It should be clear to the reader, from what was said in chapter 1, that ACO algorithms can easily be adapted to solve routing problems following the shortest-path/distance-vector paradigm.

6.1.2   The Communication Network Model

Before we can describe the AntNet algorithm, it is necessary to accurately define the problem we are going to consider. In particular, we need to define the network architecture and protocols, as well as the characteristics of the input data traffic. In turn, this also defines the characteristics of the network simulator that is used for the experiments.

In this chapter, we focus on irregular topology packet-switched data networks with an IP-like network layer (in the ISO-OSI terminology (Tanenbaum, 1996)) and a very simple transport layer. In particular, we focus on wide area networks (WANs), of which the Internet is a noteworthy instance. In WANs, *hierarchical* organization schemes are adopted. Roughly speaking, subnetworks are seen as single host nodes connected to interface nodes called gateways. Gateways perform fairly sophisticated network layer tasks, including routing. Groups of gateways, connected by an arbitrary topology, define logical areas. Inside each area, all the gateways are at the same hierarchical level and "flat" routing is performed among them. Areas communicate only by means of area border gateways. In this way, the computational complexity of the routing problem, as seen by each gateway, is much reduced, at the cost of an increase in the complexity of the design and management of the routing protocols.

The instances of the communication networks that we consider in the following can be mapped on directed weighted graphs with n processing/forwarding nodes. All the links between pairs of nodes are viewed as bit pipes characterized by a bandwidth (bit/s) and a transmission delay (s). Every node is of type store-and-forward and has a buffer space where the incoming and outgoing packets are stored. This buffer is a shared resource among all the queues attached to every incoming and outgoing link of the node. All the traveling packets are subdivided into two classes: data and routing packets. Additionally, there are two priority levels in queues. Usually, data packets are served in the low-priority queues, while routing packets are served in the high-priority queues. The workload is defined in terms of applications whose arrival rate is given by a probabilistic model. By application (or session, or connection in the following), we mean a process sending data packets from an origin node to a destination node. The number of packets to send, their sizes, and the intervals between them are assigned according to some defined stochastic process. We do not make any distinction among nodes, which act at the same time as hosts (session endpoints) and gateways/routers (forwarding elements). The adopted workload model incorporates a simple flow control mechanism implemented by using a fixed production window for the session's packets generation. The window determines the maximum number of data packets that can be waiting to be sent. Once sent, a packet is considered to be acknowledged. This means that the transport layer neither manages error control, nor packet sequencing, nor acknowledgments and retransmissions. (This choice, which is the same as in the "Simple_Traffic" model in the MaRS network simulator (Alaettinoğlu, Shankar, Dussa-Zieger, & Matta, 1992), can be seen as a very basic form of file transfer protocol (FTP).)

For each incoming packet, the node's routing component uses the information stored in the local routing table to choose the outgoing link to be used to forward the packet toward its destination node. When the link resources become available, they are reserved and the transfer is set up. The time it takes to move a packet from one node to a neighboring one depends on the packet size and on the link's transmission characteristics. If, on a packet's arrival, there is not enough buffer space to hold it, the packet is discarded. Otherwise, a service time is stochastically generated for the newly arrived packet. This time represents the delay between the packet arrival time and the time when it will be put in the buffer queue of the outgoing link the local routing component has selected for it.

Situations causing a temporary or steady alteration of the network topology or of its physical characteristics (link or node failure, adding or deleting of network components, and so on) are not taken into account in the discussed implementation, though it is easy to add them.

In order to run experiments with AntNet, a complete network simulator was developed in C++ by Gianni Di Caro (Di Caro, 2003; Di Caro & Dorigo, 1998c). It is a discrete event simulator using as its main data structure an event list, which holds the next future events. The simulation time is a continuous variable and is set by the currently scheduled event. The aim of the simulator is to closely mirror the essential features of the concurrent and distributed behavior of a generic communication network without sacrificing efficiency and flexibility in code development.

6.2   The AntNet Algorithm

AntNet, the routing algorithm we discuss in this chapter, is a direct extension of the Simple Ant Colony Optimization algorithm discussed in chapter 1. As will become clear in the following, AntNet is even closer to the real ants' behavior that inspired the development of the ACO metaheuristic than the ACO algorithms for $\mathcal{NP}$-hard problems that we discussed in previous chapters.

Informally, the AntNet algorithm and its main characteristics can be summarized as follows.

- At regular intervals, and concurrently with the data traffic, from each network node artificial ants are asynchronously launched toward destination nodes selected according to the traffic distribution [see equation (6.2)].

- Artificial ants act concurrently and independently, and communicate in an indirect way (i.e., stigmergically; see chapter 7, section 7.3), through the pheromones they read and write locally on the nodes.

- Each artificial ant searches for a minimum cost path joining its source and destination node.

- Each artificial ant moves step by step toward its destination node. At each intermediate node a greedy stochastic policy is applied to choose the next node to move to. The policy makes use of (1) node-local artificial pheromones, (2) node-local problem-dependent heuristic information, and (3) the ant's memory.

- While moving, the artificial ants collect information about the time length, the congestion status, and the node identifiers of the followed path.

- Once they have arrived at the destination, the artificial ants go back to their source nodes by moving along the same path as before but in the opposite direction.

- During this backward travel, node-local models of the network status and the pheromones stored on each visited node are modified by the artificial ants as a function of the path they followed and of its goodness.

- Once they have returned to their source node, the artificial ants are deleted from the system.

In the following subsections the above scheme is explained, all its components are described and discussed, and a more detailed description of the algorithm is given.

6.2.1 AntNet: Data Structures

In AntNet, artificial ants move on the construction graph $G_C = (C, L)$, with the constraint of never using the set of links that do not belong to the network graph (see also chapter 2, section 2.2.1). In practice, therefore, artificial ants move on the network graph.

Like all ACO algorithms, AntNet exploits artificial pheromone trails. These are maintained in an artificial pheromone matrix $\mathcal{T}_i$ associated with each node i of the data network. The elements τ_{ijd}'s of $\mathcal{T}_i$ indicate the learned desirability for an ant in node i and with destination d to move to node j. In AntNet pheromones have three indices because the considered problem consists of the solution of many, $n(n-1)/2$, minimum cost paths problems simultaneously. Therefore, an ant on a node i can in principle have any of the remaining $n-1$ nodes as destination. Hence the notation τ_{ijd}, in which different pheromones are associated with different destination nodes (this notation differs from the one used in chapter 3 in which pheromones do not correspond to specific destinations and are therefore denoted by τ_{ij}'s).

Another specificity of AntNet, shared with ACO algorithms within the hyper-cube framework (see chapter 3, section 3.4.3), is that τ_{ijd}'s are normalized to 1:

$$\sum_{j \in \mathcal{N}_i} \tau_{ijd} = 1, \quad d \in [1, n] \text{ and } \forall i,$$

where $\mathcal{N}_i$ is the set of neighbors of node i, and $n = |C|$.

Additionally, AntNet maintains at each node i a simple parametric statistical model $\mathcal{M}_i$ of the traffic situation over the network as seen by node i. This local model is used to evaluate the paths produced by the artificial ants. In fact, unlike the typical situation found in applications of ACO to $\mathcal{NP}$-hard problems, in network routing it is rather difficult to evaluate the quality of a path having as sole information the time it took for the artificial ant to traverse it: this is because the time it takes to go from a source to a destination node depends not only on the routing decisions but also on the network traffic. The model $\mathcal{M}_i(\mu_{id}, \sigma_{id}^2, \mathcal{W}_{id})$ is adaptive and described by the sample mean μ_{id} and the variance σ_{id}^2 computed over the trip times experienced by the artificial ants, and by a moving observation window $\mathcal{W}_{id}$ used to store the best value $\mathcal{W}_{best_{id}}$ of the artificial ants' trip time. For each destination d in the network, the estimated mean and variance, μ_{id} and σ_{id}^2, give a representation of the expected time to go from node i to node d and of its stability. To compute these statistics AntNet uses the following exponential models:

$$\mu_{id} \leftarrow \mu_{id} + \varsigma(o_{i \rightarrow d} - \mu_{id}),$$

$$\sigma_{id}^2 \leftarrow \sigma_{id}^2 + \varsigma((o_{i \rightarrow d} - \mu_{id})^2 - \sigma_{id}^2), \tag{6.1}$$

where $o_{i \rightarrow d}$ is the new observed agent's trip time from node i to destination d. The factor ς (read: *varsigma*) weighs the number of most recent samples that will really affect the average. The weight of the k-th sample used to estimate the value of μ_{id} after j samplings, with $j > k$, is: $\varsigma(1 - \varsigma)^{j-k}$. In this way, for example, if $\varsigma = 0.1$, approximately only the latest fifty observations will really influence the estimate, for $\varsigma = 0.05$, the latest 100, and so on. Therefore, the number of effective observations is approximately $5/\varsigma$.

As we said, $\mathcal{W}_{id}$ is used to store the value $W_{best_{id}}$ of the best ants' trip time from node i toward destination d as observed in the last w samples. The value $W_{best_{id}}$ represents a short-term memory expressing an estimate of the optimal time to go to node d from the current node. After each new sample, the length w of the window is incremented modulus w_{max}, where w_{max} is the maximum allowed size of the observation window and is set to $w_{max} = 5c/\varsigma$, with $c \leq 1$, so that, when $c = 1$, the value $W_{best_{id}}$ and the exponential estimates refer to the same set of observations.

In this way, the long-term exponential mean and the short-term windowing are referring to a comparable set of observations.

$\mathcal{T}$ and $\mathcal{M}$, illustrated in figure 6.1, can be seen as memories local to nodes capturing different aspects of the network dynamics. The model $\mathcal{M}$ maintains absolute distance/time estimates to all the nodes, while the pheromone matrix gives relative goodness measures for each link-destination pair under the current routing policy implemented over all the network.

6.2.2   AntNet: The Algorithm

AntNet is conveniently described in terms of two sets of artificial ants, called in the following *forward* and *backward* ants. Ants in each set possess the same structure, but they are differently situated in the environment; that is, they can sense different inputs and they can produce different, independent outputs. Ants communicate in an indirect way, according to the stigmergy paradigm, through the information they concurrently read and write on the network nodes they visit.

The AntNet algorithm, whose high-level description in pseudo-code is given in figure 6.2, can be described as being composed of two main phases: solution construction, and data structures update. These are described in the following.

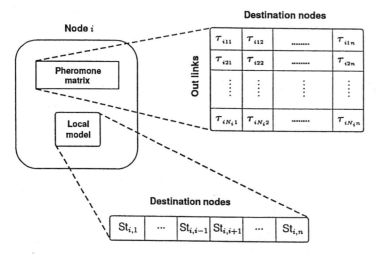

Figure 6.1
Data structures used by the artificial ants in AntNet for the case of a node i with $N_i = |\mathcal{N}_i|$ neighbors and a network with n nodes. The pheromone matrix $\mathcal{T}_i = [\tau_{ijd}]$ is isomorphic to the routing table used by the ants. The structure $\mathcal{M}_i = [\mathrm{St}_{i,d}]$, $d = 1, \ldots n$, $d \neq i$, containing the statistics $\mu_{id}, \sigma_{id}^2, \mathcal{W}_{id}$ about the local traffic, plays the role of a local adaptive model for the expected delay toward each possible destination.

Solution Construction

At regular intervals Δt from every network node s, a forward ant $F_{s \rightarrow d}$ is launched toward a destination node d to discover a feasible, low-cost path to that node and to investigate the load status of the network along the path. Forward ants share the same queues as data packets, so that they experience the same traffic load. Destinations are locally selected according to the data traffic patterns generated by the local workload: if f_{sd} is a measure (in bits or in the number of packets) of the data flow $s \rightarrow d$, then the probability of creating at node s a forward ant with node d as destination is

$$p_{sd} = \frac{f_{sd}}{\sum_{i=1}^{n} f_{si}}. \tag{6.2}$$

In this way, ants adapt their exploration activity to the varying data traffic distribution.

While traveling toward their destination nodes, the forward ants keep memory of their paths and of the traffic conditions found. The identifier of every visited node i and the time elapsed since the launching time to arrive at this i-th node are stored in a memory stack $S_{s \rightarrow d}(i)$. The ant builds a path performing the following steps:

procedure AntNet(t, t_{end}, Δt)

    input t % current time

    input t_{end} % time length of the simulation

    input Δt % time interval between ants generation

    foreach $i \in C$ **do**   % concurrent activity over the network

       $\mathcal{M} \leftarrow$ InitLocalTrafficModel

       $\mathcal{T} \leftarrow$ InitNodeRoutingTable

       while $t \leq t_{end}$ **do**

          in_parallel   % concurrent activity on each node

            if $(t \bmod \Delta t) = 0$ **then**

               $destination \leftarrow$ SelectDestination($traffic\_distribution\_at\_source$)

               LaunchForwardAnt($source$, $destination$)

            end-if

            foreach (ActiveForwardAnt[$source$, $current$, $destination$]) **do**

               while ($current \neq destination$) **do**

                  $next\_hop \leftarrow$ SelectLink($current$, $destination$, $link\_queues$, $\mathcal{T}$)

                  PutAntOnLinkQueue($current$, $next\_hop$)

                  WaitOnDataLinkQueue($current$, $next\_hop$)

                  CrossLink($current$, $next\_hop$)

                  Memorize($next\_hop$, $elapsed\_time$)

                  $current \leftarrow next\_hop$

               end-while

               LaunchBackwardAnt($destination$, $source$, $memory\_data$)

            end-foreach

            foreach (ActiveBackwardAnt[$source$, $current$, $destination$]) **do**

               while ($current \neq destination$) **do**

                  $next\_hop \leftarrow$ PopMemory

                  WaitOnHighPriorityLinkQueue($current$, $next\_hop$)

                  CrossLink($current$, $next\_hop$)

                  $from \leftarrow current$

                  $current \leftarrow next\_hop$

                  UpdateLocalTrafficModel($\mathcal{M}$, $current$, $from$, $source$, $memory\_data$)

                  $r \leftarrow$ GetNewPheromone($\mathcal{M}$, $current$, $from$, $source$, $memory\_data$)

                  UpdateLocalRoutingTable($\mathcal{T}$, $current$, $source$, r)

               end-while

            end-foreach

          end-in_parallel

       end-while

    end-foreach

end-procedure

1. At each node i, each forward ant headed toward a destination d selects the node j to move to, choosing among the neighbors it did not already visit, or over all the neighbors in case all of them had previously been visited. The neighbor j is selected with a probability P_{ijd} computed as the normalized sum of the pheromone τ_{ijd} with a heuristic value η_{ij} taking into account the state (the length) of the j-th link queue of the current node i:

$$P_{ijd} = \frac{\tau_{ijd} + \alpha \eta_{ij}}{1 + \alpha(|\mathcal{N}_i| - 1)}. \tag{6.3}$$

The heuristic value η_{ij} is a $[0, 1]$ normalized value function of the length q_{ij} (in bits waiting to be sent) of the queue on the link connecting the node i with its neighbor j:

$$\eta_{ij} = 1 - \frac{q_{ij}}{\sum_{l=1}^{|\mathcal{N}_i|} q_{il}}. \tag{6.4}$$

The value of α weighs the importance of the heuristic value with respect to the pheromone values stored in the pheromone matrix $\mathcal{T}$ (similar to what is done in the ANTS algorithm; see chapter 3, section 3.4.2). The value η_{ij} reflects the instantaneous state of the node's queues and, assuming that the queue's consuming process is almost stationary or slowly varying, η_{ij} gives a quantitative measure associated with the queue waiting time. The pheromone values, on the other hand, are the outcome of a continual learning process and capture both the current and the past status of the whole network as seen by the local node. Correcting these values with the values of η allows the system to be more "reactive," and at the same time it avoids following all the network fluctuations. An ant's decisions are therefore taken on the basis of a combination of a long-term learning process and an instantaneous heuristic prediction.

2. If a cycle is detected, that is, if an ant returns to an already visited node, the cycle's nodes are removed and all the memory about them is deleted. If the cycle lasted longer than the lifetime of the ant before entering the cycle, that is, if the cycle is greater than half the ant's age, the ant is deleted. In fact, in this case the agent wasted a lot of time, probably because of a wrong sequence of decisions and not because of congestion states. Therefore, the agent is carrying an old and misleading

Figure 6.2
AntNet's high-level description in pseudo-code. All the described actions take place in a completely distributed and concurrent way over the network nodes (while, in the text, AntNet has been described from an individual ant's perspective). All the constructs at the same level of indentation inside the context of the statement `in_parallel` are executed concurrently. The processes of data generation and forwarding are not described, but they can be thought as acting concurrently with the ants.

memory of the network's state and it could be counterproductive to use it to update the pheromone trails (see below).

3. When the destination node d is reached, the agent $F_{s \to d}$ generates another agent (backward ant) $B_{d \to s}$, transfers to it all of its memory, and is deleted. A forward ant is also deleted if its lifetime becomes greater than a value *max_life* before it reaches its destination node, where *max_life* is a parameter of the algorithm.

4. The backward ant takes the same path as that of its corresponding forward ant, but in the opposite direction. Backward ants do not share the same link queues as data packets; they use higher-priority queues reserved for routing packets, because their task is to quickly propagate to the pheromone matrices the information accumulated by the forward ants.

Data Structures Update
Arriving at a node i coming from a neighbor node, the backward ant updates the two main data structures of the node, the local model of the traffic $\mathcal{M}_i$ and the pheromone matrix $\mathcal{T}_i$, for all the entries corresponding to the (forward ant) destination node d. With some precautions, updates are performed also on the entries corresponding to every node $d' \in S_{i \to d}$, $d' \neq d$ on the "subpaths" followed by ant $F_{s \to d}$ after visiting the current node i. In fact, if the elapsed trip time of a subpath is statistically "good" (i.e., less than $\mu_{id} + I(\mu_{id}, \sigma_{id})$, where I is an estimate of a confidence interval for μ_{id}), then the time value is used to update the corresponding statistics and the pheromone matrix. On the contrary, trip times of subpaths that are not deemed good, in the same statistical sense as defined above, are not used because they might give a wrong estimate of the time to go toward the subdestination node. In fact, all the forward ant routing decisions were made only as a function of the destination node. In this perspective, subpaths are side effects, and they are potentially suboptimal because of local variations in the traffic load. Obviously, in the case of a good subpath, it can be used: the ant discovered, at zero cost, an additional good route. In the following, we describe the way $\mathcal{M}_i$ and $\mathcal{T}_i$ are updated with respect to a generic "destination" node $d' \in S_{i \to d}$. A simple example of the way AntNet's ants update $\mathcal{M}_i$ and $\mathcal{T}_i$ is given in figure 6.3.

- $\mathcal{M}_i$ is updated with the values stored in the backward ant's memory. The time elapsed to arrive (for the forward ant) to the destination node d' starting from the current node is used to update, according to equation (6.1), the mean and variance estimates, $\mu_{id'}$ and $\sigma_{id'}^2$, as well as the best value over the observation window $\mathcal{W}_{id'}$. In this way, a parametric model of the traveling time from node i to destination d' is maintained. The mean value of this time and its dispersion can vary strongly, depending on the traffic conditions: a poor time (path) under low traffic load can be

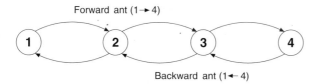

Figure 6.3
Example of the way AntNet's ants update node data structures. The forward ant, $F_{1 \to 4}$, moves along the path $1 \to 2 \to 3 \to 4$ and, arrived at node 4, launches the backward ant $B_{4 \to 1}$ which travels in the opposite direction. At each node i, $i = 3, \ldots, 1$, the backward ant uses its memory contents $S_{1 \to 4}(i)$ to update the values for $\mathcal{M}_i(\mu_{i4}, \sigma_{i4}^2, \mathcal{W}_{i4})$, and, in case of good subpaths, to update also the values for $\mathcal{M}_i(\mu_{id'}, \sigma_{id'}^2, \mathcal{W}_{id'})$, $d' = i+1, \ldots, 3$. At the same time, the pheromone matrix is updated by incrementing the pheromone τ_{ij4}, $j = i+1$, of the last node j the ant $B_{4 \to 1}$ came from, and decrementing the values of the pheromones of the other neighbors (not shown in the figure). The increment will be a function of the trip time experienced by the forward ant going from node i to destination node 4. As for $\mathcal{M}_i$, if the trip time associated by the forward ant with the subpaths to the other nodes $d' = i+1, \ldots, 3$ is statistically good, then also the corresponding pheromone matrix entries are updated. Adapted from Di Caro & Dorigo (1998c).

a very good one under heavy traffic load. The statistical model has to be able to capture this variability and to follow in a robust way the fluctuations of the traffic. This model plays a critical role in the pheromone matrix updating process, as explained in the following.

- The pheromone matrix $\mathcal{T}_i$ is updated by incrementing the pheromone $\tau_{ifd'}$ (i.e., the pheromone suggesting to choose neighbor f when destination is d') and decrementing, by normalization, the other pheromones $\tau_{ijd'}$, $j \in \mathcal{N}_i$, $j \neq f$. The way the pheromones are updated depends on a measure of goodness associated with the trip time $T_{i \to d'}$ experienced by the forward ant, and is given below. This time represents the only available explicit feedback signal to score paths. It gives a clear indication about the goodness of the followed path because it is proportional to its length from a physical point of view (number of hops, transmission capacity of the used links, processing speed of the crossed nodes) and from a traffic congestion point of view (because the forward ants share the same queues as data packets). Note that the time measure T cannot be associated with an exact error measure, given that the "optimal" trip times are not known, because they depend on the whole network load status. In fact, when the network is in a congested state, all the trip times will score poorly with respect to the times observed in low load situations. Nevertheless, a path with a high trip time should be scored as a good path if its trip time is significantly lower than the other trip times observed in the same congested situation. Therefore, T can only be used as a reinforcement signal. This gives rise to a credit assignment problem typical of the reinforcement learning field (Bertsekas & Tsitsiklis, 1996; Kaelbling, Littman, & Moore, 1996).

The reinforcement $r \equiv r(T, \mathcal{M}_i)$, $0 < r \leq 1$, is used to update the pheromones. It is computed by taking into account some average of the values observed so far and of their dispersion to score the goodness of the trip time T, such that the smaller the T, the higher the r (the exact definition of r is discussed in the next subsection). The value r is used by the backward ant $B_{d \to s}$ moving from node f to node i to increase the pheromone values $\tau_{ifd'}$. The pheromone $\tau_{ifd'}$ is increased by r as follows:

$$\tau_{ifd'} \leftarrow \tau_{ifd'} + r \cdot (1 - \tau_{ifd'}). \tag{6.5}$$

In this way, given a same value r, small pheromone values are increased proportionally more than large pheromone values, favoring in this way a quick exploitation of new, and good, discovered paths.

Pheromones $\tau_{ijd'}$ for destination d' of the other neighboring nodes j, $j \in \mathcal{N}_i$, $j \neq f$, evaporate implicitly by normalization. That is, their values are reduced so that the sum of pheromones on links exiting from node i will remain 1:

$$\tau_{ijd'} \leftarrow \tau_{ijd'} - r \cdot \tau_{ijd'}, \quad j \in \mathcal{N}_i, \, j \neq f. \tag{6.6}$$

It is important to remark that every discovered path increases its selection probability. In this way, not only does the (explicit) assigned value r play a role but also the (implicit) ant's arrival rate. In this respect, AntNet is closer to real ants' behavior than the other ACO algorithms for $\mathcal{NP}$-hard problems we have studied in the previous chapters: in fact, it exploits the *differential path length effect* described in chapter 1. This strategy is based on trusting paths that receive either high reinforcements, independent of their frequency, or low and frequent reinforcements. In fact, for any traffic load condition, a path receives one or more high reinforcements only if it is much better than previously explored paths. On the other hand, during a transient phase after a sudden increase in network load all paths will likely have high traversing times with respect to those learned by the model $\mathcal{M}$ in the preceding, low-congestion, situation. Therefore, in this case good paths can only be differentiated by the frequency of ants' arrivals. Assigning always a positive, but low, reinforcement value in the case of paths with high traversal time allows the implementation of the above mechanism based on the frequency of the reinforcements, while, at the same time, it avoids giving excessive credit to paths with high traversal time due to their poor quality.

6.2.3   How to Evaluate the Quality of an Ant's Trip

The value r is a critical quantity that has to be assigned after considering three main aspects: (1) paths should receive an increment in their selection probability proportional to their goodness, (2) the goodness is a relative measure, which depends on the traffic conditions that can be estimated by means of the models $\mathcal{M}_i$, and (3) it is important not to follow all the traffic fluctuations. This last aspect is particularly important. Uncontrolled oscillations in the routing tables are one of the main problems in shortest-path routing (Wang & Crowcroft, 1992). It is very important to be able to set the best trade-off between stability and adaptivity.

Several ways to assign the r values, trying to take into account the above three requirements, have been investigated:

- The simplest way is to set $r = constant$: independently of the ant's "experiment outcomes," the discovered paths are all rewarded in the same way. In this simple but meaningful case the core of the algorithm is based on the capability of "real" ants to discover shortest paths via stigmergic communication mediated by pheromone trails. In other words, what is at work is the differential path length effect: ants traveling along faster paths will arrive at a higher rate than other ants, hence their paths will receive a higher cumulative reward. The obvious problem with this approach lies in the fact that, although ants following longer paths arrive delayed, they will nevertheless have the same effect on the pheromone matrices as the ants that followed shorter paths.

- A more elaborate approach is to define r as a function of the ant's trip time T and of the parameters of the local statistical model $\mathcal{M}_i$. The following functional form gave good results, and was used in the experiments reported later in this chapter:

$$r = c_1 \left(\frac{W_{best}}{T} \right) + c_2 \left(\frac{I_{sup} - I_{inf}}{(I_{sup} - I_{inf}) + (T - I_{inf})} \right). \tag{6.7}$$

I_{sup} and I_{inf} are estimates of the limits of an approximate confidence interval for μ. I_{inf} is set to W_{best}, while $I_{sup} = \mu + z(\sigma/\sqrt{w})$, with $z = 1/\sqrt{(1 - v)}$ where v gives the selected confidence level. This expression is obtained using the Tchebycheff inequality that allows the definition of a confidence interval for a random variable following any distribution (Papoulis, 2001). Although usually, for specific probability densities, the Tchebycheff bound is not very tight, here its use is justified by the fact that only a raw estimate of the confidence interval is needed and that in this way there is no need to make any assumption on the distribution of μ.

The first term in equation (6.7) simply evaluates the ratio between the best trip time observed over the current observation window and the current trip time. The second term evaluates how far the value T is from I_{inf} in relation to the extension of the confidence interval, that is, considering the stability in the latest trip times. Note that the denominator of this term could go to zero, when $T = I_{sup} = I_{inf}$. In this case the whole term is set to zero. The coefficients c_1 and c_2 are parameters which weigh the importance of each term.

The value r obtained from equation (6.7) is finally transformed by means of a squash function $s(x)$:

$$r = \frac{s(r)}{s(1)}, \tag{6.8}$$

where

$$s(x) = \left(1 + exp\left(\frac{a}{x|\mathcal{N}_i|}\right)\right)^{-1}, \quad x \in (0, 1], \, a \in R^+. \tag{6.9}$$

Squashing the r-values allows the system to be more sensitive in rewarding good (high) values of r, while having the tendency to saturate the rewards for bad (near to zero) r-values: the scale is compressed for lower values and expanded in the upper part. In such a way an emphasis is put on good results.

6.3   The Experimental Settings

In this section we describe the test bed used to compare AntNet with some of the best-known routing algorithms. Note that, because the functioning of a data network is governed by many components which may interact in nonlinear and unpredictable ways, the choice of a meaningful test bed is not an easy task: the approach followed is to define a limited set of classes of tunable components. These are: the topology and the physical properties of the network, the traffic patterns, the metrics chosen for performance evaluation, the competing routing algorithms chosen, and their parameter values. In the following, for each class the choices are explained.

6.3.1   Topology and Physical Properties of the Net

The experiments presented in section 6.4 were run on models based on two real-world network instances: the US National Science Foundation network, NSFnet, and the Japanese NTT company backbone, NTTnet.

▪ *NSFnet* is the old USA T1 backbone (1987). NSFnet is a WAN composed of fourteen nodes and twenty-one bidirectional links with a bandwidth of 1.5 Mbit/s. Its topology is shown in figure 6.4. Propagation delays range from 4 to 20 ms. NSFnet is a well-balanced network, where a network is said to be well-balanced if the distribution of the shortest paths between all the pairs of nodes has a small variance.

Figure 6.4
NSFnet. Each arc in the graph represents a pair of directed links. Link bandwidth is 1.5 Mbit/s; propagation delays range from 4 to 20 ms.

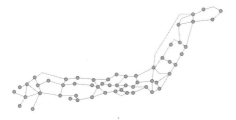

Figure 6.5
NTTnet. Each arc in the graph represents a pair of directed links. Link bandwidth is 6 Mbit/s, propagation delays range from 1 to 5 ms.

▪ *NTTnet* is a network modeled on the NTT (Nippon Telephone and Telegraph company) fiberoptic corporate backbone at the end of the '90s. It is a 57-node, 162 bidirectional links network. Link bandwidth is of 6 Mbit/s, while propagation delays range from 1 to 5 ms. Its topology is shown in figure 6.5. NTTnet is not a well-balanced network.

All the networks are simulated with link-fault and node-fault probabilities set to zero, local node buffers of 1 Gbit capacity, and data packet maximum time to live (TTL) set to 15 seconds.

6.3.2 Traffic Patterns

Traffic is defined in terms of open sessions between pairs of different nodes. Traffic patterns can show a huge variety of forms, depending on the characteristics of each session and on their distribution from geographic and temporal points of view. Each session is characterized by the number of transmitted packets, and by their size and interarrival time distributions. Sessions over a network can be characterized by their interarrival time distribution and by their geographic distribution. The latter is controlled by the probability assigned to each node to be selected as a session start or endpoint.

In the experiments three basic patterns for the temporal distribution of the sessions and three for their spatial distribution were considered.

Temporal Distributions

- *Poisson* (P): for each node a Poisson process regulates the arrival of new sessions (i.e., session interarrival times follow a negative exponential distribution).

- *Fixed* (F): at the beginning of the simulation, for each node a fixed number of one-to-all sessions is set up and left constant for the whole simulation.

- *Temporary* (TMPHS): a temporary, heavy-load traffic condition is generated, turning on some nodes that act like hot spots (see below).

Spatial Distributions

- *Uniform* (U): the assigned temporal characteristics for session arrivals are set to be identical in all the network nodes.

- *Random* (R): the assigned temporal characteristics for session arrivals are set randomly over the network nodes.

- *Hot spots* (HS): some nodes behave as hot spots, concentrating a high rate of input/output traffic. A fixed number of sessions are opened from the hot spots to all the other nodes.

General traffic patterns are obtained combining the above temporal and spatial characteristics. Therefore, for example, UP traffic means that on each node an identical Poisson process regulates the arrival of new sessions, while in the RP case the characteristics of the Poisson process are different for each node, and UP-HS means that a hot spots traffic model is superimposed on a UP traffic.

The bit streams generated by each session were chosen to have a time-varying bit rate (called *generic variable bit rate*, GVBR, in the following). The term GVBR is a broad generalization of the term *varying bit rate*, VBR, normally used to designate a bit stream with a variable bit rate but with known average characteristics and expected/admitted fluctuations. Here, a GVBR session generates packets whose sizes and interarrival times are variable and follow a negative exponential distribution. The information about these characteristics is never directly used by the routing algorithms, as in IP-based networks.

The values used in the experiments to shape traffic patterns are "reasonable" values for session generations and data packet production, taking into consideration network usage and computing power at the time the experiments were carried out (Di Caro & Dorigo, 1998c). The mean of the packet size distribution was set to 4096 bits in all the experiments. Basic temporal and spatial distributions are chosen to be representative of a wide class of possible situations that can be arbitrarily composed to generate a meaningful subset of real traffic patterns.

6.3.3 Metrics for Performance Evaluation

The metrics used for performance evaluation are *throughput* (correctly delivered bit/s) and *delay distribution* for data packets (s). These are the standard metrics for performance evaluation, when considering only sessions with equal costs, benefits, and priority and without the possibility of requests for special services like real time. Simulation results for throughput are reported as average values without an associated measure of variance. The intertrial variability is in fact always very low, within a few percentage points of the average value. Simulation results concerning packet delays are reported either using the whole empirical distribution or the 90th percentile, which allows comparison of algorithms on the basis of the upper value of delay they were able to keep 90% of the correctly delivered packets. In fact, packet delays can be spread over a wide range of values. This is an intrinsic characteristic of data networks: packet delays can range from very low values for sessions open between adjacent nodes connected by fast links, to much higher values in the case of sessions involving nodes very far apart connected by many slow links. Because of this, very often the empirical distribution of packet delays cannot be meaningfully parameterized in terms of mean and variance, and the 90th percentile statistic, or still better, the whole empirical distribution, is much more meaningful.

6.3.4 Competing Routing Algorithms and Their Parameters

AntNet performance was compared with state-of-the-art routing algorithms taken from the telecommunications and machine learning literature. The algorithms were reimplemented to make them as efficient as possible. They belong to the various possible combinations of static and adaptive, distance-vector, and link-state classes, and are listed below:

OSPF (static, link-state) is an implementation of the current Interior Gateway Protocol (IGP) of Internet (Moy, 1998). It is essentially a static shortest path algorithm.

SPF (adaptive, link-state) is the prototype of link-state algorithms with a dynamic metric for link cost evaluations. A similar algorithm was implemented in the second version of ARPANET (McQuillan, Richer, & Rosen, 1980) and in its successive revisions (Khanna & Zinky, 1989).

Table 6.1
Routing packets size for the implemented algorithms (except for the Daemon algorithm, which does not generate routing packets)

	AntNet	OSPF & SPF	BF	Q-R & PQ-R
Packet size (byte)	$24 + 8H$	$64 + 8|\mathcal{N}_i|$	$24 + 12n$	12

H is the incremental number of hops made by the forward ant, $|\mathcal{N}_i|$ is the number of neighbors of node i, and n is the number of network nodes. The values assigned to these parameters are either the same as used in previous simulation works (Alaettinoğlu et al., 1992) or were chosen on the basis of heuristic evaluations (e.g., the size of forward ants was set to be the same size as that of a BF packet plus 8 bytes for each hop to store the information about the node address and the elapsed time).

BF (adaptive, distance-vector) is the asynchronous distributed Bellman-Ford algorithm with dynamic metrics (Bertsekas & Gallager, 1992; Shankar et al., 1992).

Q-R (adaptive, distance-vector) is the Q-Routing algorithm proposed by Boyan & Littman (1994) (an online asynchronous version of the Bellman-Ford algorithm).

PQ-R (adaptive, distance-vector) is the Predictive Q-Routing algorithm of Choi & Yeung (1996).

Daemon (adaptive, optimal routing) is an approximation of an ideal algorithm defining an empirical upper bound on the achievable performance. The algorithm exploits a "daemon" able to read in every instant the state of all the queues in the network and then calculates instantaneous "real" costs for all the links and assigns paths on the basis of a network-wide shortest-paths recalculation for every packet hop.

All the algorithms used have a collection of parameters to be set. Common parameters are routing packet size and elaboration time. Settings for these parameters are shown in table 6.1.

Concerning the other main parameters, specific for each algorithm, for the AntNet competitors either the best settings available in the literature were used or the parameters were tuned as much as possible to obtain better results. For OSPF, SPF, and BF, the length of the time interval between consecutive routing information broadcasts and the length of the time window to average link costs are the same, and they are set to 0.8 or 3.0 seconds, depending on the experiment for SPF and BF, and to 30 seconds for OSPF. For Q-R and PQ-R the transmission of routing information is totally data-driven. The learning and adaptation rate used were the same as those used by the algorithms' authors (Boyan & Littman, 1994; Choi & Yeung, 1996).

Concerning AntNet, the algorithm is very robust to internal parameter settings. The parameter set was not fine-tuned and the same set of values was used in all the

Box 6.1
Parameter Settings for AntNet

In this box we report "good" values for AntNet's parameters. "Good" means that the value of these parameters was not optimized experimentally, so that it is to be expected that AntNet performance can be slightly increased by their careful optimization. Nevertheless, AntNet's performance was found to be rather robust with respect to limited variations in these parameter values.

- $\varsigma = 0.005$: exponential mean coefficient found in equation (6.1).

- $\Delta t = 0.3$ second: time interval between two consecutive ant generations.

- $\alpha = 0.45$: relative weight of heuristic information with respect to pheromones, found in equation (6.3). In all the experiments that were run it was observed that the use of the heuristic value is a very effective mechanism: depending on the characteristics of the problem, the best value to assign to the weight α can vary, but if α ranges between 0.2 and 0.5, performance doesn't change appreciably. For lower values, the effect of η_{ij} is vanishing, while for higher values the resulting routing tables oscillate and, in both cases, performance degrades.

- $max\_life = 15$: number of hops after which an ant is removed from the system.

- $w_{max} = 5(c/\varsigma)$, with $c = 0.3$: max length of the observation windows (see section 6.2.1).

- $c_1 = 0.7$, $c_2 = 0.3$: constants found in equation (6.7), to compute the value r used to update pheromones. Experiments have shown that c_2 should not be too big (i.e., smaller than 0.35), otherwise performance starts to degrade appreciably. The behavior of the algorithm is quite stable for c_2 values in the range 0.15 to 0.35, but setting c_2 below 0.15 slightly degrades performance.

- $I_{inf} = W_{best}$, $I_{sup} = \mu + z(\sigma/\sqrt{w})$, with $z = 1.7$: values found in equation (6.7).

- $a = 10$: constant found in equation (6.9).

different experiments presented in the next section. The settings for all parameters used by AntNet are summarized in box 6.1.

6.4   Results

In this section we compare AntNet with the competing routing algorithms described in section 6.3.4. The performance of the algorithms was studied for increasing traffic load and for temporary saturation conditions. In the experiments reported here, the saturating input traffic, whose level is a function of the routing algorithm used, was determined using AntNet as routing algorithm.

All reported data are averaged over ten trials lasting 1000 virtual seconds of simulation time, which was found to be a time interval long enough to make effects due to transients negligible and to get enough statistical data to evaluate the behavior of the routing algorithm. Before being fed with data traffic, the algorithms are given 500 preliminary simulation seconds with no data traffic to build initial routing tables. In this way, each algorithm builds the routing tables according to its own "vision" about minimum cost paths in relation to the physical characteristics of the network.

Parameter values for traffic characteristics are given in the figure captions with the following meaning: *MSIA* is the mean of the session interarrival time distribution for the Poisson (P) case, *MPIA* is the mean of the packet interarrival time distribution, *HS* is the number of hot-spot nodes, and *MPIA-HS* is the equivalent of MPIA for the hot-spot sessions. As we said (see section 6.3.2), the shape of the session bit streams is of the GVBR type.

It should be noted that when using AntNet, data packets are routed in a probabilistic way. This has been observed to improve AntNet performance, in some cases even by 30% to 40%, which means that the way the routing tables are built in AntNet is well matched with a probabilistic distribution of the data packets over all the good paths. Data packets are prevented from choosing links with very low probability by remapping the elements of the routing table P by means of a power function $f(x) = x^\delta, \delta > 1$, which emphasizes high probability values and reduces lower ones. This value was set to $\delta = 1.2$ in the experiments. Differently, the use of probabilistic data routing was found not to improve the performance of the algorithms used for comparison. Therefore, in all the other algorithms the routing of data was done deterministically by choosing at each hop the best neighbor among those indicated in the routing table.

Results for *throughput* and *packet delays* for all the considered network topologies are described in the two following subsections. Results concerning the *network resources utilization* are reported in section 6.4.3.

6.4.1   NSFnet

Experiments on NSFnet were run using UP, RP, UP-HS, and TMPHS-UP traffic patterns. In all the cases considered, differences in throughput were found to be of minor importance with respect to those shown by packet delays. For each of the UP, RP, and UP-HS cases, three distinct groups of ten trial experiments were run, gradually increasing the generated workload (in terms of reducing the session interarrival time). This amounts, as explained above, to studying the behavior of the algorithms when moving the traffic load toward a saturation region.

In the UP case, differences in throughput (figure 6.6a) were found to be small: the best performing algorithms were BF and SPF, which attained performances only about 10% inferior to that of Daemon and of the same amount better than those of AntNet, Q-R, and PQ-R, while OSPF behaved slightly better than the last-named. Concerning delays (figure 6.6b), the results were rather different: OSPF, Q-R, and PQ-R performed poorly, while BF and SPF had a performance on the order of 50% worse than that obtained by AntNet and 65% worse than Daemon.

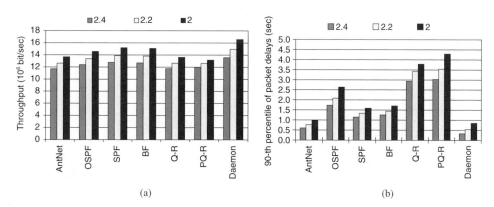

Figure 6.6
NSFnet: comparison of algorithms for increasing load for UP traffic. The load is increased reducing the MSIA value from 2.4 to 2.0 seconds (MPIA = 0.005 second). Statistics are computed over ten trials: (a) average throughput; (b) 90th percentile of the packet delays empirical distribution.

In the RP case (figure 6.7a), throughputs generated by AntNet, SPF, and BF were very similar, although AntNet presented a slightly better performance. OSPF and PQ-R behaved only slightly worse, while Q-R was the worst algorithm. Daemon was able to obtain only slightly better results than AntNet. Again, looking at packet delay results (figure 6.7b) OSPF, Q-R, and PQ-R performed very badly, while SPF showed results a bit better than those of BF but approximately 40% worse than those of AntNet. Daemon was in this case far better, which indicates that the test bed was very difficult.

For the case of UP-HS load, throughputs (figure 6.8a) for AntNet, SPF, BF, Q-R, and Daemon were found to be very similar, while OSPF and PQ-R gave much worse results. Again (figure 6.8b), packet delay results for OSPF, Q-R and PQ-R were much worse than those of the other algorithms (they were so much worse that they did not fit the scale chosen to highlight the differences between the other algorithms). AntNet was once again the best-performing algorithm (except, as usual, for Daemon). In this case, differences with SPF were found to be around 20%, and about 40% with respect to BF. Daemon performed about 50% better than AntNet and scaled much better than AntNet, which, again, indicates that the test bed was rather difficult.

The last graph for NSFnet shows how the algorithms behave in the case of a TMPHS-UP situation (figure 6.9). At time $t = 400$ four hot spots were turned on and superimposed on the existing light UP traffic. The transient was kept on for 120 seconds. In this case, only one, typical, situation is reported in detail to show how the different algorithms reacted. Reported values are the "instantaneous" values for

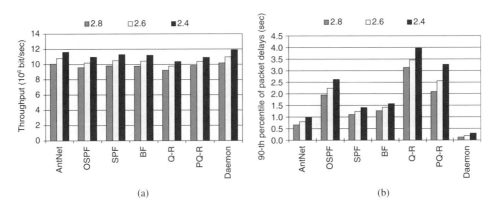

(a)   (b)

Figure 6.7
NSFnet: comparison of algorithms for increasing load for RP traffic. The load is increased reducing the MSIA value from 2.8 to 2.4 seconds (MPIA = 0.005 second). Statistics are computed over ten trials: (a) average throughput; (b) 90th percentile of the packet delays empirical distribution.

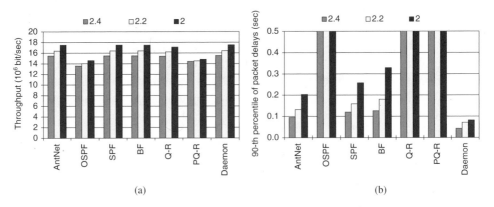

(a)   (b)

Figure 6.8
NSFnet: comparison of algorithms for increasing load for UP-HS traffic. The load is increased reducing the MSIA value from 2.4 to 2.0 seconds (MPIA = 0.3 second, HS = 4, MPIA-HS = 0.04 second). Statistics are computed over ten trials: (a) average throughput; (b) 90th percentile of the packet delays empirical distribution.

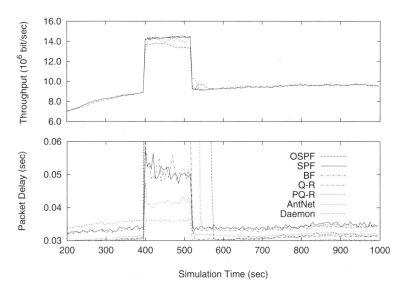

Figure 6.9
NSFnet: comparison of algorithms for transient saturation conditions with TMPHS-UP traffic (MSIA = 3.0 seconds, MPIA = 0.3 second, HS = 4, MPIA-HS = 0.04). Statistics are computed over ten trials: (up) average throughput; (down) packet delays averaged over 5-second moving windows. Reprinted by permission from Di Caro & Dorigo (1998c), © Morgan Kaufmann Publisher.

throughput and packet delays computed as the average over 5-second moving windows. All algorithms had a similar very good performance as far as throughput is concerned, except for OSPF and PQ-R, which lost a small percentage of the packets during the transitory period. The graph of packet delays confirms previous results: SPF and BF have a similar behavior, about 20% worse than AntNet and 45% worse than Daemon. The other three algorithms show a big out-of-scale jump not being able to properly dump the sudden load increase.

6.4.2   NTTnet

The same set of experiments run on the NSFnet was repeated on the NTTnet. In this case the results are even sharper than those obtained with NSFnet: AntNet performance is much better than that of all its competitors.

For the UP, RP, and UP-HS cases, differences in throughput are not significant (figures 6.10a, 6.11a, and 6.12a). All the algorithms, except the OSPF, practically behave in the same way as the Daemon algorithm. Concerning packet delays (figures 6.10b, 6.11b, and 6.12b), differences between AntNet and each of its competitors are

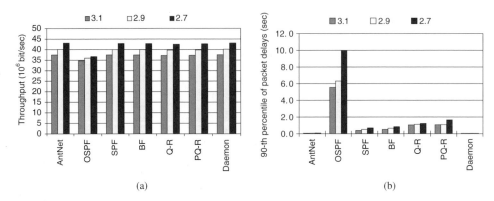

(a)   (b)

Figure 6.10
NTTnet: comparison of algorithms for increasing load for UP traffic. The load is increased reducing the MSIA value from 3.1 to 2.7 seconds (MPIA = 0.005 second). Statistics are computed over ten trials: (a) average throughput; (b) 90th percentile of the packet delays empirical distribution.

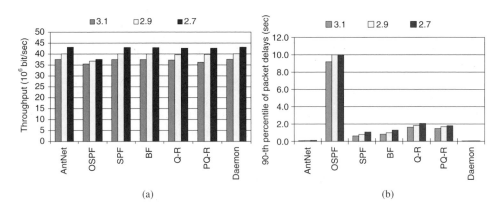

(a)   (b)

Figure 6.11
NTTnet: comparison of algorithms for increasing load for RP traffic. The load is increased reducing the MSIA value from 3.1 to 2.7 seconds (MPIA = 0.005 second). Statistics are computed over ten trials: (a) average throughput; (b) 90th percentile of the packet delays empirical distribution.

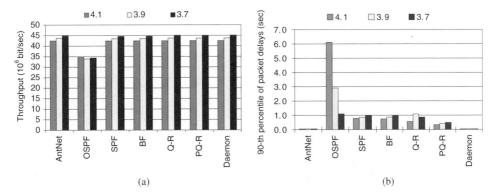

Figure 6.12
NTTnet: comparison of algorithms for increasing load for UP-HS traffic. The load is increased reducing the MSIA value from 4.1 to 3.7 seconds (MPIA = 0.3 second, HS = 4, MPIA-HS = 0.05 second). Statistics are computed over ten trials: (a) average throughput; (b) 90th percentile of the packet delays empirical distribution.

at least one order of magnitude in favor of AntNet. AntNet keeps delays at low values, very close to those obtained by Daemon, whereas SPF, BF, Q-R, and PQ-R perform poorly and OSPF completely collapses.

Note that in the UP-HS case, OSPF, which is the worst algorithm in this case, shows an interesting behavior. The increase in the generated data throughput determines a decrease or a very slow increase in the delivered throughput while delays decrease (see figure 6.12). In this case the load was too high for the algorithm and the balance between the two, conflicting, objectives, throughput and packet delays, showed an inverse dynamics: having a lot of packet losses made it possible for the surviving packets to obtain lower trip delays.

The TMPHS-UP experiment (figure 6.13), concerning sudden load variation, confirms the previous results. OSPF is not able to follow properly the variation both for throughput and delays. All the other algorithms were able to follow the sudden increase in the input throughput, but only AntNet (and Daemon) show a very regular behavior. Differences in packet delays are striking. AntNet performance is very close to that obtained by Daemon (the curves are practically superimposed at the scale used in the figure). Among the other algorithms, SPF and BF are the best, although their response is rather irregular and, in any case, much worse than AntNet's. OSPF and Q-R are out-of-scale and show a very delayed recovering curve. PQ-R, after a huge jump, which takes the graph out-of-scale in the first 40 seconds after hot spots are switched on, shows a trend approaching that of BF and SPF.

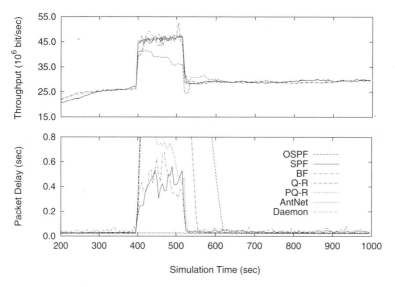

Figure 6.13
NTTnet: comparison of algorithms for transient saturation conditions with TMPHS-UP traffic (MSIA = 4.0 second, MPIA = 0.3 second, HS = 4, MPIA-HS = 0.05). Statistics are computed over ten trials: (up) average throughput; (down) packet delays averaged over 5-second moving windows. Reprinted by permission from Di Caro & Dorigo (1998c), © Morgan Kaufmann Publishers.

6.4.3   Routing Overhead

Table 6.2 reports results concerning the overhead generated by the routing packets. For each algorithm, the network load generated by the routing packets is reported as the ratio between the bandwidth occupied by the routing packets and the total available network bandwidth. Each row in the table refers to one of the experiments discussed in the two previous subsections. Routing overhead is computed for the experiment with the heaviest load in the increasing load series.

All data are scaled by a factor of 10^{-3}. The data in the table show that the routing overhead is negligible for all the algorithms with respect to the available bandwidth. Among the adaptive algorithms, BF shows the lowest overhead, closely followed by SPF. AntNet generates a slightly bigger consumption of network resources, but this is widely compensated by the higher performance it provides. Q-R and PQ-R produce an overhead a bit higher than that of AntNet. The routing load caused by the different algorithms is a function of many factors, specific to each algorithm. Q-R and PQ-R are data-driven algorithms: if the number of data packets or the length of the followed paths (because of topology or bad routing) grows, so will the number of

Table 6.2
Routing overhead: ratio between the bandwidth occupied by the routing packets and the total available network bandwidth

	AntNet	OSPF	SPF	BF	Q-R	PQ-R
NSFnet—UP	2.39	0.15	0.86	1.17	6.96	9.93
NSFnet—RP	2.60	0.15	1.07	1.17	5.26	7.74
NSFnet—UP-HS	1.63	0.15	1.14	1.17	7.66	8.46
NTTnet—UP	2.85	0.14	3.68	1.39	3.72	6.77
NTTnet—RP	4.41	0.14	3.02	1.18	3.36	6.37
NTTnet—UP-HS	3.81	0.14	4.56	1.39	3.09	4.81

All data are scaled by a factor of 10^{-3}. Adapted from Di Caro & Dorigo (1998c).

generated routing packets. BF, SPF, and OSPF have a more predictable behavior: the generated overhead is mainly a function of the topologic properties of the network and of the generation rate of the routing information packets. AntNet produces a routing overhead function of the ants' generation rate and of the length of the paths they travel.

The ant traffic can be roughly characterized as a collection of additional traffic sources, one for each network node, producing very small packets (and related acknowledgment packets) at a constant bit rate with destinations matching the input data traffic. On average, ants will travel over rather "short" paths and their size will grow by only 8 bytes at each hop. Therefore, each "ant routing traffic source" represents a very light additional traffic source with respect to network resources when the ant launching rate is not excessively high. In figure 6.14, the sensitivity of AntNet with respect to the ant launching rate is reported for a sample case of a UP data traffic model on NSFnet (previously studied in figure 6.6). The interval Δt between two consecutive ant generations is progressively decreased (Δt is the same for all nodes). Δt values are sampled at constant intervals over a logarithmic scale ranging from about 0.006 to 25 seconds. The lower, dashed, curve interpolates the generated routing overhead expressed, as before, as the fraction of the available network bandwidth used by routing packets. The upper, solid, curve plots the data for the obtained power normalized to its highest value observed during the trials, where the power is defined as the ratio between the delivered throughput and the 90th percentile of the packet delay distribution. The value used for delivered throughput is the throughput value at time 1000 averaged over ten trials, while for packet delay the 90th percentile of the empirical distribution was used.

In figure 6.14, it can be seen how an excessively small Δt causes an excessive growth of the routing overhead, with consequent reduction of the algorithm power. Similarly, when Δt is too big, the power slowly diminishes and tends toward a

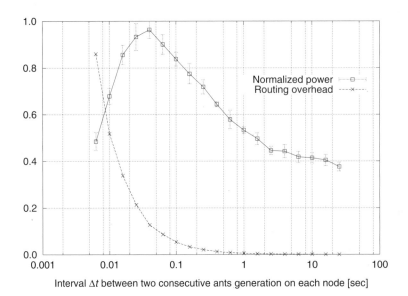

Interval Δ*t* between two consecutive ants generation on each node [sec]

Figure 6.14
AntNet normalized power versus normalized routing overhead as a function of the interval Δ*t* between two consecutive ants generation. Power is defined as the ratio between delivered throughput and the 90th percentile of the distribution of packet delays, and it is normalized to its highest value observed during the trials. Routing overhead is normalized by taking the ratio between the bandwidth used by the artificial ants and the available bandwidth over the whole network. Adapted from Di Caro & Dorigo (1998c).

plateau because the number of ants is not enough to generate and maintain up-to-date statistics of the network status. In the middle of these two extreme regions a wide range of Δ*t* intervals gives rise to similar, very good power values, while, at the same time, the routing overhead quickly falls toward negligible values. It should be noted that the value for Δ*t* used in the experiments, Δ*t* = 0.3, is not optimal. The reason for using a suboptimal parameter is that the analysis, whose results are shown in figure 6.14, was performed only after all the experiments were run, and the already good results obtained with the suboptimal parameter value did not motivate the authors to re-run the experiments.

6.5   AntNet and Stigmergy

In AntNet, the continual online adaptation of pheromone matrices (and therefore of the corresponding routing tables) is the emerging result of a collective learning process. In fact, each forward-backward ant pair is complex enough to find a good route and to adapt the pheromone matrices for a single-source destination path, but it cannot solve the global routing optimization problem. It is the interaction between the ants that determines the emergence of a global effective behavior from the point of view of network performance. Ants cooperate in their problem-solving activity by communicating in an indirect and noncoordinated asynchronous way. Each ant acts independently. Good routes are discovered by applying a policy that is a function of the information accessed through the network nodes visited, and the information collected about the route is eventually released on the same nodes. Therefore, communication among artificial ants is mediated in an explicit and implicit way by the "environment," that is, by the node's data structures and by the traffic patterns recursively generated by the data packets' utilization of the routing tables. In other words, ants exploit stigmergic communication (see chapter 1, section 1.4, for a definition of stigmergy). The stigmergic communication paradigm matches well the intrinsically distributed nature of the routing problem.

Cooperation among artificial ants goes on at two levels: (1) by modifications of the pheromone matrices, and (2) by modifications of local models that determine the way the ants' performance is evaluated. The way pheromone matrices are modified depends, among others, on the value of the reinforcement r. As we have seen in section 6.2.3, in AntNet this value is set to be a function of the ant's trip time and of the node-local statistical models [according to equations (6.7), (6.8), and (6.9)]. It is interesting, however, to note that reasonably good results are obtained when setting the value r to a constant. Results of experiments run with this strategy are presented in figure 6.15. These results suggest that the "implicit" component of the algorithm, based on the ant arrival rate (differential path length effect), plays a very important role. Of course, to compete with state-of-the-art algorithms, the available information about path costs has to be used.

As shown in the previous section, the results obtained with the above stigmergic model of computation are excellent. In terms of throughput and average delay, AntNet performs better than both classic and recently proposed routing algorithms on a wide range of experimental conditions (see Di Caro & Dorigo, 1998a,b,c,e,f, for further experimental results).

Finally, it is interesting to remark that the used stigmergy paradigm makes AntNet's artificial ants very flexible from a software engineering point of view. In this perspective, once the interface with the node's data structure is defined, the internal policy of the ants can be transparently updated. Also, the ants could be upgraded to become richer mobile agents that carry out multiple concurrent tasks as, for example, collecting information for distributed network management or for Web data-mining tasks (see Di Caro, 2003, for the first results in this direction).

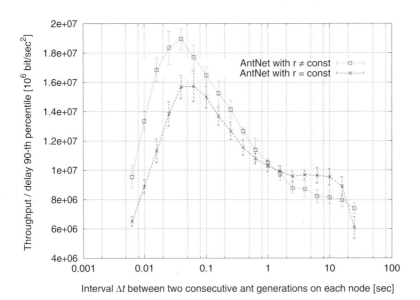

Figure 6.15
AntNet power for constant ($r = const$) and quality-based ($r \neq const$) pheromone updates as a function of the interval Δt between two consecutive ants generation (ants' launching rate). Power is defined as the ratio between delivered throughput and the 90th percentile of the distribution of packet delays.

6.6   AntNet, Monte Carlo Simulation, and Reinforcement Learning

The structure of AntNet allows one to draw some parallels with both parallel Monte Carlo simulation and with some well-known reinforcement learning (RL) algorithms. This is what is discussed in the rest of this section.

6.6.1   AntNet as an Online Monte Carlo System with Biased Exploration

The AntNet routing system can be seen as a collection of mobile agents collecting data about network status by concurrently performing online Monte Carlo simulations (Rubinstein, 1981; Streltsov & Vakili, 1996). In Monte Carlo methods, repeated experiments with stochastic transition components are run to collect data about the statistics of interest. Similarly, in AntNet, ants explore the network by performing random experiments (i.e., building paths from source to destination nodes using a stochastic policy dependent on the past and current network states), and collect online information on network status. A built-in variance reduction effect is determined by the way ants' destinations are assigned, biased by the most frequently observed data destinations, and by the way the ants' policy makes use of current and past traffic information. In this way, the explored paths match the most interesting paths from a data traffic point of view, which results in a very efficient variance reduction effect in the stochastic sampling of the paths. Unlike the usual offline Monte Carlo systems, in AntNet the state space sampling is performed online, that is, the sampling of the statistics and the controlling of the nonstationary traffic process are performed concurrently.

This way of exploring the network concurrently with data traffic is very different from what happens in the other algorithms where there is either no exploration at all (OSPF, SPF, and BF), or exploration is both tightly coupled to data traffic and of a local nature (Q-R and PQ-R). As shown in section 6.4.3, the extra traffic associated with the exploration is negligible for a wide range of values which allow very good performance.

6.6.2 AntNet and Reinforcement Learning

The characteristics of the routing problem allow one to interpret it as a distributed, stochastic time-varying RL problem (Sutton & Barto, 1998). This fact, as well as the structure of AntNet, makes it natural to draw some parallels between AntNet and classic RL approaches.

A first way to relate the structure of AntNet to that of an RL algorithm is connected to the way the outcomes of the experiments, the trip times $T_{k \to d}$, are processed. The transformation from the raw values $T_{k \to d}$ to the more refined reinforcements r are reminiscent of what happens in *actor-critic* systems (Barto, Sutton, & Anderson, 1983): the raw feedback signal from the environment is processed by a *critic* module, which is learning a model (the node's component $\mathcal{M}$) of the underlying process, and then is fed to the *actor* module that implements the policy (the pheromone matrix $\mathcal{T}$) and updates it according to the *critic* signal which consists of an evaluation of the policy followed by the ants. In our case, the *critic* is both adaptive, to take into account the variability of the traffic process, and rather simple, to meet computational requirements.

Another way of seeing AntNet as a classic RL system is related to its interpretation as a parallel replicated Monte Carlo system, as discussed in the previous subsection. In fact, as was shown first by Singh & Sutton (1996), a first-visit Monte Carlo simulation system (only the first visit to a state is used to estimate its value during a trial) is equivalent to a batch temporal difference (TD) method (Sutton, 1988) with replacing traces and decay parameter $\lambda = 1$. Although AntNet is a first-visit Monte Carlo simulation system, there are some important differences with the type of Monte Carlo used by Singh and Sutton (and in other RL works), mainly due to differences in the considered class of problems. In AntNet, outcomes of experiments are used both to update local models able to capture the variability of the whole network status (only partially observable) and to generate a sequence of stochastic policies. On the contrary, in the Monte Carlo system considered by Singh and Sutton, outcomes of the experiments are used to compute maximum-likelihood estimates of the expected mean and variance of the states' returns (i.e., the total reward following a visit of a state) of a Markov chain.

In spite of these differences, the weak parallel with $TD(\lambda)$ methods is rather interesting, and allows highlighting an important difference between AntNet and its competitors (and general TD methods): in AntNet there is no backchaining of the information from one state (i.e., a triple [current node, destination node, next hop node]) to its predecessors. Each state is rewarded only on the basis of the ant's trip time information strictly relevant to it. This approach is completely different from that followed by Q-R, PQ-R, and BF, which are TD methods, and, from a different perspective, by SPF. In fact, these algorithms build the distance estimates at each node by using the predictions made at other nodes. In particular, Q-R and PQ-R, which propagate the estimation information only one step back, are precisely distributed versions of the $TD(0)$ class of algorithms. They could be transformed into generic $TD(\lambda)$, $0 < \lambda \leq 1$, by transmitting backward to all the previously visited nodes the information collected by the routing packet generated after each data hop. Of course, this would greatly increase the routing traffic generated, because it has to be done after each hop of each data packet, making the approach at least very costly, if feasible at all.

In general, using temporal difference methods in the context of routing presents an important problem: the key condition of the method, the *self-consistency between the estimates of successive states*, may not be strictly satisfied in the general case. Here, by self-consistency between the estimates of successive states, we mean, for example, that the prediction made at node k about the time to go to the destination node d should be additively related to the prediction for the same destination from each one of k's neighbors, each neighbor being one of the ways to go to d. The lack of self-consistency in routing applications is due to the fact that (1) the dynamics at each node is related in a highly nonlinear way to the dynamics of all its neighbors, (2) the traffic process evolves concurrently over all the nodes, and (3) there is a recursive interaction between the traffic patterns and the control actions (i.e., the modifications of the pheromone matrices). This aspect can explain in part the poor performance of the pure $TD(0)$ algorithms Q-R and PQ-R.

6.7   Bibliographical Remarks

This chapter is strongly based on the work presented in Di Caro & Dorigo (1998c). Additional results obtained with AntNet on different network topologies can be found in a number of publications by Di Caro & Dorigo (1998a,b,e,f). A recent extension of AntNet in the direction of allowing network resources reservation is AntNet++. AntNet++ is a multiagent architecture for distributed learning and control in networks providing at the same time several types of services (e.g., best-effort and resource reservation at the same time) and is described in Dr. Di Caro's doctoral thesis (Di Caro, in preparation).

Schoonderwoerd and colleagues (1996) were the first to consider routing as a possible application domain for ACO algorithms. Their ant-based control (ABC) approach, which was applied to routing in telephone networks, differs from AntNet in many respects. The main differences are a direct consequence of the different network model they considered, which has the following characteristics: (1) connection links potentially carry an infinite number of full-duplex, fixed bandwidth channels, and (2) transmission nodes are crossbar switches with limited connectivity (i.e., there is no necessity for queue management in the nodes). In such a model, bottlenecks are put on the nodes, and the congestion degree of a network can be expressed in terms of connections still available at each switch. As a result, the network is cost-symmetric: the congestion status over available paths is completely bidirectional. The path $(n_0, n_1, n_2, \ldots, n_k)$ connecting nodes n_0 and n_k will exhibit the same level of congestion in both directions because the congestion depends only on the state of the nodes in the path. Moreover, dealing with telephone networks, each call occupies exactly one physical channel across the path. Therefore, "calls" are not multiplexed over the links, but they can be accepted or refused, depending on the possibility of reserving a physical circuit connecting the caller and the receiver. All these modeling assumptions make the problem of Schoonderwoerd et al. very different from the cost-asymmetric routing problem for data networks discussed in this chapter. This difference is reflected in many algorithmic differences between ABC and AntNet, the most important of which is that in ABC ants update pheromone trails after each step, without waiting for the completion of an ant's trip, as done in AntNet. This choice, which makes ABC behavior closer to real ants' behavior and which is reminiscent of the pheromone trail updating strategy implemented in the ant-density variant of AS (see chapter 3, section 3.3.1), was made possible by the cost-symmetry assumption made by the authors. Other differences are that ABC does not use local models to score the ants' trip times, nor local heuristic information and ant-private memory to improve the ants' decision policies. Also, it does not recover from cycles and does not use the information contained in all the ant subpaths.

Subramanian and colleagues (1997) have proposed an ant-based algorithm for packet-switched nets. Their algorithm is a straightforward extension of ABC, obtained by adding so-called *uniform ants*, an additional exploration mechanism that should avoid a rapid suboptimal convergence of the algorithm. A major limitation of these authors' work is that, although the algorithm they propose is based on the same cost-symmetry hypothesis as ABC, they apply it to packet-switched networks where this requirement is very often not met.

Heusse, Snyers, Guérin, & Kuntz (1998) have proposed the Cooperative Asymmetric Forward (CAF) model for routing in networks with asymmetric costs. CAF is an ant-based approach that is intended to build routing tables that permit use of a collection of paths between each pair of nodes. Unlike what happens in ABC and AntNet, in CAF routing tables are based on cost estimates to destination nodes. Depending on the objective pursued, CAF may use different metrics. Originally it was studied using the delay metric, but it was later more deeply studied using the load metric in the context of connection-oriented networks (Heusse & Kermarrec, 2000). CAF is strongly focused on convergence speed, and a complete presentation of this technique, which involves a large amount of cooperation between two types of agents, may be found in Heusse (2001).

Recently there has been a surge in interest concerning the use of ACO, and in particular of AntNet-like, algorithms for routing in mobile ad hoc networks (MANETs). Preliminary results, presented, for example, in Fujita, Saito, Matsui, & Matsuo (2002), Güneş, Sorges, & Bouazizi (2002), Güneş & Spaniol (2002), Baras & Mehta (2003), and Heissenbüttel & Braun (2003), are very promising and suggest that routing problems in these highly dynamic types of networks are a possible novel area for the successful application of ACO.

6.8   Things to Remember

- Network routing is a difficult problem because of its stochastic and time-varying nature.

- The distributed nature of network routing is well matched by the multiagent nature of ACO algorithms.

- AntNet is an ACO algorithm especially designed to solve routing problems in data networks. Its main differences with classic ACO algorithms applied to $\mathcal{NP}$-hard problems are (1) its use of the network graph as the construction graph; (2) its asynchronous nature (ants do not move synchronously on the graph as in $\mathcal{NP}$-hard applications), which allows the exploitation of the differential path length effect observed in real ants; (3) the extra machinery required to evaluate meaningfully the quality of the paths produced by the ants; and (4) the fact that it is used to solve online problems.

- In the simulation conditions and on the problems described in section 6.3, AntNet reaches a performance that is comparable to, or better than that of state-of-the-art algorithms such as OSPF, SPF, adaptive Bellman-Ford, Q-Routing, and PQ-Routing:

· Under *low load conditions*, all the algorithms tested have similar performance. In this case, also considering the huge variability in the possible traffic patterns, it is very hard to assess whether an algorithm is significantly better than another or not.

· Under *high, near-saturation loads*, all the tested algorithms are able to deliver the input throughput in a quite similar way, that is, in most of cases all the generated traffic is routed without big losses. On the contrary, the study of packet delay distributions has shown remarkable differences among the algorithms, in favor of AntNet.

· Under *saturation*, packet losses or packet delays, or both, become too big, causing the network operations to slow down. Therefore, saturation has to be only a temporary situation. If it is not, structural changes to the network characteristics, like adding new and faster connection lines, rather than improvements of the routing algorithm, should be in order. For these reasons, the responsiveness of the algorithms to traffic loads causing only a temporary saturation was studied. Here also, AntNet had a better performance than the competing algorithms.

6.9   Computer Exercises

Exercise 6.1   Reimplement the AntNet algorithm using as network traffic simulator a public domain software such as, for example, OMNeT++ or NS2. OMNeT++ is available at whale.hit.bme.hu/omnetpp/; NS2 is available at www.isi.edu/nsnam/ns/.

Exercise 6.2   In AntNet, at each intermediate node visited while building a path to their destination node, forward ants wait in line in the data packet queues. Although this allows them to simulate exactly the behavior of data packets, it causes delays in the subsequent propagation of the collected information (to be done by the corresponding backward ants). A possibility of avoiding this inherent delay would be to let forward ants use the same high-priority queues used by backward ants and to let backward ants make use of the current status of the local-link queues, that is, the number of bits waiting to be sent, to estimate the time that would have been required for a forward ant to cross the link at the current moment using the data queues. In this way, forward ants are much quicker in building a path from source to destination, and at the same time the backward ants update the local models and the routing tables with more up-to-date information.

Implement a variant of AntNet in which artificial ants use the above-mentioned estimates to evaluate the quality of their paths and compare it to the standard AntNet. Do you expect an increase or a decrease in performance? Are the results you obtain a function of the degree of variability of the data traffic?

Hint: You can find a discussion of this extension of AntNet in Di Caro & Dorigo (1998f), where it is called AntNet-CO, and in Di Caro (2003), where it is called AntNet-FA.

Exercise 6.3 Investigate the behavior of AntNet when each artificial ant can have a different value for the parameter α [α, found in equation (6.3), weighs the importance of the heuristic values with respect to the pheromone values].

Exercise 6.4 Study the behavior of AntNet when changing the values of the parameters listed in box 6.1.

Exercise 6.5 In AntNet, whenever an ant uses a link, the associated pheromone is incremented. Try to implement a version of AntNet in which ants can also cause a decrease in pheromones (i.e., negative updates are possible). The idea is that if an ant generates a very bad path, it could be sensible to decrease the probability of choosing it by future ants. Compare this version of AntNet with the standard one.

Exercise 6.6 Test the behavior of AntNet in the absence of data traffic. Check that it converges to routing tables implementing shortest paths among all node pairs. Does convergence to shortest paths depend on the way the reinforcement r is computed?

7 Conclusions and Prospects for the Future

Go to the ant, thou sluggard; consider her ways, and be wise. Which having no guide, overseer, or ruler, Provideth her meat in the summer, and gathereth her food in the harvest.
—Proverbs 6: 6–8

At the time of completing this monograph on ant colony optimization (early summer 2003), it was 13 years since the first ideas that led to ACO were developed at the Politecnico di Milano in Milan, Italy, and just 4 years since ACO was formalized as a metaheuristic. In this short time span many things have happened and ACO is now a well-recognized member of the family of metaheuristic methods for discrete optimization problems.

In this final chapter we briefly summarize what we know about ACO and we give a short overview of the current main research trends. We conclude by putting ACO in the context of the wider research field of *ant algorithms*.

7.1 What Do We Know about ACO?

What we have learned in the first 13 years of life of ACO is a lot. Even more is what we still need to learn and discover. In this section we briefly summarize our current knowledge of the ACO metaheuristic, and in the next section we overview what are, in our opinion, the most promising current research trends.

7.1.1 Theoretical Developments

To repeat the epigraph at the beginning of chapter 4: *In theory, there is no difference between theory and practice. But in practice, there is a difference!* Apart from being amusing, this aphorism contains much wisdom. It is true, in fact, that the theory developed for ACO (the same is true for other metaheuristics, though) has little use in practical terms. Nevertheless, what we have learned about theory will hopefully be useful for better understanding the working of our algorithms and, maybe, for designing better-performing ones in the future.

Summarizing, what we know about the theory aspects of ACO is the following:

- *Convergence proofs.* We know that some of the best-performing ACO algorithms ($\mathcal{MMAS}$ and ACS), both with and without local search, converge in value. As explained in chapter 4, convergence in value means that they will find, sooner or later, the optimal solution. This is a rather weak result, since it also applies to random search, and you can force the same property on any ACO algorithm by adding, for example, a procedure called every constant number of steps and that generates a random solution. The interesting point is, however, that the proof applies to two algorithms, $\mathcal{MMAS}$ and ACS, that were not designed to converge, and that at the same time have good performance over many different combinatorial optimization problems.

We also know that it is possible to force an ACO algorithm to converge in solution (i.e., generate over and over the same, optimal solution). This result can be obtained by letting the pheromones evaporate very slowly, so that the optimal solution has probability 1 of being generated before it might become impossible to generate it. This result has only theoretical interest. In fact, in optimization we are interested in generating the optimal solution once, and the fact that the algorithm generates it over and over has no practical interest.

• *Model-based search framework.*   An interesting question when a new algorithm or a new metaheuristic is proposed is its relation to other already existing algorithms or metaheuristics. By putting ACO in the framework of model-based search, a first, rather general answer to this question has been given: ACO algorithms have some general characteristics in common with such different algorithms as population based incremental learning (Baluja & Caruana, 1995), mutual-information–maximizing input clustering (MIMIC) (De Bonet et al., 1997), cross-entropy (Rubinstein, 1999), stochastic gradient descent (Robbins & Monroe, 1951), and estimation of distribution algorithms (Larrañaga & Lozano, 2001).

7.1.2   Experimental Results and Real-World Applications

As we have seen in chapter 5, ACO algorithms have been tested on a large number of academic problems. These include problems related to the traveling salesman, as well as assignment, scheduling, subset, and constraint satisfaction problems. For many of these, world-class performance has been achieved. For example, ACO algorithms are, at the time of writing, state-of-the-art (i.e., their performance is comparable to, or better than, that of the best existing methods other than ACO) for the sequential ordering problem (Gambardella & Dorigo, 2000), the vehicle routing problem with time window constraints (Gambardella et al., 1999), the quadratic assignment problem (Maniezzo, 1999; Stützle & Hoos, 2000), the group shop scheduling problem (Blum, 2003a), the arc-weighted *l*-cardinality tree problem (Blum & Blesa, 2003), and the shortest common supersequence problem (Michel & Middendorf, 1999). Additionally, very good performance has been obtained by AntNet (see chapter 6) on network routing problems (Di Caro & Dorigo, 1998c).

This success with academic problems has raised the attention of a number of companies that have started to use ACO algorithms for real-world applications. Among the first to exploit algorithms based on the ACO metaheuristic is EuroBios (www.eurobios.com). They have applied ACO to a number of different scheduling problems such as a continuous two-stage flow shop problem with finite reservoirs. The modeled problem included various real-world constraints such as setup times, capacity restrictions, resource compatibilities, and maintenance calendars. Another company that has played, and still plays, a very important role in promoting the real-world application of ACO is AntOptima (www.antoptima.com). AntOptima's researchers have developed a set of tools for the solution of vehicle routing problems whose optimization algorithms are based on ACO. Particularly successful products based on these tools are (1) DYVOIL, for the management and optimization of heating oil distribution with a nonhomogeneous fleet of trucks, used for the first time by Pina Petroli in Switzerland, and (2) ANTROUTE, for the routing of hundreds of vehicles of Migros, the leading Swiss supermarket chain. Still another vehicle routing application was developed by BiosGroup for the French company Air Liquide. Other interesting real-world applications are those of Gravel, Price, & Gagné (2002), who have applied ACO to an industrial scheduling problem in an aluminum casting center, and by Bautista & Pereira (2002), who successfully applied ACO to solve an assembly line balancing problem with multiobjective function and constraints between tasks for a bike assembly line.

7.2   Current Trends in ACO

Today, several hundred papers have been written on the applications of ACO. It is a true metaheuristic, with dozens of application areas. While both the performance of ACO algorithms and our theoretical understanding of their working have significantly increased, as shown in previous chapters, there are several areas in which until now only preliminary steps have been taken and where much more research will have to be done.

One of these research areas is the extension of ACO algorithms to more complex optimization problems that include (1) *dynamic problems*, in which the instance data, such as objective function values, decision parameters, or constraints, may change while solving the problem; (2) *stochastic problems*, in which one has only probabilistic information about objective function value(s), decision variable values, or constraint boundaries, due to uncertainty, noise, approximation, or other factors; and (3) *multiple objective problems*, in which a multiple objective function evaluates competing criteria of solution quality.

Active research directions in ACO include also the effective parallelization of ACO algorithms and, on a more theoretical level, the understanding and characterization of the behavior of ACO algorithms while solving a problem.

7.2.1   Dynamic Optimization Problems

A dynamic problem is a problem defined as a function of some quantities whose value is set by the dynamics of an underlying system. In other words, some of the characteristics of the problem change over time. A paradigmatic example is network routing, as discussed in chapter 6 (it should be noted that the network routing problems discussed in chapter 6 are both dynamic and stochastic: dynamic because traffic changes over time, and stochastic because the value assumed by traffic at different temporal instants is a stochastic variable).

Another dynamic problem that has been considered in the literature on ACO is the dynamic traveling salesman problem. In the dynamic TSP, cities may be deleted or added over time (see chapter 2, section 2.3.6). Guntsch & Middendorf (2001) and Guntsch, Middendorf, & Schmeck (2001) consider the case in which a certain percentage of the cities are deleted and replaced by new cities. The problem is then how to recompute quickly a good tour for the new TSP problem. Guntsch and colleagues propose three strategies. The first consists of a simple restart of the algorithm after all pheromone trails are reinitialized. The second and third strategies are based on the hypothesis that, at least for dynamic TSPs in which the percentage of cities replaced is not too high, it is useful to exploit the information contained in the pheromone trails. One of the two strategies reinitializes the pheromone trails exploiting heuristic information, while the other makes use of pheromone information. The experimental results, conducted on two instances from the TSPLIB (Reinelt, 1991), show that, if given enough computation time, the simple restart strategy performs better. Otherwise, if the time between two insertions/deletions is short, then the simple restart strategy has not enough time to find a new solution and the two restart strategies that partially reuse the pheromone information perform better.

Guntsch & Middendorf (2002b) have also proposed the use of *population-based* ACO for both dynamic TSPs and dynamic QAPs, where the dynamic QAP they consider is a QAP where in each fixed number of iterations some percentage of the locations is deleted and replaced by new locations. Population-based ACO differs from standard ACO as described in chapter 3 in that it maintains a population of *pop* solutions, used for updating the pheromone matrix. The algorithm works as follows: At the start of the algorithm, the population is empty. For the first *pop* iterations the iteration-best solution is added to the growing population without any constraints and no solution leaves the population. Whenever a solution enters the population, the pheromone matrix (which is initialized uniformly with a value τ_0) is updated by adding a constant quantity of pheromone $\Delta\tau$ to each element of the pheromone matrix which was used to build the solution. Beginning with iteration *pop* + 1, the iteration-best solution becomes a candidate for insertion in the population. Guntsch and Middendorf propose a few mechanisms to decide whether to insert the iteration-best solution or not. These are based on simple heuristics such as removing the oldest solution in the population, or removing the worst one, or removing solutions with a probability inversely proportional to their quality, and other combinations thereof. When a solution is removed from the population, the pheromone matrix is updated by removing a constant quantity of pheromone $\Delta\tau$ from each element of the pheromone matrix which was originally used to build the removed solution.

The interesting point of using population-based ACO for dynamic problems is that, because all the information necessary to generate the pheromone matrix is maintained in the population, in case the problem instance dynamically changes, it is easy to apply a repair operator to the solutions in the population and then to regenerate the pheromone matrix using the repaired solutions. In Guntsch & Middendorf (2002a) some preliminary tests of this idea are run on two problem instances, one TSP and one QAP. Comparisons with a restart algorithm, that is, an algorithm that does not make any repair, but simply restarts after each modification in the problem instance, showed that the population-based ACO approach is competitive either when the changes to the problem are large (i.e., many cities/locations are substituted so that the new optimal solution is very different from the old one), or when the time interval between two changes is short, so that the restart algorithm has not enough time to find new good solutions.

Last, Eyckelhof & Snoek (2002) have considered another type of dynamic TSP in which the number of cities remains constant and what changes is the distance between some pairs of cities (this is intended to represent sudden changes in traffic between selected locations). Their preliminary experimental results show that AS, as well as the few extensions they propose, work reasonably well on some simple test problems.

7.2.2   Stochastic Optimization Problems

By stochastic optimization we mean here those optimization problems for which some of the variables used to define them have a stochastic nature. This could be the problem components, as defined in chapter 2, section 2.2.1, which can have some probability of being part of the problem or not, or the values taken by some of the variables describing the problem, or the value returned by the objective function.

To the best of our knowledge, the only stochastic optimization problems to which ACO has been applied are (1) network routing, which was already discussed at length in chapter 6 and which is also a dynamic optimization problem, and (2) the probabilistic TSP (PTSP).

The first application of ACO to the PTSP was done by Bianchi, Gambardella, & Dorigo (2002a,b). In the PTSP, a TSP problem in which each city has a given probability of requiring a visit, the goal is to find an a priori tour of minimal expected length over all the cities, with the strategy of visiting a random subset of cities in the same order as they appear in the a priori tour. The ACO algorithm chosen by Bianchi et al. to solve the PTSP was ACS. In fact, they implemented two versions of ACS: the standard one (see chapter 3, section 3.4.1) and pACS, which differs from ACS only in the way the objective function is computed. In ACS, it is computed in the standard way for the case in which each city has probability 1 to require a visit, whereas in pACS the probabilities with which cities require a visit are taken into account. In practice, pACS uses the exact objective function, which can be computed in $\mathcal{O}(n^2)$ (Jaillet, 1985, 1988), while ACS uses an approximation of the objective function, which can be computed in $\mathcal{O}(n)$.

pACS was first experimentally shown to outperform some problem-specific heuristics, and it was then compared with ACS. The experimental results, which were run on homogeneous instances of the PTSP (i.e., all cities have the same probability of requiring a visit), show that pACS is the best among the two algorithms except for probabilities close to 1, in which case ACS is more efficient than pACS. This is due to the fact that the computation of the exact objective function is CPU–time-consuming, and this overhead is not justified in those cases in which the approximate objective function, which can be computed much faster, is close enough to the exact one.

More recently, Branke & Guntsch (2003) have considered two ways of improving the performance of ACO for the PTSP: they experimentally show that a coarse and fast-to-compute approximation of the exact objective function and the use of problem-specific heuristics to guide the ants during tour construction improve the algorithm performance.

7.2.3   Multiobjective Optimization Problems

Many problems from real-world applications require the evaluation of multiple, often conflicting, objectives. In such problems, which are called multiobjective optimization problems (MOOPs), the goal becomes to find a solution that gives the best compromise between the various objectives.

The selection of a compromise solution has to take into account the preferences of the decision maker. There are different ways to determine such compromise solutions. Under some mild assumptions on the preferences of the decision maker

(Steuer, 1986), compromise solutions belong to the set of efficient (or Pareto-optimal) solutions. A solution is called efficient if it is not dominated by any other solution, and the *Pareto-optimal set* is the set that contains all the efficient solutions. Hence, one possibility to solve MOOPs is to find the Pareto-optimal set, or at least a good approximation of it. The solutions in the set can then be given to the decision maker, who will choose among them according to personal criteria.

Differently, if the decision maker can give weights or priorities to the objectives before solving the problem, then the MOOP can be transformed into a single objective problem. In fact, in the first case the different objectives can be combined in a single objective given by their weighted sum, while in the second case the different solutions can be ordered according to priorities and compared lexicographically.

The first applications of ACO to the solution of multiobjective optimization problems are based on prioritized objectives. One such approach is the two-colony approach of Gambardella et al. (1999a) to the vehicle routing problem with time window constraints, which was presented in chapter 5, section 5.1.2.

A multicolony approach was also proposed by Mariano & Morales (1999) for the design of water irrigation networks. Their approach differs from that of Gambardella et al. in that (1) the first colony constructs only a partial solution to the problem that is completed to a full solution by the second colony; and (2) the solutions used to update the pheromone trails are all the nondominated solutions found after the second colony has completed the solution construction.

Other applications of ACO to MOOPs using prioritized objectives are those of T'kindt, Monmarché, Tercinet, & Laügt (2002) and Gravel et al. (2002). T'kindt et al. applied $\mathcal{MMAS}$ to a biobjective two-machine permutation flow shop problem. Computational results showed that $\mathcal{MMAS}$ yields excellent performance from a solution quality point of view. Gravel et al. applied an ACO algorithm to a four-objectives problem arising in a real-world scheduling problem for a aluminum casting center. They used the objectives to construct the heuristic information to be used by the ACO algorithm. For the pheromone update, only the most important objective was taken into account.

Doerner, Hartl, & Reimann (2001, 2003) used two cooperative ant colonies for solving biobjective transportation problems. They combined the two objective functions into a single one using a weighted sum approach. However, they used two colonies, which exploit different heuristic information for the solution construction; each of the heuristic information used takes into account one of the objectives. From an algorithmic point of view, the two approaches presented in Doerner, Hartl & Reimann (2001, 2003) differ mainly in the way information is exchanged between the colonies and in the way these colonies interact. Computational results suggest that the multiple-colony approach leads to improved performance when compared to the use of a single colony with single heuristic information.

Few ACO approaches exist that try to approximate the set of Pareto-optimal solutions. Doerner, Gutjahr, Hartl, Strauss, & Stummer (2003) apply an extension of ACS to a portfolio optimization problem. In their approach, for each of the objectives there is one pheromone matrix. An ant constructs a solution based on a weighted combination of the pheromone matrices; the weights used by each ant are chosen randomly when the ant is generated and kept fixed over its lifetime. After all ants have finished their solution construction, the pheromone matrices for each objective are updated by allowing the two ants with the best solutions for the corresponding objective to deposit pheromone. Experimental results obtained on instances with five and ten objectives showed that the proposed ACS algorithm performed better than the Pareto-simulated annealing proposed in Czyzak & Jaszkiewicz (1998) and the nondominated sorting genetic algorithm (Deb, 2001).

Iredi, Merkle, & Middendorf (2001) applied ACO algorithms to a biobjective single-machine total tardiness scheduling problem with changeover costs c_{ij} when switching from a job i to a job j. They used a multicolony ACO algorithm, in which each of the multiple colonies specializes in a different region of the Pareto front. Each colony uses two pheromone matrices, one corresponding to the total tardiness criterion and one to the changeover costs. For the solution construction, an ant uses a weighted combination of the pheromones and of the heuristic information with respect to the two criteria. After all ants of all colonies have completed their solution, the set of all nondominated solutions from all colonies is determined; only ants in this set are allowed to deposit pheromone. Experimental tests were done considering various possibilities for defining the region of the Pareto front to which the colonies specialize and the strategies for the pheromone update. This work was extended by Guntsch & Middendorf (2002b), who adapted population-based ACO to multiobjective problems and applied it to the same problem treated in the Iredi et al. paper as well as to a variant of this problem with four objectives (Guntsch & Middendorf, 2003).

7.2.4   Parallelization

Even when using metaheuristics, the solution of real-world optimization problems may require long computation times. Parallel implementations of ACO algorithms, for running on distributed (parallel) computing hardware, are therefore desirable.

ACO is inherently a distributed methodology which makes use of many individual and local procedures, so it is particularly suited to parallelization. Although a number of parallel versions of ACO have been implemented and tested in limited settings (see chapter 3, section 3.5), it is still an open question as to how to implement efficient parallel versions of ACO, and what type of performance improvement can be obtained over sequential versions.

An interesting research direction would also be to develop and test truly distributed ACO algorithms running on parallel hardware. In particular, ACO software running on Beowulf-style clusters of PCs (Sterling, Salmon, Becker, & Savarese, 1999) and GRID computing systems would be very useful to allow experimentation with real-world problems presenting multiobjective, dynamic, and stochastic characteristics, as discussed earlier.

Finally, a recent work by Merkle & Middendorf (2002b) has opened the way to implementations of ACO algorithms on *run-time reconfigurable processor arrays*.

7.2.5   Understanding ACO's Behavior

ACO algorithms are complex systems whose behavior is determined by the interaction of many components such as parameters, macroscopic algorithm components (e.g., the form of the probabilistic rule used by ants to build solutions, or the type of pheromone update rule used), and problem characteristics. Because of this, it is very difficult to predict their performance when they are applied to the solution of a novel problem.

Recently, researchers have started to try to understand ACO algorithm behavior by two typical approaches of science: (1) the study of the complex system under consideration in controlled and simplified experimental conditions, and (2) the study of the conditions under which the performance of the studied system degrades. Contributions along these two lines of research are briefly discussed in the following.

Study of ACO in Controlled and Simplified Experimental Conditions
The analysis of ACO algorithm behavior on simple problems is interesting because the behavior of the algorithm is not obscured by factors due to the complexity of the problem itself. A first such analysis was presented in chapter 1 (see also Dorigo & Stützle, 2001), where Simple-ACO was applied to the problem of finding the shortest path in a graph. The experimental results show that many algorithm components, which are essential to more advanced ACO models applied to challenging tasks, are also important for efficiently finding shortest paths.

In a similar vein, Merkle & Middendorf (2003b) apply ACO to the linear assignment problem, a permutation problem that is solvable in polynomial time (Papadimitriou & Steiglitz, 1982). By varying the cost matrix, they are able to generate classes of instances that differ in the number and structure of the optimal solutions. They tested three different ways of using the pheromones for the construction of a permutation and explored three different ways of filling the permutation (forward construction, backward construction, and assigning the positions in a random order). The experiments enabled the identification of situations in which particular construction rules fail to achieve good behavior and the explaination of why this can be the case.

Merkle & Middendorf (2002c) also proposed a deterministic model of ACO algorithms based on the ants' expected behavior and used it to model the dynamics of an ACO algorithm that uses iteration-best update when applied to a special type of permutation problem that consists of several, independent subproblems. They studied the behavior of the ACO model analytically and performed a fixed point analysis of the pheromone matrices, showing that the position of the fixed points in the state space of the system has a strong influence on the algorithm's optimization behavior.

Finally, Blum & Dorigo (2003, 2004) experimentally and theoretically studied the behaviour of AS applied to unconstrained binary problems, that is, binary problems for which the values of different decision variables are independent of each other. They were able to prove that, in this setting, the expected quality of the solutions generated by AS increases monotonically over time. Although their result cannot be transferred to the application of AS to constrained problems, in Blum & Dorigo (2003) they give empirical evidence that it holds for one of the most studied constrained problems: the TSP.

Study of the Conditions under which ACO Algorithm Performance Degrades

The study of problems in which a stochastic algorithm's performance degrades is an important research direction to understand an algorithm's behavior. Work in this direction has recently been done by Blum, Sampels, & Zlochin (2002), who have shown analytically that the expected quality of the solutions found by AS on particular instances of the arc-weighted l-cardinality tree problem (see chapter 5, section 5.4.3) may decrease over time. In their example this was the case because for the particular instance chosen there are two equally good competing solutions, and a third, bad solution is taking profit from this. However, in this particular case, the use of different update rules like iteration-best update would not lead to such behavior. This type of analysis is extended by Blum & Sampels (2002b) to ACO algorithms for the group shop scheduling problem (see also chapter 5, section 5.3.2). They show experimentally that for particular choices in the definition of the pheromone trails, the average quality of the solutions returned by ACO algorithms may decrease for a number of iterations, even if iteration-best update is used. A more detailed analysis showed that this effect becomes stronger when the problem becomes more constrained. Hence, the problem constraints, together with the chosen meaning given to the pheromone trails, determine how strong this detrimental effect, which Blum and Sampels call model bias, is.

7.3   Ant Algorithms

The ideas presented in this book are part of a growing discipline known collectively as *ant algorithms*. Ant algorithms have been defined in Dorigo et al. (2000a) and Dorigo (2001) as multiagent systems inspired by the observation of real ant colony behavior exploiting *stigmergy*.

Stigmergy, defined in chapter 1, section 1.4, plays an important role in ant algorithms research because the implementation of ant algorithms is made possible by the use of so-called *stigmergic variables*, that is, variables that contain the information used by the artificial ants to communicate indirectly. In some cases, as in the foraging behavior discussed in chapter 1 and at the base of ACO, the stigmergic variable is a specifically defined variable used by ants to adaptively change the way they build solutions to a considered problem.

But ant foraging is not the only social insect behavior that has inspired computer scientists and roboticists. Other examples, that we shall discuss only briefly here, are brood sorting and division of labor. In these cases, as discussed in the following, the stigmergic variable is one of the problem variables: a change in its value determines not only a change in the way a solution to the problem is built but also a direct change in the solution of the problem itself. A more comprehensive discussion of ant algorithms and stigmergy can be found in Bonabeau, Dorigo, & Theraulaz (1999, 2000) and in Dorigo et al. (2000a). In the following subsections we briefly describe some of the current directions in ant algorithms research.

7.3.1   Other Models Inspired by Foraging and Path Marking

As we know by now, the foraging behavior of ant colonies is at the basis of the ACO metaheuristic. But foraging, and more generally path-marking, behaviors have also inspired other types of algorithms. For example, Wagner, Lindenbaum, & Bruckstein proposed two algorithms for exploring a graph called, respectively, *Edge Ant Walk* (Wagner, Lindenbaum, & Bruckstein, 1996) and *Vertex Ant Walk* (Wagner, Lindenbaum, & Bruckstein, 1998) in which one or more artificial ants walk along the arcs of the graph, lay a pheromone trail on the visited arcs (or nodes), and use the pheromone trails deposited by previous ants to direct their exploration. Although the general idea behind the algorithm is similar to the one that inspired ACO, the actual implementation is very different. In the work of Wagner et al., pheromone trail is the stigmergic variable and is used as a kind of distributed memory that directs the ants toward unexplored areas of the search space. In fact, their goal is to cover the graph, that is, to visit all the nodes, without knowing the graph topology. They were able to prove a number of theoretical results, for example, concerning the time complexity for covering a generic graph. Also, they recently extended their algorithms (Wagner, Lindenbaum, & Bruckstein, 2000) so that they can be applied to dynamically changing graphs. A possible and promising application of this work is Internet search, where the problem is to keep track of the hundreds of thousands of pages added every day (Lawrence & Giles, 1998) (as well as of those that disappear).

7.3.2   Models Inspired by Brood Sorting

Brood sorting is an activity which can be observed in many ant species (e.g., in *Leptothorax unifasciatus* [Franks & Sendova-Franks, 1992], in *Lasius niger* [Chrétien, 1996], and in *Pheidole pallidula* [Deneubourg, Goss, Franks, Sendova-Franks, Detrain, & Chrétien, 1991]), which compactly cluster their smaller eggs and microlarvae at the center of the nest brood area and the largest larvae at the periphery of the brood cluster. Deneubourg et al. (1991) proposed a model of this phenomenon in which an ant picks up and drops an item according to the number of similar surrounding items. For example, if an ant carries a small egg, it will, with high probability, drop the small egg in a region populated by small eggs. On the contrary, if an ant is unloaded and finds a large larva surrounded by small eggs, it will, with high probability, pick up the larva. In all other situations the probability with which an ant picks up or drops an item is set to a small value.

Lumer & Faieta (1994) and Kuntz, Snyers, & Layzell (1999) have applied this model to the following clustering problem. Given are a set of points in an *n*-dimensional space and a metric *d* which measures the distance between pairs of points. The points must be projected onto the plane in such a way that if any two projected points in the plane are neighbors, their corresponding points in the *n*-dimensional space are neighbors under the metric *d*. The initial projection is random and the artificial ants then perform random walks on the plane and pick up or drop projected data items using rules from the model of Deneubourg et al. (1991). The results obtained are promising: they are qualitatively equivalent to those obtained by classic techniques like spectral decomposition or stress minimization (Kuntz et al., 1999), but at a lower computational cost. Recently, Handl & Meyer (2002) extended Lumer and Faieta's algorithm and proposed an application to the classification of Web documents and to their visualization in the form of topic maps (Fabrikant, 2000).

The model of Deneubourg et al. (1991) has also inspired a number of researchers in collective robotics who have implemented robotic systems capable of building clusters of objects without the need for any centralized control (Beckers, Holland, & Deneubourg, 1994; Martinoli & Mondada, 1998). Holland & Melhuish (1999) have extended the model of Deneubourg et al. so that it can be used by a colony of robots to sort objects.

In all these applications the stigmergic variable is represented by the physical distribution of the items: different configurations of the items determine different actions by the artificial agents.

7.3.3   Models Inspired by Division of Labor

In ant colonies individual workers tend to specialize in certain tasks (Robinson, 1992). Nevertheless, ants can adapt their behavior to the circumstances: a soldier ant can become a forager, a nurse ant a guard, and so on. Such a combination of specialization and flexibility in task allocation is appealing for multiagent optimization and control, particularly in resource or task allocation problems that require continuous adaptation to changing conditions. Robinson (1992) developed a threshold model in which workers with low response thresholds respond to lower levels of stimuli than do workers with high response thresholds. In this model the stimuli play the role of stigmergic variables.

A response threshold model of division of labor in which task performance reduces the intensity of stimuli has been used to solve dynamic task-scheduling problems (Bonabeau et al., 1999; Bonabeau, Sobkowski, Theraulaz, & Deneubourg, 1997a). When workers with low thresholds perform their normal tasks, the task-associated stimuli never reach the thresholds of the high-threshold workers. But if, for any reason, the intensity of task-associated stimuli increases, high-threshold workers engage in task performance. Bonabeau et al. (1999) and Campos, Bonabeau, Theraulaz, & Deneubourg (2000) present an application of these ideas to the problem of choosing a paint booth for trucks coming out of an assembly line in a truck factory. In this system each paint booth is considered an insect-like agent that, although specialized in one color, can, if needed, change its color (though it is expensive). The ant algorithm minimizes the number of booth setups (i.e., paint changeovers). This and similar scheduling and task allocation problems were also recently investigated by Nouyan (2002) and Cicirello & Smith (2001, 2003).

The threshold model was also used by Krieger & Billeter (2000) and Krieger, Billeter, & Keller (2000) to organize a group of robots. They designed a group of Khepera robots (miniature mobile robots aimed at "desktop" experiments [Mondada, Franzi, & Ienne, 1993]) to collectively perform an object-retrieval task.

In one of the experiments they performed, the objects were spread in the environment and the robots' task was to take them back to their "nest" where they were dropped in a basket. The available "energy" of the group of robots decreased regularly with time, but was increased when pucks were dropped into the basket. More energy was consumed during retrieval trips than when robots were immobile in the nest. Each robot had a threshold for the retrieval task: when the energy of the colony went below the threshold of a robot, the robot left the nest to look for objects in the environment. Krieger and Billeter's experiment has shown the viability of the threshold-based ant algorithm in a rather simple environment. Further experimentation is necessary to test the methodology on more complex tasks.

7.3.4   Models Inspired by Cooperative Transport

The behavior of ant colonies has also inspired roboticists interested in the design of distributed control algorithms for groups of robots (Martinoli & Mondada, 1998). An example of a task that has been used as a benchmark for ant algorithms applied to distributed robotics problems is cooperative box pushing (Kube & Zhang, 1994). In several ant species, when it is impossible for a single ant to retrieve a large item, nest mates are recruited to help through direct contact or chemical marking (Franks, 1986; Moffett, 1988; Sudd, 1965), implementing in this way a form of cooperative transport. The ants move around the item they want to carry, changing position and alignment until they succeed in carrying it toward the nest. An ant algorithm which reproduces the behavior of real ants in a group of robots whose task is to push a box toward a goal has been implemented and tested by Kube & Zhang (1994). Another example of application of ant algorithms is the related problem of pulling an object. This has been achieved (Dorigo, Trianni, Şahin, Labella, Gross, Baldassare, Nolfi; Deneubourg, Mondada, Floreano, & Gambardella, 2003) within the Swarm-bots project (www.swarm-bots.org), a project dedicated to the study of ant algorithms for autonomous robotics applications.

Appendix: Sources of Information about the ACO Field

There are a number of sources for information about the ACO field. The most important ones are listed in the following.

- *Webpages*

  - www.aco-metaheuristic.org: These are the official webpages dedicated to collecting information about ACO.

  - www.metaheuristics.org: These are the webpages of the "Metaheuristics Network" project. This European Union–funded project is dedicated to the theoretical analysis and experimental comparison of metaheuristics.

- *Software.* Software, distributed under the GNU license, is available at: www.aco-metaheuristic.org/aco-code/

- *Popular press.* ACO is often covered by the popular press. Pointers to popularization articles can be found at: www.aco-metaheuristic.org/aco-in-the-press.html

- *Mailing list.* A moderated mailing list dedicated to the exchange of information related to ACO is accessible at: www.aco-metaheuristic.org/mailing-list.html

- *Conferences and journals*

  - "ANTS 2004—Fourth International Workshop on Ant Colony Optimization and Swarm Intelligence." The ANTS biannual series of workshops (see iridia.ulb.ac.be/~ants), held for the first time in 1998, is the oldest conference in the ACO and swarm intelligence fields.

  - "From Worker to Colony: International Workshop on the Mathematics and Algorithms of Social Insects." This workshop was held for the first time in Cambridge, UK, in 2001, and the second workshop took place at the Georgia Institute of Technology, Atlanta, in December 2003.

  - Special sessions or special tracks on ACO are organized in many conferences. Examples are the IEEE Congress on Evolutionary Computation (CEC) and the Genetic and Evolutionary Computation (GECCO) series of conferences.

  - Papers on ACO can regularly be found in many other conferences such as "Parallel Problem Solving from Nature" conferences, INFORMS meetings, ECCO conferences, the Metaheuristics International Conference, the European Workshop on Evolutionary Computation in Combinatorial Optimization, and many others, and in many journals, such as *Artificial Life*; *Evolutionary Computation*; *IEEE Transactions on Systems, Man, and Cybernetics*; *IEEE Transactions on Evolutionary Computation*; *INFORMS Journal on Computing*; *Journal of Operations Research Society*; *European Journal of Operational Research*; and so on.

References

Aardal, K. I., van Hoesel, S. P. M., Koster, A. M. C. A., Mannino, C., & Sassano, A. (2001). Models and solution techniques for the frequency assignment problem. Technical report 01-40, Konrad-Zuse-Zentrum für Informationstechnik, Berlin.

Aarts, E. H. L., Korst, J. H. M., & van Laarhoven, P. J. M. (1997). Simulated annealing. In E. H. L. Aarts & J. K. Lenstra (Eds.), *Local Search in Combinatorial Optimization* (pp. 91–120). Chichester, UK, John Wiley & Sons.

Aarts, E. H. L., & Lenstra, J. K. (Eds.). (1997). *Local Search in Combinatorial Optimization*. Chichester, UK, John Wiley & Sons.

Abdul-Razaq, T. S., Potts, C. N., & Wassenhove, L. N. V. (1990). A survey of algorithms for the single machine total weighted tardiness scheduling problem. *Discrete Applied Mathematics, 26*(2), 235–253.

Alaettinoğlu, C., Shankar, A. U., Dussa-Zieger, K., & Matta, I. (1992). Design and implementation of MaRS: A routing testbed. Technical report UMIACS-TR-92-103, CS-TR-2964, Institute for Advanced Computer Studies and Department of Computer Science, University of Maryland, College Park.

Anstreicher, K. M., Brixius, N. W., Goux, J.-P., & Linderoth, J. (2002). Solving large quadratic assignment problems on computational grids. *Mathematical Programming, 91*(3), 563–588.

Applegate, D., Bixby, R., Chvátal, V., & Cook, W. (1995). Finding cuts in the TSP. Technical report 95-05, DIMACS Center, Rutgers University, Piscataway, NJ.

Applegate, D., Bixby, R., Chvátal, V., & Cook, W. (1998). On the solution of traveling salesman problems. *Documenta Mathematica, Extra Volume ICM III*, 645–656.

Applegate, D., Bixby, R., Chvátal, V., & Cook, W. (1999). Finding tours in the TSP. Technical report 99885, Forschungsinstitut für Diskrete Mathematik, University of Bonn, Germany.

Applegate, D., Cook, W., & Rohe, A. (2003). Chained Lin-Kernighan for large traveling salesman problems. *INFORMS Journal on Computing, 15*(1), 82–92.

Ausiello, G., Crescenzi, P., Gambosi, G., Kann, V., Marchetti-Spaccamela, A., & Protasi, M. (1999). *Complexity and Approximation—Combinatorial Optimization Problems and Their Approximability Properties*. Berlin, Springer-Verlag.

Bacchus, F., Chen, X., van Beek, P., & Walsh, T. (2002). Binary vs non-binary constraints. *Artificial Intelligence, 140*(1–2), 1–37.

Baird, L. C., & Moore, A. W. (1999). Gradient descent for general reinforcement learning. In M. Kearns, S. Solla, & D. Cohn (Eds.), *Advances in Neural Information Processing Systems, 11* (pp. 968–974). Cambridge, MA, MIT Press.

Balas, E., & Vazacopoulos, A. (1998). Guided local search with shifting bottleneck for job shop scheduling. *Management Science, 44*(2), 262–275.

Baluja, S., & Caruana, R. (1995). Removing the genetics from the standard genetic algorithm. In A. Prieditis & S. Russell (Eds.), *Proceedings of the Twelfth International Conference on Machine Learning (ML-95)* (pp. 38–46). Palo Alto, CA, Morgan Kaufmann.

Baras, J. S., & Mehta, H. (2003). A probabilistic emergent routing algorithm for mobile ad hoc networks. In *Proceedings of WiOpt03: Modeling and Optimization in Mobile, Ad Hoc and Wireless Networks. Sophia-Antipolis, France, INRIA*.

Barto, A. G., Sutton, R. S., & Anderson, C. W. (1983). Neuronlike adaptive elements that can solve difficult learning control problems. *IEEE Transactions on Systems, Man, and Cybernetics, 13*, 834–846.

Battiti, R., & Protasi, M. (2001). Reactive local search for the maximum clique problem. *Algorithmica, 29*(4), 610–637.

Battiti, R., & Tecchiolli, G. (1994). The reactive tabu search. *ORSA Journal on Computing, 6*(2), 126–140.

Bauer, A., Bullnheimer, B., Hartl, R. F., & Strauss, C. (2000). Minimizing total tardiness on a single machine using ant colony optimization. *Central European Journal for Operations Research and Economics, 8*(2), 125–141.

Baum, E. B. (1986). Towards practical "neural" computation for combinatorial optimization problems. In J. S. Denker (Ed.), *Neural Networks for Computing*, vol. 151 (pp. 53–58). New York, American Institute of Physics Conference Proceedings.

Bautista, J., & Pereira, J. (2002). Ant algorithms for assembly line balancing. In M. Dorigo, G. Di Caro, & M. Sampels (Eds.), *Proceedings of ANTS 2002—From Ant Colonies to Artificial Ants: Third International Workshop on Ant Algorithms*, vol. 2463 of *Lecture Notes in Computer Science* (pp. 65–75). Berlin, Springer-Verlag.

Baxter, J. (1981). Local optima avoidance in depot location. *Journal of the Operational Research Society*, 32, 815–819.

Beckers, R., Deneubourg, J.-L., & Goss, S. (1993). Modulation of trail laying in the ant *Lasius niger* (hymenoptera: Formicidae) and its role in the collective selection of a food source. *Journal of Insect Behavior*, 6(6), 751–759.

Beckers, R., Holland, O. E., & Deneubourg, J.-L. (1994). From local actions to global tasks: Stigmergy and collective robotics. In R. Brooks & P. Maes (Eds.), *Artificial Life IV* (pp. 181–189). Cambridge, MA, MIT Press.

Bellman, R. (1958). On a routing problem. *Quarterly of Applied Mathematics*, 16(1), 87–90.

Bellman, R., Esogbue, A. O., & Nabeshima, I. (1982). *Mathematical Aspects of Scheduling and Applications*. New York, Pergamon Press.

Bentley, J. L. (1992). Fast algorithms for geometric traveling salesman problems. *ORSA Journal on Computing*, 4(4), 387–411.

Berger, B., & Leight, T. (1998). Protein folding in the hydrophobic-hydrophilic (hp) model is NP-complete. *Journal of Computational Biology*, 5(1), 27–40.

Bertsekas, D. (1995a). *Dynamic Programming and Optimal Control*. Belmont, MA, Athena Scientific.

Bertsekas, D. (1995b). *Nonlinear Programming*. Belmont, MA, Athena Scientific.

Bertsekas, D., & Gallager, R. (1992). *Data Networks*. Englewood Cliffs, NJ, Prentice Hall.

Bertsekas, D., & Tsitsiklis, J. (1996). *Neuro-Dynamic Programming*. Belmont, MA, Athena Scientific.

Bianchi, L., Gambardella, L. M., & Dorigo, M. (2002a). An ant colony optimization approach to the probabilistic traveling salesman problem. In J. J. Merelo, P. Adamidis, H.-G. Beyer, J.-L. Fernández-Villacanas, & H.-P. Schwefel (Eds.), *Proceedings of PPSN-VII, Seventh International Conference on Parallel Problem Solving from Nature*, vol. 2439 of *Lecture Notes in Computer Science* (pp. 883–892). Berlin, Springer-Verlag.

Bianchi, L., Gambardella, L. M., & Dorigo, M. (2002b). Solving the homogeneous probabilistic traveling salesman problem by the ACO metaheuristic. In M. Dorigo, G. Di Caro, & M. Sampels (Eds.), *Proceedings of ANTS 2002—From Ant Colonies to Artificial Ants: Third International Workshop on Ant Algorithms*, vol. 2463 of *Lecture Notes in Computer Science* (pp. 176–187). Berlin, Springer-Verlag.

Birattari, M., Di Caro, G., & Dorigo, M. (2002a). Toward the formal foundation of ant programming. In M. Dorigo, G. Di Caro, & M. Sampels (Eds.), *Proceedings of ANTS 2002—From Ant Colonies to Artificial Ants: Third International Workshop on Ant Algorithms*, vol. 2463 of *Lecture Notes in Computer Science* (pp. 188–201). Berlin, Springer-Verlag.

Birattari, M., Stützle, T., Paquete, L., & Varrentrapp, K. (2002b). A racing algorithm for configuring metaheuristics. In W. B. Langdon, E. Cantú-Paz, K. Mathias, R. Roy, D. Davis, R. Poli, K. Balakrishnan, V. Honavar, G. Rudolph, J. Wegener, L. Bull, M. A. Potter, A. C. Schultz, J. F. Miller, E. Burke, & N. Jonoska (Eds.), *Proceedings of the Genetic and Evolutionary Computation Conference (GECCO-2002)* (pp. 11–18). San Francisco, Morgan Kaufmann.

Bland, R. G., & Shallcross, D. F. (1989). Large traveling salesman problems arising from experiments in X-ray crystallography: A preliminary report on computation. *Operations Research Letters*, 8, 125–128.

Blum, C. (2002a). ACO applied to group shop scheduling: A case study on intensification and diversification. In M. Dorigo, G. Di Caro, & M. Sampels (Eds.), *Proceedings of ANTS 2002—From Ant Colonies to*

Artificial Ants: Third International Workshop on Ant Algorithms, vol. 2463 of *Lecture Notes in Computer Science* (pp. 14–27). Berlin, Springer-Verlag.

Blum, C. (2002b). *Metaheuristics for Group Shop Scheduling*. DEA thesis, Université Libre de Bruxelles, Brussels.

Blum, C. (2003a). An ant colony optimization algorithm to tackle shop scheduling problems. Technical report TR/IRIDIA/2003-1, IRIDIA, Université Libre de Bruxelles, Brussels.

Blum, C. (2003b). Beam-ACO. Hybridizing ant colony optimization with beam search. An application to open shop scheduling. Technical report TR/IRIDIA/2003-17, IRIDIA, Université Libre de Bruxelles, Brussels.

Blum, C., & Blesa, M. J. (2003). Metaheuristics for the edge-weighted k-cardinality tree problem. Technical report LSI-03-1-R, Departament de Llenguatges i Sistemes Informátics, Universitat Politécnica de Catalunya, Barcelona, Spain.

Blum, C., & Dorigo, M. (2003). Deception in ant colony optimization. Part I: Definition and examples. Technical report TR/IRIDIA/2003-18, IRIDIA, Université Libre de Bruxelles, Brussels.

Blum, C., & Dorigo, M. (2004). The hyper-cube framework for ant colony optimization. *IEEE Transactions on Systems, Man, and Cybernetics–Part B*, to appear.

Blum, C., & Roli, A. (2003). Metaheuristics in combinatorial optimization: Overview and conceptual comparison. *ACM Computing Surveys, 35*(3), 268–308.

Blum, C., Roli, A., & Dorigo, M. (2001). HC–ACO: The hyper-cube framework for Ant Colony Optimization. In *Proceedings of MIC'2001—Metaheuristics International Conference*, vol. 2 (pp. 399–403).

Blum, C., & Sampels, M. (2002a). Ant colony optimization for FOP shop scheduling: A case study on different pheromone representations. In D. B. Fogel, M. A. El-Sharkawi, X. Yao, G. Greenwood, H. Iba, P. Marrow, & M. Shackleton (Eds.), *Proceedings of the 2002 Congress on Evolutionary Computation (CEC'02)* (pp. 1558–1563). Piscataway, NJ, IEEE Press.

Blum, C., & Sampels, M. (2002b). When model bias is stronger than selection pressure. In J. J. Merelo, P. Adamidis, H.-G. Beyer, J.-L. Fernández-Villacañas, & H.-P. Schwefel (Eds.), *Proceedings of PPSN-VII, Seventh International Conference on Parallel Problem Solving from Nature*, vol. 2439 in *Lecture Notes in Computer Science* (pp. 893–902). Berlin, Springer-Verlag.

Blum, C., Sampels, M., & Zlochin, M. (2002). On a particularity in model-based search. In W. B. Langdon, E. Cantú-Paz, K. Mathias, R. Roy, D. Davis, R. Poli, K. Balakrishnan, V. Honavar, G. Rudolph, J. Wegener, L. Bull, M. A. Potter, A. C. Schultz, J. F. Miller, E. Burke, & N. Jonoska (Eds.), *Proceedings of the Genetic and Evolutionary Computation Conference (GECCO-2002)* (pp. 35–42). San Francisco, Morgan Kaufmann.

Boese, K. D., Kahng, A. B., & Muddu, S. (1994). A new adaptive multi-start technique for combinatorial global optimization. *Operations Research Letters, 16*, 101–113.

Bolondi, M., & Bondanza, M. (1993). Parallelizzazione di un algoritmo per la risoluzione del problema del commesso viaggiatore. Master's thesis, Dipartimento di Elettronica, Politecnico di Milano, Italy.

Bonabeau, E., Dorigo, M., & Theraulaz, G. (1999). *Swarm Intelligence: From Natural to Artificial Systems*. New York, Oxford University Press.

Bonabeau, E., Dorigo, M., & Theraulaz, G. (2000). Inspiration for optimization from social insect behavior. *Nature, 406*, 39–42.

Bonabeau, E., Henaux, F., Guérin, S., Snyers, D., Kuntz, P., & Theraulaz, G. (1998). Routing in telecommunication networks with "smart" ant-like agents. In S. Albayrak & F. Garijo (Eds.), *Proceedings of IATA'98, Second International Workshop on Intelligent Agents for Telecommunication Applications*, vol. 1437 of *Lecture Notes in Artificial Intelligence* (pp. 60–72). Berlin, Springer-Verlag.

Bonabeau, E., Sobkowski, A., Theraulaz, G., & Deneubourg, J.-L. (1997a). Adaptive task allocation inspired by a model of division of labor in social insects. In D. Lundha, B. Olsson, & A. Narayanan (Eds.), *Bio-Computation and Emergent Computing* (pp. 36–45). Singapore, World Scientific Publishing.

Bonabeau, E., Theraulaz, G., Deneubourg, J.-L., Aron, S., & Camazine, S. (1997b). Self-organization in social insects. *Tree, 12*(5), 188–193.

Borndörfer, R., Eisenblätter, A., Grötschel, M., & Martin, A. (1998a). The orientation model for frequency assignment problems. Technical report 98-01, Konrad Zuse Zentrum für Informationstechnik, Berlin.

Borndörfer, R., Ferreira, C., & Martin, A. (1998b). Decomposing matrices into blocks. *SIAM Journal on Optimization, 9*(1), 236–269.

Boyan, J., & Littman, M. (1994). Packet routing in dynamically changing networks: A reinforcement learning approach. In J. Cowan, G. Tesauro, & J. Alspector (Eds.), *Advances in Neural Information Processing Systems 6 (NIPS6)* (pp. 671–678). San Francisco, Morgan Kaufmann.

Branke, J., & Guntsch, M. (2003). New ideas for applying ant colony optimization to the probabilistic TSP. In G. R. Raidl, J.-A. Meyer, M. Middendorf, S. Cagnoni, J. J. R. Cardalda, D. W. Corne, J. Gottlieb, A. Guillot, E. Hart, C. G. Johnson, & E. Marchiori (Eds.), *Applications of Evolutionary Computing, Proceedings of EvoWorkshops 2003*, vol. 2611 of *Lecture Notes in Computer Science* (pp. 165–175). Berlin, Springer-Verlag.

Branke, J., Middendorf, M., & Schneider, F. (1998). Improved heuristics and a genetic algorithm for finding short supersequences. *OR Spektrum, 20*(1), 39–46.

Bräysy, O. (2003). A reactive variable neighborhood search for the vehicle routing problem with time windows. *INFORMS Journal on Computing, 15*(4), 347–368.

Brelaz, D. (1979). New methods to color the vertices of a graph. *Communications of the ACM, 22*, 251–256.

Brixius, N. W., & Anstreicher, K. M. (2001). The Steinberg wiring problem. Technical report, College of Business Administration, University of Iowa, Iowa City.

Brucker, P. (1998). *Scheduling Algorithms*. Berlin, Springer-Verlag.

Brucker, P., Drexl, A., Möhring, R., Neumann, K., & Pesch, E. (1999). Resource-constrained project scheduling: Notation, classification, models, and methods. *European Journal of Operational Research, 112*(1), 3–41.

Brucker, P., Hurink, J., & Werner, F. (1996). Improving local search heuristics for some scheduling problems—Part I. *Discrete Applied Mathematics, 65*(1–3), 97–122.

Bruinsma, O. H. (1979). *An Analysis of Building Behaviour of the Termite* Macrotemes subhyalinus. PhD thesis, Lanbouwhogeschool te Wageningen, Netherlands.

Bullnheimer, B., Hartl, R. F., & Strauss, C. (1997). A new rank based version of the Ant System—A computational study. Technical report, Institute of Management Science, University of Vienna, Austria.

Bullnheimer, B., Hartl, R. F., & Strauss, C. (1999a). Applying the Ant System to the vehicle routing problem. In S. Voss, S. Martello, I. H. Osman, & C. Roucairol (Eds.), *Meta-Heuristics: Advances and Trends in Local Search Paradigms for Optimization* (pp. 285–296). Dordrecht, Netherlands, Kluwer Academic Publishers.

Bullnheimer, B., Hartl, R. F., & Strauss, C. (1999b). An improved ant system algorithm for the vehicle routing problem. *Annals of Operations Research, 89*, 319–328.

Bullnheimer, B., Hartl, R. F., & Strauss, C. (1999c). A new rank-based version of the Ant System: A computational study. *Central European Journal for Operations Research and Economics, 7*(1), 25–38.

Bullnheimer, B., Kotsis, G., & Strauss, C. (1998). Parallelization strategies for the Ant System. In R. D. Leone, A. Murli, P. Pardalos, & G. Toraldo (Eds.), *High Performance Algorithms and Software in Nonlinear Optimization*, No. 24 in Kluwer Series of Applied Optmization (pp. 87–100). Dordrecht, Netherlands, Kluwer Academic Publishers.

Burkard, R. E., & Offermann, J. (1977). Entwurf von Schreibmaschinentastaturen mittels quadratischer Zuordnungsprobleme. *Zeitschrift für Operations Research, 21*, B121–B132.

Camazine, S., Deneubourg, J.-L., Franks, N. R., Sneyd, J., Theraulaz, G., & Bonabeau, E. (Eds.). (2001). *Self-Organization in Biological Systems*. Princeton, NJ, Princeton University Press.

Campos, M., Bonabeau, E., Theraulaz, G., & Deneubourg, J.-L. (2000). Dynamic scheduling and division of labor in social insects. *Adaptive Behavior, 8*(3), 83–96.

Cantú-Paz, E. (2000). *Efficient and Accurate Parallel Genetic Algorithms*. Boston, Kluwer Academic Publishers.

Caprara, A., Fischetti, M., & Toth, P. (1999). A heuristic method for the set covering problem. *Operations Research, 47*(5), 730–743.

Casillas, J., Cordón, O., & Herrera, F. (2000). Learning cooperative fuzzy linguistic rules using ant colony algorithms. Technical report DECSAI-00-01-19, Department of Computer Science and Artificial Intelligence, University of Granada, Granada, Spain.

Casillas, J., Cordón, O., & Herrera, F. (2002). COR: A methodology to improve ad hoc data-driven linguistic rule learning methods by inducing cooperation among rules. *IEEE Transactions on Systems, Man, and Cybernetics, 32*(4), 526–537.

Cerný, V. (1985). A thermodynamical approach to the traveling salesman problem. *Journal of Optimization Theory and Applications, 45*(1), 41–51.

Chandra, R., Dagum, L., Kohr, D., Maydan, D., McDonald, J., & Menon, R. (2000). *Parallel Programming in OpenMP*. San Francisco, Morgan Kaufmann.

Choi, S., & Yeung, D.-Y. (1996). Predictive Q-routing: A memory-based reinforcement learning approach to adaptive traffic control. In D. Touretzky, M. Mozer, & M. Hasselmo (Eds.), *Advances in Neural Information Processing Systems 8 (NIPS8)* (pp. 945–951). Cambridge, MA, MIT Press.

Chrétien, L. (1996). *Organisation spatiale du matériel provenant de l'excavation du nid chez* Messor barbarus *et des cadavres d'ouvrières chez* Lasius niger *(Hymenopterae: Formicidae)*. PhD thesis, Université Libre de Bruxelles, Brussels.

Christofides, N. (1976). Worst-case analysis of a new heuristic for the travelling salesman problem. Technical report 388, Graduate School of Industrial Administration, Carnegie Mellon University, Pittsburgh.

Cicirello, V. A., & Smith, S. F. (2001). Ant colony control for autonomous decentralized shop floor routing. In *Proceedings of the 5th International Symposium on Autonomous Decentralized Systems* (pp. 383–390). Los Alamitos, CA, IEEE Computer Society Press.

Cicirello, V. A., & Smith, S. F. (2003). Wasp-like agents for distributed factory coordination. *Autonomous Agents and Multi-Agent Systems*, to appear.

Clark, P., & Boswell, R. (1991). Rule induction with CN2: Some recent improvements. In *Proceedings of the European Working Session on Learning (EWSL-91)*, vol. 482 of *Lecture Notes in Artificial Intelligence* (pp. 151–163). Berlin, Springer-Verlag.

Clark, P., & Niblett, T. (1989). The CN2 induction algorithm. *Machine Learning, 3*(4), 261–283.

Clarke, G., & Wright, J. W. (1964). Scheduling of vehicles from a central depot to a number of delivery points. *Operations Research, 12*, 568–581.

Coffman, E. G., Jr., Garey, M. R., & Johnson, D. S. (1997). Approximation algorithms for bin packing: A survey. In D. Hochbaum (Ed.), *Approximation Algorithms for NP-Hard Problems* (pp. 46–93). Boston, PWS Publishing.

Colorni, A., Dorigo, M., & Maniezzo, V. (1992a). Distributed optimization by ant colonies. In F. J. Varela & P. Bourgine (Eds.), *Proceedings of the First European Conference on Artificial Life* (pp. 134–142). Cambridge, MA, MIT Press.

Colorni, A., Dorigo, M., & Maniezzo, V. (1992b). An investigation of some properties of an ant algorithm. In R. Männer & B. Manderick (Eds.), *Proceedings of PPSN-II, Second International Conference on Parallel Problem Solving from Nature* (pp. 509–520). Amsterdam, Elsevier.

Colorni, A., Dorigo, M., Maniezzo, V., & Trubian, M. (1994). Ant System for job-shop scheduling. *JORBEL—Belgian Journal of Operations Research, Statistics and Computer Science, 34*(1), 39–53.

Congram, R. K., Potts, C. N., & de Velde, S. L. V. (2002). An iterated dynasearch algorithm for the single-machine total weighted tardiness scheduling problem. *INFORMS Journal on Computing, 14*(1), 52–67.

Cook, W. J., Cunningham, W. H., Pulleyblank, W. R., & Schrijver, A. (1998). *Combinatorial Optimization*. New York, John Wiley & Sons.

Cooper, G. F., & Herskovits, E. (1992). A Bayesian method for the induction of probabilistic networks from data. *Machine Learning*, *9*(4), 309–348.

Cordón, O., de Viana, I. F., & Herrera, F. (2002). Analysis of the best-worst Ant System and its variants on the TSP. *Mathware and Soft Computing*, *9*(2–3), 177–192.

Cordón, O., de Viana, I. F., Herrera, F., & Moreno, L. (2000). A new ACO model integrating evolutionary computation concepts: The best-worst Ant System. In M. Dorigo, M. Middendorf, & T. Stützle (Eds.), *Abstract Proceedings of ANTS 2000—From Ant Colonies to Artificial Ants: Second International Workshop on Ant Algorithms* (pp. 22–29). Brussels, IRIDIA, Université Libre de Bruxelles.

Cordón, O., & Herrera, F. (2000). A proposal for improving the accuracy of linguistic modeling. *IEEE Transactions on Fuzzy Systems*, *8*(3), 335–344.

Cordone, R., & Maffioli, F. (2001). Coloured Ant System and local search to design local telecommunication networks. In E. J. W. Boers, J. Gottlieb, P. L. Lanzi, R. E. Smith, S. Cagnoni, E. Hart, G. R. Raidl, & H. Tijink (Eds.), *Applications of Evolutionary Computing: Proceedings of EvoWorkshops 2001*, vol. 2037 of *Lecture Notes in Computer Science* (pp. 60–69). Berlin, Springer-Verlag.

Cordone, R., & Maffioli, F. (2003). On the complexity of graph tree partition problems. *Discrete Applied Mathematics*, *134*(1–3), 51–65.

Corne, D., Dorigo, M., & Glover, F. (Eds.). (1999). *New Ideas in Optimization*. London, McGraw Hill.

Costa, D., & Hertz, A. (1997). Ants can colour graphs. *Journal of the Operational Research Society*, *48*, 295–305.

Crauwels, H. A. J., Potts, C. N., & Wassenhove, L. N. V. (1998). Local search heuristics for the single machine total weighted tardiness scheduling problem. *INFORMS Journal on Computing*, *10*(3), 341–350.

Crescenzi, P., Goldman, D., Papadimitriou, C. H., Piccolboni, A., & Yannakakis, M. (1998). On the complexity of protein folding. *Journal of Computational Biology*, *5*(3), 423–466.

Croes, G. A. (1958). A method for solving traveling salesman problems. *Operations Research*, *6*, 791–812.

Czyzak, P., & Jaszkiewicz, A. (1998). Pareto simulated annealing—A metaheuristic technique for multiple objective combinatorial optimization. *Journal of Multi-Criteria Decision Analysis*, *7*, 34–47.

Dantzig, G. B., Fulkerson, D. R., & Johnson, S. M. (1954). Solution of a large-scale traveling salesman problem. *Operations Research*, *2*, 393–410.

Davenport, A., Tsang, E., Wang, C. J., & Zhu, K. (1994). GENET: A connectionist architecture for solving constraint satisfaction problems by iterative improvement. In *Proceedings of the 14th National Conference on Artificial Intelligence* (pp. 325–330). Menlo Park, CA, AAAI Press/MIT Press.

Dawid, H., Doerner, K., Hartl, R. F., & Reimann, M. (2002). Ant systems to solve operational problems. In H. Dawid, K. Doerner, G. Dorffner, T. Fent, M. Feurstein, R. F. Hartl, A. Mild, M. Natter, M. Reimann, & A. Taudes (Eds.), *Quantitative Models of Learning Organizations* (pp. 65–82). Vienna, Springer-Verlag.

De Bonet, J. S., Isbell, C. L., & Viola, P. (1997). MIMIC: Finding optima by estimating probability densities. In M. C. Mozer, M. I. Jordan, & T. Petsche (Eds.), *Advances in Neural Information Processing Systems 9 (NIPS9)*, vol. 9 (pp. 424–431). Cambridge, MA, MIT Press.

de Campos, L. M., Fernández-Luna, J. M., Gámez, J. A., & Puerta, J. M. (2002a). Ant colony optimization for learning Bayesian networks. *International Journal of Approximate Reasoning*, *31*(3), 291–311.

de Campos, L. M., Gámez, J. A., & Puerta, J. M. (2002b). Learning Bayesian networks by ant colony optimisation: Searching in the space of orderings. *Mathware and Soft Computing*, *9*(2–3), 251–268.

de Campos, L. M., & Puerta, J. M. (2001). Stochastic local search and distributed search algorithms for learning Bayesian networks. In *III International Symposium on Adaptive Systems (ISAS): Evolutionary Computation and Probabilisitic Graphical Models* (pp. 109–115). La Habana, Cuba: Institute of Cybernetics, Mathematics and Physics.

Deb, K. (2001). *Multi-Objective Optimization Using Evolutionary Algorithms*. Chichester, UK, John Wiley & Sons.

Dechter, R. (2003). *Constraint Processing*. San Francisco, Morgan Kaufmann.

Dechter, R., Meiri, I., & Pearl, J. (1991). Temporal constraint networks. *Artificial Intelligence, 49*(1–3), 61–95.

Dechter, R., & Pearl, J. (1989). Tree clustering schemes for constraint-processing. *Artificial Intelligence, 38*(3), 353–366.

Delisle, P., Krajecki, M., Gravel, M., & Gagné, C. (2001). Parallel implementation of an ant colony optimization metaheuristic with OpenMP. In *Proceedings of the 3rd European Workshop on OpenMP (EWOMP'01)*, Barcelona, Spain.

Dell'Amico, M., Maffioli, F., & Martello, S. (Eds.). (1997). *Annotated Bibliographies in Combinatorial Optimization*. Chichester, UK, John Wiley & Sons.

den Besten, M. (2000). Ants for the single machine total weighted tardiness problem. Master's thesis, University of Amsterdam.

den Besten, M. L., Stützle, T., & Dorigo, M. (2000). Ant colony optimization for the total weighted tardiness problem. In M. Schoenauer, K. Deb, G. Rudolph, X. Yao, E. Lutton, J. J. Merelo, & H.-P. Schwefel (Eds.), *Proceedings of PPSN-VI, Sixth International Conference on Parallel Problem Solving from Nature*, vol. 1917 of *Lecture Notes in Computer Science* (pp. 611–620). Berlin, Springer-Verlag.

Deneubourg, J.-L. (2002). Personal communication. Université Libre de Bruxelles, Brussels.

Deneubourg, J.-L., Aron, S., Goss, S., & Pasteels, J.-M. (1990). The self-organizing exploratory pattern of the Argentine ant. *Journal of Insect Behavior, 3*, 159–168.

Deneubourg, J.-L., Goss, S., Franks, N., Sendova-Franks, A., Detrain, C., & Chrétien, L. (1991). The dynamics of collective sorting: Robot-like ants and ant-like robots. In J.-A. Meyer & S. W. Wilson (Eds.), *Proceedings of the First International Conference on Simulation of Adaptive Behavior: From Animals to Animats* (pp. 356–363). Cambridge, MA, MIT Press.

Di Caro, G. (in preparation). *Systems of Ant-like Agents for Adaptive Network Control and Combinatorial Optimization*. PhD thesis, Université Libre de Bruxelles, Brussels.

Di Caro, G., & Dorigo, M. (1997). AntNet: A mobile agents approach to adaptive routing. Technical report IRIDIA/97-12, IRIDIA, Université Libre de Bruxelles, Brussels.

Di Caro, G., & Dorigo, M. (1998a). An adaptive multi-agent routing algorithm inspired by ants behavior. In K. A. Hawick & H. A. James (Eds.), *Proceedings of PART98—5th Annual Australasian Conference on Parallel and Real-Time Systems* (pp. 261–272). Singapore, Springer-Verlag.

Di Caro, G., & Dorigo, M. (1998b). Ant colonies for adaptive routing in packet-switched communications networks. In A. E. Eiben, T. Bäck, M. Schoenauer, & H.-P. Schwefel (Eds.), *Proceedings of PPSN-V, Fifth International Conference on Parallel Problem Solving from Nature*, vol. 1498 of *Lecture Notes in Computer Science* (pp. 673–682). Berlin, Springer-Verlag.

Di Caro, G., & Dorigo, M. (1998c). AntNet: Distributed stigmergetic control for communications networks. *Journal of Artificial Intelligence Research, 9*, 317–365.

Di Caro, G., & Dorigo, M. (1998d). Extending AntNet for best-effort quality-of-service routing. Unpublished presentation at *ANTS'98—From Ant Colonies to Artificial Ants: First International Workshop on Ant Colony Optimization*, Brussels.

Di Caro, G., & Dorigo, M. (1998e). Mobile agents for adaptive routing. In H. El-Rewini (Ed.), *Proceedings of the 31st International Conference on System Sciences (HICSS-31)* (pp. 74–83). Los Alamitos, CA, IEEE Computer Society Press.

Di Caro, G., & Dorigo, M. (1998f). Two ant colony algorithms for best-effort routing in datagram networks. In Y. Pan, S. G. Akl, & K. Li (Eds.), *Proceedings of the Tenth IASTED International Conference on Parallel and Distributed Computing and Systems (PDCS'98)* (pp. 541–546). Anaheim, CA, IASTED/ACTA Press.

Dickey, J. W., & Hopkins, J. W. (1972). Campus building arrangement using TOPAZ. *Transportation Science*, *6*, 59–68.

Dijkstra, E. W. (1959). A note on two problems in connection with graphs. *Numerische Mathematik*, *1*, 269–271.

Doerner, K., Gutjahr, W. J., Hartl, R. F., Strauss, C., & Stummer, C. (2003). Pareto ant colony optimization: A metaheuristic approach to multiobjective portfolio selection. *Annals of Operations Research*, to appear.

Doerner, K., Hartl, R. F., & Reimann, M. (2001). Cooperative ant colonies for optimizing resource allocation in transportation. In E. J. W. Boers, J. Gottlieb, P. L. Lanzi, R. E. Smith, S. Cagnoni, E. Hart, G. R. Raidl, & H. Tijink (Eds.), *Applications of Evolutionary Computing: Proceedings of EvoWorkshops 2001*, vol. 2037 of *Lecture Notes in Computer Science* (pp. 70–79). Berlin, Springer-Verlag.

Doerner, K., Hartl, R. F., & Reimann, M. (2003). Competants for problem solving: The case of full truckload transportation. *Central European Journal for Operations Research and Economics*, *11*(2), 115–141.

Dorigo, M. (1992). *Optimization, Learning and Natural Algorithms* [in Italian]. PhD thesis, Dipartimento di Elettronica, Politecnico di Milano, Milan.

Dorigo, M. (2001). Ant algorithms solve difficult optimization problems. In J. Kelemen (Ed.), *Proceedings of the Sixth European Conference on Artificial Life*, vol. 2159 of *Lecture Notes in Artificial Intelligence* (pp. 11–22). Berlin, Springer-Verlag.

Dorigo, M., Bonabeau, E., & Theraulaz, G. (2000a). Ant algorithms and stigmergy. *Future Generation Computer Systems*, *16*(8), 851–871.

Dorigo, M., & Di Caro, G. (1999a). Ant colony optimization: A new meta-heuristic. In P. J. Angeline, Z. Michalewicz, M. Schoenauer, X. Yao, & A. Zalzala (Eds.), *Proceedings of the 1999 Congress on Evolutionary Computation (CEC'99)* (pp. 1470–1477). Piscataway, NJ, IEEE Press.

Dorigo, M., & Di Caro, G. (1999b). The ant colony optimization meta-heuristic. In D. Corne, M. Dorigo, & F. Glover (Eds.), *New Ideas in Optimization* (pp. 11–32). London, McGraw Hill.

Dorigo, M., Di Caro, G., & Gambardella, L. M. (1999). Ant algorithms for discrete optimization. *Artificial Life*, *5*(2), 137–172.

Dorigo, M., Di Caro, G., & Sampels, M. (Eds.). (2002a). *Proceedings of ANTS 2002—From Ant Colonies to Artificial Ants: Third International Workshop on Ant Algorithms*, vol. 2463 of *Lecture Notes in Computer Science*. Berlin, Springer-Verlag.

Dorigo, M., & Gambardella, L. M. (1996). A study of some properties of Ant-Q. In H. Voigt, W. Ebeling, I. Rechenberg, & H. Schwefel (Eds.), *Proceedings of PPSN-IV, Fourth International Conference on Parallel Problem Solving from Nature*, vol. 1141 of *Lecture Notes in Computer Science* (pp. 656–665). Berlin, Springer-Verlag.

Dorigo, M., & Gambardella, L. M. (1997a). Ant colonies for the traveling salesman problem. *BioSystems*, *43*(2), 73–81.

Dorigo, M., & Gambardella, L. M. (1997b). Ant Colony System: A cooperative learning approach to the traveling salesman problem. *IEEE Transactions on Evolutionary Computation*, *1*(1), 53–66.

Dorigo, M., Gambardella, L. M., Middendorf, M., & Stützle, T. (Eds.). (2002b). Special section on "Ant Colony Optimization." *IEEE Transactions on Evolutionary Computation*, *6*(4), 317–365.

Dorigo, M., Maniezzo, V., & Colorni, A. (1991a). Positive feedback as a search strategy. Technical report 91-016, Dipartimento di Elettronica, Politecnico di Milano, Milan.

Dorigo, M., Maniezzo, V., & Colorni, A. (1991b). The Ant System: An autocatalytic optimizing process. Technical report 91-016 revised, Dipartimento di Elettronica, Politecnico di Milano, Milan.

Dorigo, M., Maniezzo, V., & Colorni, A. (1996). Ant System: Optimization by a colony of cooperating agents. *IEEE Transactions on Systems, Man, and Cybernetics—Part B*, *26*(1), 29–41.

Dorigo, M., Middendorf, M., & Stützle, T. (Eds.). (2000b). *Abstract Proceedings of ANTS 2000—From Ant Colonies to Artificial Ants: Second International Workshop on Ant Algorithms*. Brussels, IRIDIA, Université Libre de Bruxelles.

Dorigo, M., & Stützle, T. (2001). An experimental study of the simple ant colony optimization algorithm. In N. Mastorakis (Ed.), *2001 WSES International Conference on Evolutionary Computation (EC'01)* (pp. 253–258). WSES Press.

Dorigo, M., & Stützle, T. (2002). The ant colony optimization metaheuristic: Algorithms, applications and advances. In F. Glover & G. Kochenberger (Eds.), *Handbook of Metaheuristics*, vol. 57 of *International Series in Operations Research & Management Science* (pp. 251–285). Norwell, MA, Kluwer Academic Publishers.

Dorigo, M., Stützle, T., & Di Caro, G. (Eds.). (2000c). Special issue on "Ant Algorithms." *Future Generation Computer Systems*, *16*, 851–956.

Dorigo, M., Trianni, V., Şahin, E., Labella, T., Gross, R., Baldassarre, G., Nolfi, S., Deneubourg, J.-L., Mondada, F., Floreano, D., & Gambardella, L. M. (2003). Evolving self-organizing behaviors for a *Swarm-bot*. Technical report IRIDIA/2003-11, IRIDIA, Université Libre de Bruxelles, Brussels.

Dorigo, M., Zlochin, M., Meuleau, N., & Birattari, M. (2002c). Updating ACO pheromones using stochastic gradient ascent and cross-entropy methods. In S. Cagnoni, J. Gottlieb, E. Hart, M. Middendorf, & G. R. Raidl (Eds.), *Applications of Evolutionary Computing, Proceedings of EvoWorkshops 2002*, vol. 2279 of *Lecture Notes in Computer Science* (pp. 21–30). Berlin, Springer-Verlag.

Dorne, R., & Hao, J. (1999). Tabu search for graph coloring, t-colorings and set t-colorings. In S. Voss, S. Martello, I. Osman, & C. Roucairol (Eds.), *Meta-heuristics: Advances and Trends in Local Search Paradigms for Optimization* (pp. 77–92). Boston, Kluwer Academic Publishers.

Elmaghraby, S. E. (1977). *Activity Networks*. New York, John Wiley & Sons.

Elshafei, A. N. (1977). Hospital layout as a quadratic assignment problem. *Operations Research Quarterly*, *28*, 167–179.

Eyckelhof, C. J., & Snoek, M. (2002). Ant systems for a dynamic TSP: Ants caught in a traffic jam. In M. Dorigo, G. Di Caro, & M. Sampels (Eds.), *Proceedings of ANTS 2002—From Ant Colonies to Artificial Ants: Third International Workshop on Ant Algorithms*, vol. 2463 of *Lecture Notes in Computer Science* (pp. 88–99). Berlin, Springer-Verlag.

Fabrikant, S. I. (2000). *Spatial Metaphors for Browsing Large Data Archives*. PhD thesis, Department of Geography, University of Colorado at Boulder.

Faigle, U., & Kern, W. (1992). Some convergence results for probabilistic tabu search. *ORSA Journal on Computing*, *4*(1), 32–37.

Falkenauer, E. (1996). A hybrid grouping genetic algorithm for bin packing. *Journal of Heuristics*, *2*(1), 5–30.

Fang, H.-L., Ross, P., & Corne, D. (1994). A promising hybrid GA/heuristic approach for open-shop scheduling problems. In A. G. Cohn (Ed.), *Proceedings of the 11th European Conference on Artificial Intelligence* (pp. 590–594). Chichester, John Wiley & Sons.

Fenet, S., & Solnon, C. (2003). Searching for maximum cliques with ant colony optimization. In G. R. Raidl, J.-A. Meyer, M. Middendorf, S. Cagnoni, J. J. R. Cardalda, D. W. Corne, J. Gottlieb, A. Guillot, E. Hart, C. G. Johnson, & E. Marchiori (Eds.), *Applications of Evolutionary Computing, Proceedings of EvoWorkshops 2003*, vol. 2611 of *Lecture Notes in Computer Science* (pp. 236–245). Berlin, Springer-Verlag.

Feo, T. A., & Resende, M. G. C. (1989). A probabilistic heuristic for a computationally difficult set covering problem. *Operations Research Letters*, *8*, 67–71.

Feo, T. A., & Resende, M. G. C. (1995). Greedy randomized adaptive search procedures. *Journal of Global Optimization*, *6*, 109–133.

Festa, P., & Resende, M. G. C. (2002). GRASP: An annotated bibliography. In P. Hansen & C. C. Ribeiro (Eds.), *Essays and Surveys on Metaheuristics* (pp. 325–367). Boston, Kluwer Academic Publishers.

Fischetti, M., Hamacher, H. W., Jörnsten, K., & Maffioli, F. (1994). Weighted k-cardinality trees: Complexity and polyhedral structure. *Networks, 24*, 11–21.

Fleurent, C., & Ferland, J. A. (1996). Genetic and hybrid algorithms for graph coloring. *Annals of Operations Research, 63*, 437–461.

Flood, M. M. (1956). The traveling-salesman problem. *Operations Research, 4*, 61–75.

Fogel, D. B. (1995). *Evolutionary Computation*. Piscataway, NJ, IEEE Press.

Fogel, L. J., Owens, A. J., & Walsh, M. J. (1966). *Artificial Intelligence through Simulated Evolution*. New York, John Wiley & Sons.

Ford, L., & Fulkerson, D. (1962). *Flows in Networks*. Princeton, NJ, Princeton University Press.

Foulds, L., Hamacher, H., & Wilson, J. (1998). Integer programming approaches to facilities layout models with forbidden areas. *Annals of Operations Research, 81*, 405–417.

Foulser, D. E., Li, M., & Yang, Q. (1992). Theory and algorithms for plan merging. *Artificial Intelligence, 57*(2–3), 143–181.

Frank, J. (1996). Weighting for Godot: Learning heuristics for GSAT. In *Proceedings of the AAAI National Conference on Artificial Intelligence* (pp. 338–343). Menlo Park, CA, AAAI Press/MIT Press.

Franks, N. R. (1986). Teams in social insects: Group retrieval of prey by army ants (*Eciton burchelli*, Hymenoptera: Formicidae). *Behavioral Ecology and Sociobiology, 18*, 425–429.

Franks, N. R., & Sendova-Franks, A. B. (1992). Brood sorting by ants: Distributing the workload over the work surface. *Behavioral Ecology and Sociobiology, 30*, 109–123.

Freuder, E. C., & Wallace, R. J. (1992). Partial constraint satisfaction. *Artificial Intelligence, 58*(1–3), 21–70.

Fujita, K., Saito, A., Matsui, T., & Matsuo, H. (2002). An adaptive ant-based routing algorithm used routing history in dynamic networks. In L. Wang, K. C. T. Furuhashi, J.-H. Kim, & X. Yao (Eds.), *4th Asia-Pacific Conference on Simulated Evolution and Learning (SEAL'02)*, vol. 1 (pp. 46–50). Orchid Country Club, Singapore, 18–22 Nov. 2002.

Gagné, C., Price, W. L., & Gravel, M. (2002). Comparing an ACO algorithm with other heuristics for the single machine scheduling problem with sequence-dependent setup times. *Journal of the Operational Research Society, 53*, 895–906.

Galinier, P., & Hao, J.-K. (1997). Tabu search for maximal constraint satisfaction problems. In G. Smolka (Ed.), *Principles and Practice of Constraint Programming—CP97*, vol. 1330 of *Lecture Notes in Computer Science* (pp. 196–208). Berlin, Springer-Verlag.

Galinier, P., & Hao, J.-K. (1999). Hybrid evolutionary algorithms for graph coloring. *Journal of Combinatorial Optimization, 3*(4), 379–397.

Gambardella, L. M., & Dorigo, M. (1995). Ant-Q: A reinforcement learning approach to the traveling salesman problem. In A. Prieditis & S. Russell (Eds.), *Proceedings of the Twelfth International Conference on Machine Learning (ML-95)* (pp. 252–260). Palo Alto, CA, Morgan Kaufmann.

Gambardella, L. M., & Dorigo, M. (1996). Solving symmetric and asymmetric TSPs by ant colonies. In T. Baeck, T. Fukuda, & Z. Michalewicz (Eds.), *Proceedings of the 1996 IEEE International Conference on Evolutionary Computation (ICEC'96)* (pp. 622–627). Piscataway, NJ, IEEE Press.

Gambardella, L. M., & Dorigo, M. (1997). HAS-SOP: An hybrid Ant System for the sequential ordering problem. Technical report IDSIA-11-97, IDSIA, Lugano, Switzerland.

Gambardella, L. M., & Dorigo, M. (2000). Ant Colony System hybridized with a new local search for the sequential ordering problem. *INFORMS Journal on Computing, 12*(3), 237–255.

Gambardella, L. M., Taillard, É. D., & Agazzi, G. (1999a). MACS-VRPTW: A multiple ant colony system for vehicle routing problems with time windows. In D. Corne, M. Dorigo, & F. Glover (Eds.), *New Ideas in Optimization* (pp. 63–76). London, McGraw Hill.

Gambardella, L. M., Taillard, E. D., & Dorigo, M. (1999b). Ant colonies for the quadratic assignment problem. *Journal of the Operational Research Society, 50*(2), 167–176.

Gámez, J. A., & Puerta, J. M. (2002). Searching the best elimination sequence in Bayesian networks by using ant colony optimization. *Pattern Recognition Letters*, *23*(1–3), 261–277.

Gamst, A. (1986). Some lower bounds for a class of frequency assignment problems. *IEEE Transactions of Vehicular Technology*, *35*(1), 8–14.

Garey, M. R., & Johnson, D. S. (1979). *Computers and Intractability: A Guide to the Theory of NP-Completeness*. San Francisco, Freeman.

Geman, S., & Geman, D. (1984). Stochastic relaxation, Gibbs distribution, and the Bayesian restoration of images. *IEEE Transactions on Pattern Analysis and Machine Intelligence*, *6*, 721–741.

Giffler, B., & Thompson, G. L. (1960). Algorithms for solving production scheduling problems. *Operations Research*, *8*, 487–503.

Gilmore, P. C. (1962). Optimal and suboptimal algorithms for the quadratic assignment problem. *Journal of the SIAM*, *10*, 305–313.

Glover, F. (1977). Heuristics for integer programming using surrogate constraints. *Decision Sciences*, *8*, 156–166.

Glover, F. (1989). Tabu search—Part I. *ORSA Journal on Computing*, *1*(3), 190–206.

Glover, F. (1990). Tabu search—Part II. *ORSA Journal on Computing*, *2*(1), 4–32.

Glover, F. (1996). Ejection chains, reference structures and alternating path methods for traveling salesman problems. *Discrete Applied Mathematics*, *65*(1–3), 223–253.

Glover, F., & Hanafi, S. (2002). Tabu search and finite convergence. *Discrete Applied Mathematics*, *119*(1–2), 3–36.

Glover, F., & Kochenberger, G. (Eds.). (2002). *Handbook of Metaheuristics*. Norwell, MA, Kluwer Academic Publishers.

Glover, F., & Laguna, M. (1997). *Tabu Search*. Boston, Kluwer Academic Publishers.

Glover, F., Laguna, M., & Martí, R. (2002). Scatter search and path relinking: Advances and applications. In F. Glover & G. Kochenberger (Eds.), *Handbook of Metaheuristics*, vol. 57 of *International Series in Operations Research & Management Science* (pp. 1–35). Norwell, MA, Kluwer Academic Publishers.

Goldberg, D. E. (1989). *Genetic Algorithms in Search, Optimization and Machine Learning*. Reading, MA, Addison-Wesley.

Golden, B. L., & Stewart, W. R. (1985). Enpirical analysis of heuristics. In E. L. Lawler, J. K. Lenstra, A. H. G. Rinnooy Kan, & D. B. Shmoys (Eds.), *The Traveling Salesman Problem* (pp. 307–360). Chichester, UK, John Wiley & Sons.

Goss, S., Aron, S., Deneubourg, J. L., & Pasteels, J. M. (1989). Self-organized shortcuts in the Argentine ant. *Naturwissenschaften*, *76*, 579–581.

Gottlieb, J., Puchta, M., & Solnon, C. (2003). A study of greedy, local search, and ant colony optimization approaches for car sequencing problems. In G. R. Raidl, J.-A. Meyer, M. Middendorf, S. Cagnoni, J. J. R. Cardalda, D. W. Corne, J. Gottlieb, A. Guillot, E. Hart, C. G. Johnson, & E. Marchiori (Eds.), *Applications of Evolutionary Computing, Proceedings of EvoWorkshops 2003*, vol. 2611 of *Lecture Notes in Computer Science* (pp. 246–257). Berlin, Springer-Verlag.

Grabowski, J., & Wodecki, M. (2001). A new very fast tabu search algorithm for the job shop problem. Technical report 21/2001, Wroclaw University of Technology, Institute of Engineering Cybernetics, Wroclaw, Poland.

Grassé, P. P. (1959). La reconstruction du nid et les coordinations interindividuelles chez *Bellicositermes natalensis* et *Cubitermes* sp. La théorie de la stigmergie: Essai d'interprétation du comportement des termites constructeurs. *Insectes Sociaux*, *6*, 41–81.

Gravel, M., Price, W. L., & Gagné, C. (2002). Scheduling continuous casting of aluminum using a multiple objective ant colony optimization metaheuristic. *European Journal of Operational Research*, *143*(1), 218–229.

Grosso, A., Della Croce, F., & Tadei, R. (2004). An enhanced dynasearch neighborhood for the single-machine total weighted tardiness scheduling problem. *Operations Research Letters*, *32*(1), 68–72.

Grötschel, M. (1981). On the symmetric travelling salesman problem: Solution of a 120-city problem. *Mathematical Programming Study*, *12*, 61–77.

Grötschel, M., & Holland, O. (1991). Solution of large-scale symmetric traveling salesman problems. *Mathematical Programming*, *51*, 141–202.

Guesgen, H., & Hertzberg, J. (1992). *A Perspective of Constraint-Based Reasoning*, vol. 597 of *Lecture Notes in Artificial Intelligence*. Berlin, Springer-Verlag.

Güneş, M., Sorges, U., & Bouazizi, I. (2002). ARA—The ant-colony based routing algorithm for MANETS. In S. Olariu (Ed.), *2002 ICPP Workshop on Ad Hoc Networks (IWAHN 2002)* (pp. 79–85). Los Alamitos, CA, IEEE Computer Society Press.

Güneş, M., & Spaniol, O. (2002). Routing algorithms for mobile multi-hop ad-hoc networks. In H. Turlakov & L. Boyanov (Eds.), *International Workshop on Next Generation Network Technologies* (pp. 10–24). Rousse, Bulgaria: Central Laboratory for Parallel Processing—Bulgarian Academy of Sciences.

Guntsch, M., & Middendorf, M. (2001). Pheromone modification strategies for ant algorithms applied to dynamic TSP. In E. J. W. Boers, J. Gottlieb, P. L. Lanzi, R. E. Smith, S. Cagnoni, E. Hart, G. R. Raidl, & H. Tijink (Eds.), *Applications of Evolutionary Computing*, vol. 2037 of *Lecture Notes in Computer Science* (pp. 213–222). Berlin, Springer-Verlag.

Guntsch, M., & Middendorf, M. (2002a). Applying population based ACO to dynamic optimization problems. In M. Dorigo, G. Di Caro, & M. Sampels (Eds.), *Proceedings of ANTS 2002—From Ant Colonies to Artificial Ants: Third International Workshop on Ant Algorithms*, vol. 2463 of *Lecture Notes in Computer Science* (pp. 111–122). Berlin, Springer-Verlag.

Guntsch, M., & Middendorf, M. (2002b). A population based approach for ACO. In S. Cagnoni, J. Gottlieb, E. Hart, M. Middendorf, & G. R. Raidl (Eds.), *Applications of Evolutionary Computing*, vol. 2279 of *Lecture Notes in Computer Science* (pp. 71–80). Berlin, Springer-Verlag.

Guntsch, M., Middendorf, M., & Schmeck, H. (2001). An ant colony optimization approach to dynamic TSP. In L. Spector, E. D. Goodman, A. Wu, W. B. Langdon, H.-M. Voigt, M. Gen, S. Sen, M. Dorigo, S. Pezeshk, M. H. Garzon, & E. Burke (Eds.), *Proceedings of the Genetic and Evolutionary Computation Conference (GECCO-2001)* (pp. 860–867). San Francisco, Morgan Kaufmann.

Guntsch, M. G., & Middendorf, M. (2003). Solving multi-criteria optimization problems with population-based ACO. In C. M. Fonseca, P. J. Fleming, E. Zitzler, K. Deb, & L. Thiele (Eds.), *Evolutionary Multi-Criterion Optimization*, vol. 2632 of *Lecture Notes in Computer Science* (pp. 464–478). Berlin, Springer-Verlag.

Gutjahr, W. J. (2000). A graph-based Ant System and its convergence. *Future Generation Computer Systems*, *16*(8), 873–888.

Gutjahr, W. J. (2002). ACO algorithms with guaranteed convergence to the optimal solution. *Information Processing Letters*, *82*(3), 145–153.

Hadji, R., Rahoual, M., Talbi, E., & Bachelet, V. (2000). Ant colonies for the set covering problem. In M. Dorigo, M. Middendorf, & T. Stützle (Eds.), *Abstract Proceedings of ANTS 2000—From Ant Colonies to Artificial Ants: Second International Workshop on Ant Algorithms* (pp. 63–66). Brussels, Université Libre de Bruxelles.

Hahn, P., & Krarup, J. (2001). A hospital facility layout problem finally solved. *Journal of Intelligent Manufacturing*, *12*(5–6), 487–496.

Hahn, P. M., Hightower, W. L., Johnson, T. A., Guignard-Spielberg, M., & Roucairol, C. (2001). Tree elaboration strategies in branch and bound algorithms for solving the quadratic assignment problem. *Yugoslavian Journal of Operational Research*, *11*(1), 41–60.

Hajek, B. (1988). Cooling schedules for optimal annealing. *Mathematics of Operations Research*, *13*(2), 311–329.

Haken, H. (1983). *Synergetics*. Berlin, Springer-Verlag.

Hamacher, H. W., & Jörnsten, K. (1993). Optimal relinquishment according to the Norwegian petrol law: A combinatorial optimization approach. Technical report 7/93, Norwegian School of Economics and Business Administration, Bergen, Norway.

Hanafi, S. (2000). On the convergence of tabu search. *Journal of Heuristics*, 7(1), 47–58.

Handl, J., & Meyer, B. (2002). Improved ant-based clustering and sorting in a document retrieval interface. In J. J. Merelo, P. Adamidis, H.-G. Beyer, J.-L. Fernández-Villacañas, & H.-P. Schwefel (Eds.), *Proceedings of PPSN-VII, Seventh International Conference on Parallel Problem Solving from Nature*, vol. 2439 in *Lecture Notes in Computer Science* (pp. 913–923). Berlin, Springer-Verlag.

Hansen, P., & Mladenović, N. (1999). An introduction to variable neighborhood search. In S. Voss, S. Martello, I. H. Osman, & C. Roucairol (Eds.), *Meta-Heuristics—Advances and Trends in Local Search Paradigms for Optimization* (pp. 433–458). Dordrecht, Netherlands, Kluwer Academic Publishers.

Hansen, P., & Ribeiro, C. (Eds.). (2001). *Essays and Surveys on Metaheuristics*. Boston, Kluwer Academic Publishers.

Hartmann, S., & Kolisch, R. (1999). Self adapting genetic algorithm with an application to project scheduling. Technical report 506, University of Kiel, Kiel, Germany.

Hartmann, S., & Kolisch, R. (2000). Experimental evaluation of state-of-the-art heuristics for resource constrained project scheduling. *European Journal of Operational Research*, 127(2), 394–407.

Haupt, R. (1989). A survey of priority rule-based scheduling. *OR Spektrum*, 11, 3–6.

Heckerman, D., Geiger, D., & Chickering, D. M. (1995). Learning Bayesian networks: The combination of knowledge and statistical data. *Machine Learning*, 20(3), 197–244.

Heissenbüttel, M., & Braun, T. (2003). Ants-based routing in large-scale mobile ad-hoc networks. In K. Irmscher & R.-P. Fähnrich (Eds.), *Proceedings of Kommunikation in verteilten Systemen (KiVS '03)* (pp. 91–99). Berlin, VDE Verlag GmbH.

Helsgaun, K. (2000). An effective implementation of the Lin-Kernighan traveling salesman heuristic. *European Journal of Operational Research*, 126(1), 106–130.

Hertz, A., Taillard, É. D., & de Werra, D. (1997). A tutorial on tabu search. In E. H. L. Aarts & J. K. Lenstra (Eds.), *Local Search in Combinatorial Optimization* (pp. 121–136). Chichester, UK, John Wiley & Sons.

Heusse, M. (2001). *Routage et équilibrage de charge par agents dans les réseaux de communication*. PhD thesis, École des Hautes Études en Sciences Sociales, Paris.

Heusse, M., & Kermarrec, Y. (2000). Adaptive routing and load balancing of ephemeral connections. In *Proceedings of the 1st IEEE European Conference on Universal Multiservice Networks ECUMN'2000* (pp. 100–108). Piscataway, NJ, IEEE Press.

Heusse, M., Snyers, D., Guérin, S., & Kuntz, P. (1998). Adaptive agent-driven routing and load balancing in communication networks. *Advances in Complex Systems*, 1(2), 237–254.

Hochbaum, D. S. (Ed.). (1997). *Approximation Algorithms for NP-Hard Problems*. Boston, PWS Publishing Company.

Holland, J. (1975). *Adaptation in Natural and Artificial Systems*. Ann Arbor, University of Michigan Press.

Holland, O., & Melhuish, C. (1999). Stigmergy, self-organization, and sorting in collective robotics. *Artificial Life*, 5(2), 173–202.

Hromkovic, J. (2003). *Algorithmics for Hard Problems*, 2nd ed. Berlin, Springer-Verlag.

Hsu, H.-P., Mehra, V., Nadler, W., & Grassberger, P. (2003). Growth algorithms for lattice heteropolymers at low temperatures. *Journal of Chemical Physics*, 118(1), 444–451.

Hurkens, C. A. J., & Tiourine, S. R. (1995). Upper and lower bounding techniques for frequency assignment problems. Technical report 95-34, Department of Mathematics and Computing Science, Eindhoven University of Technology, Netherlands.

Iredi, S., Merkle, D., & Middendorf, M. (2001). Bi-criterion optimization with multi colony ant algorithms. In E. Zitzler, K. Deb, L. Thiele, C. C. Coello, & D. Corne (Eds.), *First International Conference on*

Evolutionary Multi-Criterion Optimization (EMO'01), vol. 1993 of *Lecture Notes in Computer Science* (pp. 359–372). Berlin, Springer-Verlag.

Jacobs, L. W., & Brusco, M. J. (1995). A local search heuristic for large set covering problems. *Naval Research Logistics, 42*, 1129–1140.

Jaillet, P. (1985). *Probabilistic Traveling Salesman Problems.* PhD thesis, MIT, Cambridge, MA.

Jaillet, P. (1988). A priori solution of a travelling salesman problem in which a random subset of the customers are visited. *Operations Research, 36*(6), 929–936.

Jensen, F. V. (2001). *Bayesian Networks and Decision Graphs.* Berlin, Springer-Verlag.

Johnson, D. S., Aragon, C. R., McGeoch, L. A., & Schevon, C. (1991). Optimization by simulated annealing: An experimental evaluation: Part II, Graph coloring and number partitioning. *Operations Research, 39*(3), 378–406.

Johnson, D. S., Gutin, G., McGeoch, L. A., Yeo, A., Zhang, W., & Zverovitch, A. (2002). Experimental analysis of heuristics for the ATSP. In G. Gutin & A. Punnen (Eds.), *The Traveling Salesman Problem and Its Variations* (pp. 445–487). Norwell, MA, Kluwer Academic Publishers.

Johnson, D. S., & McGeoch, L. A. (1997). The travelling salesman problem: A case study in local optimization. In E. H. L. Aarts & J. K. Lenstra (Eds.), *Local Search in Combinatorial Optimization* (pp. 215–310). Chichester, UK, John Wiley & Sons.

Johnson, D. S., & McGeoch, L. A. (2002). Experimental analysis of heuristics for the STSP. In G. Gutin & A. Punnen (Eds.), *The Traveling Salesman Problem and Its Variations* (pp. 369–443). Norwell, MA, Kluwer Academic Publishers.

Johnson, D. S., Papadimitriou, C. H., & Yannakakis, M. (1988). How easy is local search? *Journal of Computer System Science, 37*, 79–100.

Jünger, M., Reinelt, G., & Thienel, S. (1994). Provably good solutions for the traveling salesman problem. *Zeitschrift für Operations Research, 40*, 183–217.

Kaelbling, L. P., Littman, M. L., & Moore, A. W. (1996). Reinforcement learning: A survey. *Journal of Artificial Intelligence Research, 4*, 237–285.

Karaboga, D., & Pham, D. T. (2000). *Intelligent Optimisation Techniques.* Berlin, Springer-Verlag.

Khanna, A., & Zinky, J. (1989). The revised ARPANET routing metric. *ACM SIGCOMM Computer Communication Review, 19*(4), 45–56.

Kirkpatrick, S., Gelatt, C. D., Jr., & Vecchi, M. P. (1983). Optimization by simulated annealing. *Science, 220*, 671–680.

Knox, J. (1994). Tabu search performance on the symmetric travelling salesman problem. *Computers & Operations Research, 21*(8), 867–876.

Kolisch, R., & Hartmann, S. (1999). Heuristic algorithms for solving the resource constrained project scheduling: Classification and computational analysis. In J. Weglarz (Ed.), *Handbook on Recent Advances in Project Scheduling* (pp. 197–212). Dordrecht, Netherlands, Kluwer Academic Publishers.

Krasnogor, N., Hart, W. E., Smith, J., & Pelta, D. A. (1999). Protein structure prediction with evolutionary algorithms. In W. Banzhaf, J. Daida, A. E. Eiben, M. H. Garzon, V. Honavar, M. Jakiela, & R. E. Smith (Eds.), *Proceedings of the Genetic and Evolutionary Computation Conference*, vol. 2, (pp. 1596–1601). San Francisco, Morgan Kaufmann.

Krieger, M. J. B., & Billeter, J.-B. (2000). The call of duty: Self-organised task allocation in a population of up to twelve mobile robots. *Robotics and Autonomous Systems, 30*(1–2), 65–84.

Krieger, M. J. B., Billeter, J.-B., & Keller, L. (2000). Ant-like task allocation and recruitment in cooperative robots. *Nature, 406*, 992–995.

Krüger, F., Merkle, D., & Middendorf, M. (1998). Studies on a parallel Ant System for the BSP model. Unpublished manuscript.

Kube, C. R., & Zhang, H. (1994). Collective robotics: From social insects to robots. *Adaptive Behavior, 2*, 189–218.

Kullback, S. (1959). *Information Theory and Statistics*. New York, John Wiley & Sons.

Kuntz, P., Snyers, D., & Layzell, P. (1999). A stochastic heuristic for visualizing graph clusters in a bi-dimensional space prior to partitioning. *Journal of Heuristics, 5*(3), 327–351.

Laguna, M., & Martí, R. (2003). *Scatter Search: Methodology and Implementations in C*, vol. 24 of *Operations Research/Computer Science Interface*. Boston, Kluwer Academic Publishers.

Larrañaga, P., Kuijpers, C., Poza, M., & Murga, R. (1997). Decomposing Bayesian networks by genetic algorithms. *Statistics and Computing, 7*(1), 19–34.

Larrañaga, P., & Lozano, J. A. (2001). *Estimation of Distribution Algorithms. A New Tool for Evolutionary Computation. Genetic Algorithms and Evolutionary Computation*. Dordrecht, Netherlands, Kluwer Academic Publishers.

Lau, K. F., & Dill, K. A. (1989). A lattice statistical mechanics model of the conformation and sequence space of proteins. *Macromolecules, 22*, 3986–3997.

Lawler, E. L. (1963). The quadratic assignment problem. *Management Science, 9*, 586–599.

Lawler, E. L. (1976). *Combinatorial Optimization: Networks and Matroids*. New York, Holt, Rinehart, and Winston.

Lawler, E. L. (1977). A pseudopolynomial algorithm for sequencing jobs to minimize total tardiness. *Annals of Discrete Mathematics, 1*, 331–342.

Lawler, E. L., Lenstra, J. K., Rinnooy Kan, A. H. G., & Shmoys, D. B. (1985). *The Travelling Salesman Problem*. Chichester, UK, John Wiley & Sons.

Lawrence, S., & Giles, C. L. (1998). Searching the world wide web. *Science, 280*, 98–100.

Leguizamón, G., & Michalewicz, Z. (1999). A new version of Ant System for subset problems. In P. J. Angeline, Z. Michalewicz, M. Schoenauer, X. Yao, & A. Zalzala (Eds.), *Proceedings of the 1999 Congress on Evolutionary Computation (CEC'99)* (pp. 1459–1464). Piscataway, NJ, IEEE Press.

Leguizamón, G., & Michalewicz, Z. (2000). Ant Systems for subset problems. Unpublished manuscript.

Leguizamón, G., Michalewicz, Z., & Schütz, M. (2001). A ant system for the maximum independent set problem. In *Proceedings of the VII Argentinian Congress on Computer Science*, El Calafate, Santa Cruz, Argentina, vol. 2 (pp. 1027–1040).

Leighton, F. (1979). A graph coloring algorithm for large scheduling problems. *Journal of Research of the National Bureau of Standards, 85*, 489–506.

Lenstra, J. K., Rinnooy Kan, A. H. G., & Brucker, P. (1977). Complexity of machine scheduling problems. In P. L. Hammer, E. L. Johnson, B. H. Korte, & G. L. Nemhauser (Eds.), *Studies in Integer Programming*, vol. 1 of *Annals of Discrete Mathematics* (pp. 343–362). Amsterdam, North-Holland.

Levine, J., & Ducatelle, F. (2003). Ant colony optimisation and local search for bin packing and cutting stock problems. *Journal of the Operational Research Society*, to appear.

Liang, Y.-C., & Smith, A. E. (1999). An Ant System approach to redundancy allocation. In P. J. Angeline, Z. Michalewicz, M. Schoenauer, X. Yao, & A. Zalzala (Eds.), *Proceedings of the 1999 Congress on Evolutionary Computation (CEC'99)* (pp. 1478–1484). Piscataway, NJ, IEEE Press.

Liaw, C.-F. (2000). A hybrid genetic algorithm for the open shop scheduling problem. *European Journal of Operational Research, 124*(1), 28–42.

Lin, S. (1965). Computer solutions for the traveling salesman problem. *Bell Systems Technology Journal, 44*, 2245–2269.

Lin, S., & Kernighan, B. W. (1973). An effective heuristic algorithm for the travelling salesman problem. *Operations Research, 21*, 498–516.

Liu, J. S. (2001). *Monte Carlo Strategies in Scientific Computing*. New York, Springer-Verlag.

Lokketangen, A. (2000). Satisfied ants. In M. Dorigo, M. Middendorf, & T. Stützle (Eds.), *Abstract Proceedings of ANTS 2000—From Ant Colonies to Artificial Ants: Second International Workshop on Ant Algorithms* (pp. 73–77). Université Libre de Bruxelles, Brussels.

Lourenço, H., & Serra, D. (1998). Adaptive approach heuristics for the generalized assignment problem. Technical report No. 304, Universitat Pompeu Fabra, Department of Economics and Management, Barcelona, Spain.

Lourenço, H., & Serra, D. (2002). Adaptive search heuristics for the generalized assignment problem. *Mathware and Soft Computing, 9*(2–3), 209–234.

Lourenço, H. R., Martin, O., & Stützle, T. (2002). Iterated local search. In F. Glover & G. Kochenberger (Eds.), *Handbook of Metaheuristics*, vol. 57 of *International Series in Operations Research & Management Science* (pp. 321–353). Norwell, MA, Kluwer Academic Publishers.

Lumer, E., & Faieta, B. (1994). Diversity and adaptation in populations of clustering ants. In J.-A. Meyer & S. W. Wilson (Eds.), *Proceedings of the Third International Conference on Simulation of Adaptive Behavior: From Animals to Animats* (pp. 501–508). Cambridge, MA, MIT Press.

Lundy, M., & Mees, A. (1986). Convergence of an annealing algorithm. *Mathematical Programming, 34*, 111–124.

Maniezzo, V. (1999). Exact and approximate nondeterministic tree-search procedures for the quadratic assignment problem. *INFORMS Journal on Computing, 11*(4), 358–369.

Maniezzo, V. (2000). Personal communication.

Maniezzo, V., & Carbonaro, A. (2000). An ANTS heuristic for the frequency assignment problem. *Future Generation Computer Systems, 16*(8), 927–935.

Maniezzo, V., & Colorni, A. (1999). The Ant System applied to the quadratic assignment problem. *IEEE Transactions on Data and Knowledge Engineering, 11*(5), 769–778.

Maniezzo, V., Colorni, A., & Dorigo, M. (1994). The Ant System applied to the quadratic assignment problem. Technical report IRIDIA/94-28, IRIDIA, Université Libre de Bruxelles, Brussels.

Maniezzo, V., & Milandri, M. (2002). An ant-based framework for very strongly constrained problems. In M. Dorigo, G. Di Caro, & M. Sampels (Eds.), *Proceedings of ANTS 2002—From Ant Colonies to Artificial Ants: Third International Workshop on Ant Algorithms*, vol. 2463 of *Lecture Notes in Computer Science* (pp. 222–227). Berlin, Springer-Verlag.

Marathe, M. V., Ravi, R., Ravi, S. S., Rosenkrantz, D. J., & Sundaram, R. (1996). Spanning trees short or small. *SIAM Journal on Discrete Mathematics, 9*(2), 178–200.

Marchiori, E. (2002). Genetic, iterated, and multistart local search for the maximum clique problem. In S. Cagnoni, J. Gottlieb, E. Hart, M. Middendorf, & G. R. Raidl (Eds.), *Applications of Evolutionary Computing, Proceedings of EvoWorkshops 2002*, vol. 2279 of *Lecture Notes in Computer Science* (pp. 112–121). Berlin, Springer-Verlag.

Marchiori, E., & Steenbeek, A. (2000). An evolutionary algorithm for large scale set covering problems with application to airline crew scheduling. In *Real World Applications of Evolutionary Computing*, vol. 1083 of *Lecture Notes in Computer Science* (pp. 367–381). Berlin, Springer-Verlag.

Mariano, C. E., & Morales, E. (1999). MOAQ: An Ant-Q algorithm for multiple objective optimization problems. In W. Banzhaf, J. Daida, A. E. Eiben, M. H. Garzon, V. Honavar, M. Jakiela, & R. E. Smith (Eds.), *Proceedings of the Genetic and Evolutionary Computation Conference (GECCO-1999)*, vol. 1 (pp. 894–901). San Francisco, Morgan Kaufmann.

Martello, S., & Toth, P. (1990). *Knapsack Problems, Algorithms and Computer Implementations*. Chichester, John Wiley & Sons.

Martin, O., & Otto, S. W. (1996). Combining simulated annealing with local search heuristics. *Annals of Operations Research, 63*, 57–75.

Martin, O., Otto, S. W., & Felten, E. W. (1991). Large-step Markov chains for the traveling salesman problem. *Complex Systems, 5*(3), 299–326.

Martinoli, A., & Mondada, F. (1998). Probabilistic modelling of a bio-inspired collective experiment with real robots. In T. L. R. Dillman, P. Dario, & H. Wörn (Eds.), *Proceedings of the Fourth International Symposium on Distributed Autonomous Robotic Systems (DARS-98)* (pp. 289–308). Berlin, Springer-Verlag.

McQuillan, J. M., Richer, I., & Rosen, E. C. (1980). The new routing algorithm for the ARPANET. *IEEE Transactions on Communications, 28*, 711–719.

Merkle, D., & Middendorf, M. (2000). An ant algorithm with a new pheromone evaluation rule for total tardiness problems. In S. Cagnoni, R. Poli, G. D. Smith, D. Corne, M. Oates, E. Hart, P. L. Lanzi, E. J. Willem, Y. Li, B. Paechter, & T. C. Fogarty (Eds.), *Real-World Applications of Evolutionary Computing*, vol. 1803 of *Lecture Notes in Computer Science* (pp. 287–296). Berlin, Springer-Verlag.

Merkle, D., & Middendorf, M. (2002a). Ant colony optimization with the relative pheromone evaluation method. In S. Gagnoni, J. Gottlieb, E. Hart, M. Middendorf, & G. Raidl (Eds.), *Applications of Evolutionary Computing: Proceedings of EvoWorkshops 2002*, vol. 2279 of *Lecture Notes in Computer Science* (pp. 325–333). Berlin, Springer-Verlag.

Merkle, D., & Middendorf, M. (2002b). Fast ant colony optimization on runtime reconfigurable processor arrays. *Genetic Programming and Evolvable Machines, 3*(4), 345–361.

Merkle, D., & Middendorf, M. (2002c). Modeling the dynamics of ant colony optimization. *Evolutionary Computation, 10*(3), 235–262.

Merkle, D., & Middendorf, M. (2003a). Ant colony optimization with global pheromone evaluation for scheduling a single machine. *Applied Intelligence, 18*(1), 105–111.

Merkle, D., & Middendorf, M. (2003b). On the behavior of ACO algorithms: Studies on simple problems. In M. G. C. Resende & J. P. de Sousa (Eds.), *Metaheuristics: Computer Decision-Making, Combinatorial Optimization* (pp. 465–480). Boston, Kluwer Academic Publishers.

Merkle, D., Middendorf, M., & Schmeck, H. (2000a). Ant colony optimization for resource-constrained project scheduling. In D. Whitley, D. Goldberg, E. Cantu-Paz, L. Spector, I. Parmee, & H.-G. Beyer (Eds.), *Proceedings of the Genetic and Evolutionary Computation Conference (GECCO-2000)* (pp. 893–900). San Francisco, Morgan Kaufmann.

Merkle, D., Middendorf, M., & Schmeck, H. (2000b). Pheromone evaluation in ant colony optimization. In *Proceedings of the 26th Annual Conference of the IEEE Electronics Society* (pp. 2726–2731). Piscataway, NJ, IEEE Press.

Merkle, D., Middendorf, M., & Schmeck, H. (2002). Ant colony optimization for resource-constrained project scheduling. *IEEE Transactions on Evolutionary Computation, 6*(4), 333–346.

Merz, P., & Freisleben, B. (1997). Genetic local search for the TSP: New results. In T. Bäck, Z. Michalewicz, & X. Yao (Eds.), *Proceedings of the 1997 IEEE International Conference on Evolutionary Computation (ICEC'97)* (pp. 159–164). Piscataway, NJ, IEEE Press.

Metropolis, N., Rosenbluth, A., Rosenbluth, M., Teller, A., & Teller, E. (1953). Equation of state calculations by fast computing machines. *Journal of Chemical Physics, 21*, 1087–1092.

Meuleau, N., & Dorigo, M. (2002). Ant colony optimization and stochastic gradient descent. *Artificial Life, 8*(2), 103–121.

Michalewicz, Z. (1994). *Genetic Algorithms + Data Structures = Evolution Programs*. Berlin, Springer-Verlag.

Michalewicz, Z., & Fogel, D. B. (2000). *How to Solve It: Modern Heuristics*. Berlin, Springer-Verlag.

Michel, R., & Middendorf, M. (1998). An island model based Ant System with lookahead for the shortest supersequence problem. In A. E. Eiben, T. Bäck, M. Schoenauer, & H.-P. Schwefel (Eds.), *Proceedings of PPSN-V, Fifth International Conference on Parallel Problem Solving from Nature*, vol. 1498 of *Lecture Notes in Computer Science* (pp. 692–701). Berlin, Springer-Verlag.

Michel, R., & Middendorf, M. (1999). An ACO algorithm for the shortest supersequence problem. In D. Corne, M. Dorigo, & F. Glover (Eds.), *New Ideas in Optimization* (pp. 51–61). London, McGraw Hill.

Middendorf, M., Reischle, F., & Schmeck, H. (2002). Multi colony ant algorithms. *Journal of Heuristics, 8*(3), 305–320.

Minton, S., Johnston, M., Philips, A., & Laird, P. (1992). Minimizing conflicts: A heuristic repair method for constraint satisfaction and scheduling problems. *Artificial Intelligence, 58*(1–3), 161–205.

Mitchell, M. (1996). *An Introduction to Genetic Algorithms*. Cambridge, MA, MIT Press.

Mitchell, T. (1997). *Machine Learning*. Boston, McGraw Hill.

Mockus, J., Eddy, E., Mockus, A., Mockus, L., & Reklaitis, G. V. (1997). *Bayesian Heuristic Approach to Discrete and Global Optimization*. Dordrecht, The Netherlands, Kluwer Academic Publishers.

Moffett, M. W. (1988). Cooperative food transport by an Asiatic ant. *National Geographic Research, 4*, 386–394.

Mondada, F., Franzi, E., & Ienne, P. (1993). Mobile robot miniaturization: A tool for investigation in control algorithms. In T. Yoshikawa & F. Miyazaki (Eds.), *Proceedings of the Third International Symposium on Simulation on Experimental Robotics (ISER-93)*, vol. 200 of *Lecture Notes in Control and Information Sciences* (pp. 501–513). Berlin, Springer-Verlag.

Morris, P. (1993). The breakout method for escaping from local minima. In *Proceedings of the 11th National Conference on Artificial Intelligence* (pp. 40–45). Menlo Park, CA, AAAI Press/MIT Press.

Morton, T. E., Rachamadugu, R. M., & Vepsalainen, A. (1984). Accurate myopic heuristics for tardiness scheduling. GSIA working paper 36-83-84, Carnegie Mellon University, Pittsburgh.

Moy, J. T. (1998). *OSPF Anatomy of an Internet Routing Protocol*. Boston, Addison-Wesley.

Mühlenbein, H. (1998). The equation for response to selection and its use for prediction. *Evolutionary Computation, 5*(3), 303–346.

Mühlenbein, H., & Paass, G. (1996). From recombination of genes to the estimation of distributions. In W. Ebeling, I. Rechenberg, H.-P. Schwefel, & H.-M. Voigt (Eds.), *Proceedings of PPSN-IV, Fourth International Conference on Parallel Problem Solving from Nature*, vol. 1141 of *Lecture Notes in Computer Science* (pp. 178–187). Berlin, Springer-Verlag.

Navarro Varela, G., & Sinclair, M. C. (1999). Ant colony optimisation for virtual-wavelength-path routing and wavelength allocation. In P. J. Angeline, Z. Michalewicz, M. Schoenauer, X. Yao, & A. Zalzala (Eds.), *Proceedings of the 1999 Congress on Evolutionary Computation (CEC'99)* (pp. 1809–1816). Piscataway, NJ, IEEE Press.

Nawaz, M., Enscore, E., Jr., & Ham, I. (1983). A heuristic algorithm for the *m*-machine, *n*-job flow-shop sequencing problem. *OMEGA, 11*(1), 91–95.

Nemhauser, G. L., & Wolsey, L. A. (1988). *Integer and Combinatorial Optimization*. Chichester, UK, John Wiley & Sons.

Nicolis, G., & Prigogine, I. (1977). *Self-Organisation in Non-Equilibrium Systems*. New York, John Wiley & Sons.

Nouyan, S. (2002). Agent-based approach to dynamic task allocation. In M. Dorigo, G. Di Caro, & M. Sampels (Eds.), *Proceedings of ANTS 2002—From Ant Colonies to Artificial Ants: Third International Workshop on Ant Algorithms*, vol. 2463 of *Lecture Notes in Computer Science* (pp. 28–39). Berlin, Springer-Verlag.

Nowicki, E., & Smutnicki, C. (1996a). A fast taboo search algorithm for the job-shop problem. *Management Science, 42*(2), 797–813.

Nowicki, E., & Smutnicki, C. (1996b). A fast tabu search algorithm for the permutation flow-shop problem. *European Journal of Operational Research, 91*(1), 160–175.

Nozaki, K., Ishibuchi, H., & Tanaka, H. (1997). A simple but powerful heuristic method for generating fuzzy rules from numerical data. *Fuzzy Sets and Systems, 86*, 251–270.

Nyström, M. (1999). Solving certain large instances of the quadratic assignment problem: Steinberg's examples. Technical report, Department of Computer Science, California Institute of Technology, Pasadena.

Osman, I., & Laporte, G. (1996). Metaheuristics: A bibliography. *Annals of Operations Research, 63*, 513–628.

Osman, I. H., & Kelly, J. P. (Eds.). (1996). *Meta-Heuristics: Theory and Applications*. Boston, Kluwer Academic Publishers.

Padberg, M. W., & Grötschel, M. (1985). Polyhedral computations. In E. L. Lawler, J. K. Lenstra, A. H. G. Rinnooy Kan, & D. B. Shmoys (Eds.), *The Traveling Salesman Problem* (pp. 307–360). Chichester, UK, John Wiley & Sons.

Paessens, H. (1988). The savings algorithm for the vehicle routing problem. *European Journal of Operational Research, 34*, 336–344.

Papadimitriou, C. H., & Steiglitz, K. (1982). *Combinatorial Optimization—Algorithms and Complexity.* Englewood Cliffs, NJ, Prentice Hall.

Papoulis, A. (1991). *Probability, Random Variables and Stochastic Process,* 3rd ed. New York, McGraw Hill.

Paquete, L., & Stützle, T. (2002). An experimental investigation of iterated local search for coloring graphs. In S. Cagnoni, J. Gottlieb, E. Hart, M. Middendorf, & G. R. Raidl (Eds.), *Applications of Evolutionary Computing, Proceedings of EvoWorkshops 2002,* vol. 2279 of *Lecture Notes in Computer Science* (pp. 122–131). Berlin, Springer-Verlag.

Parpinelli, R. S., Lopes, H. S., & Freitas, A. A. (2002a). An ant colony algorithm for classification rule discovery. In H. A. Abbass, R. A. Sarker, & C. S. Newton (Eds.), *Data Mining: A Heuristic Approach* (pp. 191–208). Hershey, PA, Idea Group Publishing.

Parpinelli, R. S., Lopes, H. S., & Freitas, A. A. (2002b). Data mining with an ant colony optimization algorithm. *IEEE Transactions on Evolutionary Computation, 6*(4), 321–332.

Pearl, J. (1998). *Probabilisitic Reasoning in Intelligent Systems: Networks of Plausible Inference.* San Mateo, CA, Morgan Kaufmann.

Pelikan, M., Goldberg, D. E., & Lobo, F. (1999). A survey of optimization by building and using probabilistic models. Technical report IlliGAL, 99018, University of Illinois at Urbana-Champaign, Urbana, IL.

Pfahringer, B. (1996). Multi-agent search for open shop scheduling: Adapting the Ant-Q formalism. Technical report TR-96-09, Austrian Research Institute for Artificial Intelligence, Vienna.

Pimont, S., & Solnon, C. (2000). A generic ant algorithm for solving constraint satisfaction problems. In M. Dorigo, M. Middendorf, & T. Stützle (Eds.), *Abstract proceedings of ANTS 2000—From Ant Colonies to Artificial Ants: Second International Workshop on Ant Algorithms* (pp. 100–108). Université Libre de Bruxelles, Brussels.

Pinedo, M. (1995). *Scheduling—Theory, Algorithms, and Systems.* Englewood Cliffs, NJ, Prentice Hall.

Potts, C. N., & Wassenhove, L. N. V. (1991). Single machine tardiness sequencing heuristics. *IIE Transactions, 23*, 346–354.

Quinlan, J. (1993a). *C4.5: Programs for Machine Learning.* San Francisco, Morgan Kaufmann.

Quinlan, J. (1993b). Combining instance-based and model-based learning. In *Proceedings of the Tenth International Conference on Machine Learning (ML-93)* (pp. 236–243). San Mateo, CA, Morgan Kaufmann.

Rajendran, C., & Ziegler, H. (2003). Ant-colony algorithms for permutation flowshop scheduling to minimize makespan/total flowtime of jobs. *European Journal of Operational Research,* to appear.

Rechenberg, I. (1973). *Evolutionsstrategie—Optimierung technischer Systeme nach Prinzipien der biologischen Information.* Freiburg, Germany, Fromman Verlag.

Reeves, C. (Ed.). (1995). *Modern Heuristic Techniques for Combinatorial Problems.* London, McGraw Hill.

Reimann, M., Doerner, K., & Hartl, R. F. (2002a). Insertion based ants for the vehicle routing problem with backhauls and time windows. In M. Dorigo, G. Di Caro, & M. Sampels (Eds.), *Proceedings of ANTS 2002—From Ant Colonies to Artificial Ants: Third International Workshop on Ant Algorithms,* vol. 2463 of *Lecture Notes in Computer Science* (pp. 135–148). Berlin, Springer-Verlag.

Reimann, M., Doerner, K., & Hartl, R. F. (2003). Analyzing a unified Ant System for the VRP and some of its variants. In G. R. Raidl, J.-A. Meyer, M. Middendorf, S. Cagnoni, J. J. R. Cardalda, D. W. Corne, J. Gottlieb, A. Guillot, E. Hart, C. G. Johnson, & E. Marchiori (Eds.), *Applications of Evolutionary*

Computing, Proceedings of EvoWorkshops 2003, vol. 2611 of *Lecture Notes in Computer Science* (pp. 300–310). Berlin, Springer-Verlag.

Reimann, M., Doerner, K., & Hartl, R. F. (2004). D-ants: Savings based ants divide and conquer the vehicle routing problem. *Computers & Operations Research, 31*(4), 563–591.

Reimann, M., Stummer, M., & Doerner, K. (2002b). A savings based Ant System for the vehicle routing problem. In W. B. Langdon, E. Cantú-Paz, K. Mathias, R. Roy, D. Davis, R. Poli, K. Balakrishnan, V. Honavar, G. Rudolph, J. Wegener, L. Bull, M. A. Potter, A. C. Schultz, J. F. Miller, E. Burke, & N. Jonoska (Eds.), *Proceedings of the Genetic and Evolutionary Computation Conference (GECCO-2002)* (pp. 1317–1325). San Francisco, Morgan Kaufmann.

Reinelt, G. (1991). TSPLIB—A traveling salesman problem library. *ORSA Journal on Computing, 3*, 376–384.

Reinelt, G. (1994). *The Traveling Salesman: Computational Solutions for TSP Applications*, vol. 840 of *Lecture Notes in Computer Science*. Berlin, Springer-Verlag.

Resende, M. G. C., Pitsoulis, L. S., & Pardalos, P. M. (2000). Fortran subroutines for computing approximate solutions of weighted MAX-SAT problems using GRASP. *Discrete Applied Mathematics, 100*(1–2), 95–113.

Resende, M. G. C., & Ribeiro, C. C. (2002). Greedy randomized adaptive search procedures. In F. Glover & G. Kochenberger (Eds.), *Handbook of Metaheuristics*, International Series in Operations Research & Management Science (pp. 219–249). Norwell, MA, Kluwer Academic Publishers.

Resnick, M. (1994). *Turtles, Termites, and Traffic Jams*. Cambridge, MA, MIT Press.

Robbins, H., & Monroe, H. (1951). A stochastic approximation method. *Annals of Mathematics and Statistics, 22*, 400–407.

Robinson, G. E. (1992). Regulation of division of labor in insect societies. *Annual Review of Entomology, 37*, 637–665.

Roli, A., Blum, C., & Dorigo, M. (2001). ACO for maximal constraint satisfaction problems. In *Proceedings of MIC'2001—Meta–heuristics International Conference*, vol. 1 (pp. 187–191). Porto, Portugal.

Romeo, F., & Sangiovanni-Vincentelli, A. (1991). A theoretical framework for simulated annealing. *Algorithmica, 6*(3), 302–345.

Rossi, F., Petrie, C., & Dhar, V. (1990). On the equivalence of constraint satisfaction problems. In L. Carlucci Aiello (Ed.), *Proceedings of the 9th European Conference on Artificial Intelligence* (pp. 550–556). London, Pitman Publishing.

Roy, B., & Sussmann, B. (1964). Les problèmes d'ordonnancement avec contraintes disjonctives. Technical report DS No. 9bis, SEMA, Paris.

Rubin, P. A., & Ragatz, G. L. (1995). Scheduling in a sequence dependent setup environment with genetic search. *Computers & Operations Research, 22*(1), 85–99.

Rubinstein, R. Y. (1981). *Simulation and the Monte Carlo Method*. New York, John Wiley & Sons.

Rubinstein, R. Y. (1999). The cross-entropy method for combinatorial and continuous optimization. *Methodology and Computing in Applied Probability, 1*(2), 127–190.

Rubinstein, R. Y. (2001). Combinatorial optimization via the simulated cross-entropy method. In S. I. Gass & C. M. Harris (Eds.), *Encyclopedia of Operations Research and Management Science*. Boston, Kluwer Academic Publishers.

Sadeh, N., & Fox, M. (1996). Variable and value ordering heuristics for the job shop scheduling constraint satisfaction problem. *Artificial Intelligence, 86*(1), 1–41.

Sadeh, N., Sycara, K., & Xiong, Y. (1995). Backtracking techniques for the job shop scheduling constraint satisfaction problem. *Artificial Intelligence, 76*(1–2), 455–480.

Sahni, S., & Gonzalez, T. (1976). P-complete approximation problems. *Journal of the ACM, 23*(3), 555–565.

Sait, S. M., & Youssef, H. (1999). *Iterative Computer Algorithms with Applications to Engineering*. Los Alamitos, CA, IEEE Computer Society Press.

Schoofs, L., & Naudts, B. (2000). Solving CSPs with ant colonies. In M. Dorigo, M. Middendorf, & T. Stützle (Eds.), *Abstract proceedings of ANTS 2000—From Ant Colonies to Artificial Ants: Second International Workshop on Ant Algorithms* (pp. 134–137). Université Libre de Bruxelles, Brussels.

Schoonderwoerd, R., Holland, O., & Bruten, J. (1997). Ant-like agents for load balancing in telecommunications networks. In *Proceedings of the First International Conference on Autonomous Agents* (pp. 209–216). New York, ACM Press.

Schoonderwoerd, R., Holland, O., Bruten, J., & Rothkrantz, L. (1996). Ant-based load balancing in telecommunications networks. *Adaptive Behavior*, 5(2), 169–207.

Schreiber, G. R., & Martin, O. C. (1999). Cut size statistics of graph bisection heuristics. *SIAM Journal on Optimization*, 10(1), 231–251.

Schrjiver, A. (2002). On the history of combinatorial optimization. Preprint available at www.cwi.nl/~lex/.

Schwefel, H.-P. (1981). *Numerical Optimization of Computer Models*. Chichester, UK, John Wiley & Sons.

Selman, B., & Kautz, H. (1993). Domain-independent extensions to GSAT: Solving large structured satisfiability problems. In *Proceedings of the 13th International Joint Conference on Artificial Intelligence* (pp. 290–295). San Francisco, Morgan Kaufmann.

Shang, Y., & Wah, B. W. (1998). A discrete Lagrangian-based global-search method for solving satisfiability problems. *Journal of Global Optimization*, 12(1), 61–99.

Shankar, A. U., Alaettinoğlu, C., Dussa-Zieger, K., & Matta, I. (1992). Performance comparison of routing protocols under dynamic and static file transfer connections. *ACM Computer Communication Review*, 22(5), 39–52.

Shmygelska, A., Aguirre-Hernández, R., & Hoos, H. H. (2002). An ant colony optimization algorithm for the 2D HP protein folding problem. In M. Dorigo, G. Di Caro, & M. Sampels (Eds.), *ANTS 2002*, vol. 2463 of *Lecture Notes in Computer Science* (pp. 40–52). Berlin, Springer-Verlag.

Shmygelska, A., & Hoos, H. H. (2003). An improved ant colony optimization algorithm for the 2D HP protein folding problem. In Y. Xiang, & B. Chaib-draa (Eds.), *Advances in Artificial Intelligence*, vol. 2671 of *Lecture Notes in Artificial Intelligence* (pp. 400–417). Berlin, Springer-Verlag.

Singh, S. P., & Sutton, R. S. (1996). Reinforcement learning with replacing eligibility traces. *Machine Learning*, 22(1–3), 123–158.

Smith, B. M., & Dyer, M. E. (1996). Locating the phase transition in binary constraint satisfaction problems. *Artificial Intelligence*, 81(1–2), 155–181.

Smith, D. H., Hurley, S., & Thiel, S. U. (1998). Improving heuristics for the frequency assignment problem. *European Journal of Operational Research*, 107(1), 76–86.

Socha, K., Knowles, J., & Sampels, M. (2002). A $\mathcal{MAX}$-$\mathcal{MIN}$ Ant System for the university course timetabling problem. In M. Dorigo, G. Di Caro, & M. Sampels (Eds.), *Proceedings of ANTS 2002—From Ant Colonies to Artificial Ants: Third International Workshop on Ant Algorithms*, vol. 2463 of *Lecture Notes in Computer Science* (pp. 1–13). Berlin, Springer-Verlag.

Socha, K., Sampels, M., & Manfrin, M. (2003). Ant algorithms for the university course timetabling problem with regard to the state-of-the-art. In G. R. Raidl, J.-A. Meyer, M. Middendorf, S. Cagnoni, J. J. R. Cardalda, D. W. Corne, J. Gottlieb, A. Guillot, E. Hart, C. G. Johnson, & E. Marchiori (Eds.), *Applications of Evolutionary Computing, Proceedings of EvoWorkshops 2003*, vol. 2611 of *Lecture Notes in Computer Science* (pp. 334–345). Berlin, Springer-Verlag.

Solnon, C. (2000). Solving permutation constraint satisfaction problems with artificial ants. In W. Horn (Ed.), *Proceedings of the 14th European Conference on Artificial Intelligence* (pp. 118–122). Amsterdam, IOS Press.

Solnon, C. (2002). Ants can solve constraint satisfaction problems. *IEEE Transactions on Evolutionary Computation*, 6(4), 347–357.

Steenstrup, M. E. (Ed.). (1995). *Routing in Communications Networks*. Englewood Cliffs, NJ, Prentice Hall.

Steinberg, L. (1961). The backboard wiring problem: A placement algorithm. *SIAM Review*, *3*, 37–50.

Sterling, T., Salmon, J., Becker, D. J., & Savarese, D. F. (1999). *How to Build a Beowulf*. Cambridge, MA, MIT Press.

Steuer, R. E. (1986). *Multiple Criteria Optimization: Theory, Computation and Application*. Wiley Series in Probability and Mathematical Statistics. New York, John Wiley & Sons.

Streltsov, S., & Vakili, P. (1996). Variance reduction algorithms for parallel replicated simulation of uniformized Markov chains. *Discrete Event Dynamic Systems: Theory and Applications*, *6*, 159–180.

Stützle, T. (1997a). An ant approach to the flow shop problem. Technical report AIDA-97-07, FG Intellektik, FB Informatik, TU Darmstadt, Germany.

Stützle, T. (1997b). $\mathcal{MAX}$-$\mathcal{MIN}$ Ant System for the quadratic assignment problem. Technical report AIDA-97-4, FG Intellektik, FB Informatik, TU Darmstadt, Germany.

Stützle, T. (1998a). An ant approach to the flow shop problem. In *Proceedings of the Sixth European Congress on Intelligent Techniques & Soft Computing (EUFIT'98)*, vol. 3 (pp. 1560–1564). Aachen, Germany, Verlag Mainz, Wissenschaftsverlag.

Stützle, T. (1998b). Parallelization strategies for ant colony optimization. In A. E. Eiben, T. Bäck, M. Schoenauer, & H.-P. Schwefel (Eds.), *Proceedings of PPSN-V, Fifth International Conference on Parallel Problem Solving from Nature*, vol. 1498 of *Lecture Notes in Computer Science* (pp. 722–731). Berlin, Springer-Verlag.

Stützle, T. (1999). *Local Search Algorithms for Combinatorial Problems: Analysis, Improvements, and New Applications*, vol. 220 of *DISKI*. Sankt Augustin, Germany, Infix.

Stützle, T., & Dorigo, M. (1999a). ACO algorithms for the quadratic assignment problem. In D. Corne, M. Dorigo, & F. Glover (Eds.), *New Ideas in Optimization* (pp. 33–50). London, McGraw Hill.

Stützle, T., & Dorigo, M. (1999b). ACO algorithms for the traveling salesman problem. In K. Miettinen, M. M. Mäkelä, P. Neittaanmäki, & J. Périaux (Eds.), *Evolutionary Algorithms in Engineering and Computer Science* (pp. 163–183). Chichester, UK, John Wiley & Sons.

Stützle, T., & Dorigo, M. (2002). A short convergence proof for a class of ACO algorithms. *IEEE Transactions on Evolutionary Computation*, *6*(4), 358–365.

Stützle, T., & Hoos, H. H. (1996). Improving the Ant System: A detailed report on the $\mathcal{MAX}$-$\mathcal{MIN}$ Ant System. Technical report AIDA-96-12, FG Intellektik, FB Informatik, TU Darmstadt, Germany.

Stützle, T., & Hoos, H. H. (1997). The $\mathcal{MAX}$-$\mathcal{MIN}$ Ant System and local search for the traveling salesman problem. In T. Bäck, Z. Michalewicz, & X. Yao (Eds.), *Proceedings of the 1997 IEEE International Conference on Evolutionary Computation (ICEC'97)* (pp. 309–314). Piscataway, NJ, IEEE Press.

Stützle, T., & Hoos, H. H. (1999). $\mathcal{MAX}$-$\mathcal{MIN}$ Ant System and local search for combinatorial optimization problems. In S. Voss, S. Martello, I. Osman, & C. Roucairol (Eds.), *Meta-Heuristics: Advances and Trends in Local Search Paradigms for Optimization* (pp. 137–154). Dordrecht, Netherlands, Kluwer Academic Publishers.

Stützle, T., & Hoos, H. H. (2000). $\mathcal{MAX}$-$\mathcal{MIN}$ Ant System. *Future Generation Computer Systems*, *16*(8), 889–914.

Stützle, T., & Linke, S. (2002). Experiments with variants of ant algorithms. *Mathware and Soft Computing*, *9*(2–3), 193–207.

Subramanian, D., Druschel, P., & Chen, J. (1997). Ants and reinforcement learning: A case study in routing in dynamic networks. In *Proceedings of the 15th International Joint Conference on Artificial Intelligence* (pp. 832–838). San Francisco, Morgan Kaufmann.

Sudd, J. H. (1965). The transport of prey by ants. *Behaviour*, *25*, 234–271.

Sutton, R. S. (1988). Learning to predict by the methods of temporal differences. *Machine Learning*, *3*, 9–44.

Sutton, R. S., & Barto, A. G. (1998). *Reinforcement Learning: An Introduction*. Cambridge, MA, MIT Press.

Szwarc, W., Grosso, A., & Della Croce, F. (2001). Algorithmic paradoxes of the single machine total tardiness problem. *Journal of Scheduling, 4*(2), 93–104.

Taillard, É. D. (1991). Robust taboo search for the quadratic assignment problem. *Parallel Computing, 17*, 443–455.

Taillard, É. D. (1995). Comparison of iterative searches for the quadratic assignment problem. *Location Science, 3*, 87–105.

Taillard, É. D. (1998). FANT: Fast Ant System. Technical report IDSIA-46-98, IDSIA, Lugano, Switzerland.

Taillard, É. D., Badeau, P., Gendreau, M., Guertin, F., & Potvin, J.-Y. (1997). A tabu search heuristic for the vehicle routing problem with soft time windows. *Transportation Science, 31*, 170–186.

Tan, K. C., & Narashiman, R. (1997). Minimizing tardiness on a single processor with sequence-dependent setup times: A simulated annealing approach. *OMEGA, 25*(6), 619–634.

Tanenbaum, A. (1996). *Computer Networks*. Englewood Cliffs, NJ, Prentice Hall.

Teich, T., Fischer, M., Vogel, A., & Fischer, J. (2001). A new ant colony algorithm for the job shop scheduling problem. In L. Spector, E. D. Goodman, A. Wu, W. B. Langdon, H.-M. Voigt, M. Gen, S. Sen, M. Dorigo, S. Pezeshk, M. H. Garzon, & E. Burke (Eds.), *Proceedings of the Genetic and Evolutionary Computation Conference (GECCO-2001)* (pp. 803). San Francisco, Morgan Kaufmann.

Theraulaz, G., & Bonabeau, E. (1999). A brief history of stigmergy. *Artificial Life, 5*, 97–116.

T'kindt, V., Monmarché, N., Tercinet, F., & Laügt, D. (2002). An ant colony optimization algorithm to solve a 2-machine bicriteria flowshop scheduling problem. *European Journal of Operational Research, 142*(2), 250–257.

Toth, P., & Vigo, D. (Eds.). (2001). *The Vehicle Routing Problem*. SIAM Monographs on Discrete Mathematics and Applications. Philadelphia, Society for Industrial & Applied Mathematics.

van der Put, R. (1998). Routing in the faxfactory using mobile agents. Technical report R&D-SV-98-276, KPN Research, The Netherlands.

Vasquez, M., & Hao, J.-K. (2001). A hybrid approach for the 0-1 multidimensional knapsack problem. In *Proceedings of the 17th International Joint Conference on Artificial Intelligence* (pp. 328–333). San Francisco, Morgan Kaufmann.

Vazirani, V. V. (2001). *Approximation Algorithms*. Berlin, Springer-Verlag.

Voss, S., Martello, S., Osman, I. H., & Roucairol, C. (Eds.). (1999). *Meta-Heuristics: Advances and Trends in Local Search Paradigms for Optimization*. Dordrecht, Netherlands, Kluwer Academic Publishers.

Voudouris, C. (1997). *Guided Local Search for Combinatorial Optimization Problems*. PhD thesis, Department of Computer Science, University of Essex, Colchester, UK.

Voudouris, C., & Tsang, E. (1995). Guided local search. Technical report CSM-247, Department of Computer Science, University of Essex, Colchester, UK.

Voudouris, C., & Tsang, E. P. K. (1999). Guided local search. *European Journal of Operational Research, 113*(2), 469–499.

Wagner, I. A., Lindenbaum, M., & Bruckstein, A. M. (1996). Smell as a computational resource—A lesson we can learn from the ant. In M. Y. Vardi (Ed.), *Proceedings of the Fourth Israeli Symposium on Theory of Computing and Systems (ISTCS-99)* (pp. 219–230). Los Alamitos, CA, IEEE Computer Society Press.

Wagner, I. A., Lindenbaum, M., & Bruckstein, A. M. (1998). Efficient graph search by a smell-oriented vertex process. *Annals of Mathematics and Artificial Intelligence, 24*, 211–223.

Wagner, I. A., Lindenbaum, M., & Bruckstein, A. M. (2000). ANTS: Agents, networks, trees and subgraphs. *Future Generation Computer Systems, 16*(8), 915–926.

Wallace, R. J. (1996). Analysis of heuristic methods for partial constraint satisfaction problems. In E. Freuder (Ed.), *Principles and Practice of Constraint Programming—CP'96*, vol. 1118 of *Lecture Notes in Computer Science* (pp. 482–496). Berlin, Springer-Verlag.

Wallace, R. J., & Freuder, E. C. (1996). Heuristic methods for over-constrained constraint satisfaction problems. In M. Jampel, E. Freuder, & M. Maher (Eds.), *OCS'95: Workshop on Over-Constrained Systems at CP'95*, vol. 1106 of *Lecture Notes in Computer Science* (pp. 207–216). Berlin, Springer-Verlag.

Walrand, J., & Varaiya, P. (1996). *High-performance communication networks*. San Francisco, Morgan Kaufmann.

Walters, T. (1998). Repair and brood selection in the traveling salesman problem. In A. Eiben, T. Bäck, M. Schoenauer, & H.-P. Schwefel (Eds.), *Proceedings of PPSN-V, Fifth International Conference on Parallel Problem Solving from Nature*, vol. 1498 of *Lecture Notes in Computer Science* (pp. 813–822). Berlin, Springer-Verlag.

Wang, L. X., & Mendel, J. M. (1992). Generating fuzzy rules by learning from examples. *IEEE Transactions on Systems, Man, and Cybernetics, 22*(6), 1414–1427.

Wang, Z., & Crowcroft, J. (1992). Analysis of shortest-path routing algorithms in a dynamic network environment. *ACM Computer Communication Review, 22*(2), 63–71.

Wäscher, G., & Gau, T. (1996). Heuristics for the integer one-dimensional cutting stock problem: A computational study. *OR Spektrum, 18*, 131–144.

Watkins, C. J., & Dayan, P. (1992). *Q*-Learning. *Machine Learning, 8*, 279–292.

White, T., Pagurek, B., & Oppacher, F. (1998). Connection management using adaptive mobile agents. In H. R. Arabnia (Ed.), *Proceedings of the International Conference on Parallel and Distributed Processing Techniques and Applications (PDPTA'98)* (pp. 802–809). Las Vegas, NV, CSREA Press.

Whitley, D., Gordon, S., & Mathias, K. (1994). Lamarckian evolution, the Baldwin effect and function optimization. In Y. Davidor, H. Schwefel & R. Männer (Eds.), *Proceedings of PPSN-III, Third International Conference on Parallel Problem Solving from Nature*, vol. 866 of *Lecture Notes in Computer Science* (pp. 6–15). Berlin, Springer-Verlag.

Williams, R. J. (1992). Simple statistical gradient-following algorithms for connectionist reinforcement learning. *Machine Learning, 8*(3), 229–256.

Yagiura, M., Ibaraki, T., & Glover, F. (2004). An ejection chain approach for the generalized assignment problem. *INFORMS Journal on Computing*, to appear.

Zachariasen, M., & Dam, M. (1996). Tabu search on the geometric traveling salesman problem. In I. H. Osman & J. P. Kelly (Eds.), *Meta-heuristics: Theory and Applications* (pp. 571–587). Boston, Kluwer Academic Publishers.

Zlochin, M., Birattari, M., Meuleau, N., & Dorigo, M. (2001). Combinatorial optimization using model-based search. Technical report IRIDIA/2001-15, IRIDIA, Université Libre de Bruxelles, Brussels. To appear in *Annals of Operations Research*, 2004.

Zlochin, M., & Dorigo, M. (2002). Model-based search for combinatorial optimization: A comparative study. In J. J. Merelo, P. Adamidis, H.-G. Beyer, J.-L. Fernández-Villacanas, & H.-P. Schwefel (Eds.), *Proceedings of PPSN-VII, Seventh International Conference on Parallel Problem Solving from Nature*, vol. 2439 of *Lecture Notes in Computer Science* (pp. 651–661). Berlin, Springer-Verlag.

Index